Enantiomer Separation

Enantiomer Separation

Fundamentals and Practical Methods

Edited by

Fumio Toda

Professor of Chemistry,
Department of Chemistry,
Okayama University of Science, Japan

KLUWER ACADEMIC PUBLISHERS

DORDRECHT / BOSTON / LONDON

A C.I.P. Catalogue record for this book is available from the Library of Congress.

ISBN 1-4020-2336-7 (HB)
ISBN 1-4020-2337-5 (e-book)

Published by Kluwer Academic Publishers,
P.O. Box 17, 3300 AA Dordrecht, The Netherlands.

Sold and distributed in North, Central and South America
by Kluwer Academic Publishers,
101 Philip Drive, Norwell, MA 02061, U.S.A.

In all other countries, sold and distributed
by Kluwer Academic Publishers,
P.O. Box 322, 3300 AH Dordrecht, The Netherlands.

Printed on acid-free paper

Printed in the Netherlands.

Table of Contents

PREFACE

In spite of important advances in asymmetric synthesis, chiral compounds cannot all be obtained in a pure state by asymmetric synthesis. As a result, enantiomer separation remains an important technique for obtaining optically active materials. Although asymmetric synthesis is a once-only procedure, an enantiomer separation process can be repeated until the optically pure sample is obtained.

This book discusses several new enantiomer separation methods using modern techniques developed by experts in the field. These methods consist mainly of the following three types:

1) Enantiomer separation by inclusion complexation with a chiral host compound

2) Enantiomer separation using biological methods

3) Enantiomer separation by HPLC chromatography using a column containing a chiral stationary phase.

Separation of a racemic compound has been called "optical resolution" or simply "resolution". Nowadays, the descriptions "enantiomer resolution" or "enantiomer separation" are also commonly used. Accordingly, "Enantiomer Separation" is used in the title of this book. The editor and all chapter contributors hope that this book is helpful for scientists and engineers working in this field.

Fumio Toda
Okayama, March 2004

OPTICAL RESOLUTIONS BY INCLUSION COMPLEXATION WITH A CHIRAL HOST COMPOUND

Fumio Toda
Department of Chemistry, Faculty of Science, Okayama University of Science
Ridai-cho 1-1, Okayama 700-0005, Japan

1.   Introduction

When a chiral host compound includes one enantiomer of racemic guest compound selectively, optical resolution of the guest can be accomplished. In this chapter, efficient resolutions of racemic compounds by the complexation with various artificial chiral hosts are described. All the data described in this chapter are those obtained in the author's research group.

In most cases, chiral alcohol and phenol derivatives are used as host compounds for the resolution. In these cases, guest molecules are accommodated in the complex by formation of hydrogen bond with the hydroxyl group of the host. Since the hydrogen bond is not very strong, the included guest compound can be recovered easily from the inclusion complex by distillation, recrystallization, chromatography or some other simple procedures.

This chapter consists mainly of two sections, 1) preparation of artificial chiral host compounds and 2) optical resolution of various racemic guest compounds by inclusion complexation with these hosts.

Some other resolutions by inclusion complexation with achiral host and by distillation technique are also described. In the last section, progress of the resolution of binaphthol, biphenols and related compounds is described.

Although mechanism of the precise chiral recognition between host and guest molecules in their inclusion crystal has been studied in detail by X-ray structural analysis, these X-ray structures are not shown in this chapter, since this chapter deals with practical procedures of optical resolutions.

2.   Preparation of Artificial Chiral Host Compounds

In 1986, we have found that 1,1,6,6-tetraphenylhexa-2,4-diyne-1,6-diol (**1**) includes various guest molecules in a stoichiometrical ratio and forms crystalline inclusion complexes.[1] X-ray analysis of a 1:2 inclusion complex of **1** and acetone showed that the guest molecules are accommodated in inclusion crystalline cavity by the formation of hydrogen bond with the hydroxyl groups of **1**.[2] It was also found that inclusion complexation with **1** occurs selectively, and a mixture of isomers can be separated by the selective inclusion process.[3] This suggests that racemic guest compound   can   be   separated   into enantiomers   by   inclusion

1

F. Toda (ed.), Enantiomer Separation, 1-47.
© 2004 *Kluwer Academic Publishers, Printed in the Netherlands.*

complexation with a chiral derivative of **1**. According to this idea, (*S*,*S*)-(-)-1,6-di(*o*-chlorophenyl)-1,6-diphenyl- hexa-2,4-diyne-1,6-diol (**3**) was prepared.[4] Since the optically active host **3** was found to be effective for optical resolution of various guest compounds,[4] some other chiral alcohol and phenol host compounds were also prepared. Some chiral amide hosts were also designed for optical resolution of phenol derivatives. In the section 4, miscellaneous resolutions by using some other interesting chiral host compounds are also described.

All yields of the chiral hosts obtained by the resolution are calculated based on the amount of enantiomers containing in the racemic host compounds used for the resolution.

2-1   Acetylenic Alcohols

Firstly, we found that *rac*-1-(*o*-chlorophenyl)-1-phenylprop-2-yn-1-ol (**2**) can easily be resolved by formation of a 1:1 inclusion complex with brucine (**4**)[5] or sparteine (**5**).[5,6] For example, when a solution of *rac*-**2** (16.0 g, 66 mmol) and **5** (15.4 g, 66 mmol) in acetone (50 ml) was kept at room temperature for 12 h, a 1:1 complex of (*S*)-(-)-**2** and **5** was formed (16.0 g). Pure complex obtained after two recrystallizations of the crude complex from acetone was decomposed by dil HCl to give (*S*)-(-)-**2** of 100% ee (4.2 g, 52%).[6] Inclusion complexation of *rac*-**2** with **4** also gave optically pure (*S*)-(-)-**2** in good yield. Mechanism of the precise chiral recognition of **2** with **4** in their inclusion complexes has been clarified by X-ray structural study. In both cases, the inclusion complexes are constructed by formation of hydrogen bond between the hydroxyl group of **2** and the amine nitrogen atom of the alkaloid.[5,6] Partially resolved optically impure **2** can easily be purified to pure enantiomers by complexation with achiral amine such as DABCO, *N*,*N*'-dialkylpiperazine, *N*,*N*,*N*',*N*'-tetramethyl- ethylenediamine and pyrazine.[6c,6d] Oxidative coupling reaction of (*S*)-(-)-**2** gave (*S*,*S*)-(-)-**3**.[4] Since **2** itself is also useful as a host compound for optical resolution of guests, its chiral derivatives were also prepared by the enantioselective complexation procedure with **4** or **5**. Mutual resolution between **2** and alkaloid can also be accomplished. For example, complexation of racemic sparteine (**5**) with (*S*)-(-)-**2** gave (-)-**5** of 100% ee in 38% yield.[6]

2-2. Tartaric Acid Derivatives

By using cheap chiral source, tartaric acid, some useful chiral hosts were designed.
(R,R)-(+)-*trans*-4,5-bis(hydroxydiphenylmethyl)-2,3-dimethyl-1,3-dioxacyclopentane (**8a**) was prepared by the reaction with PhMgBr of the acetal (**7**) derived from diethyl tartarate (**6**) and acetone.[7] By the same method, **9** and **10** were also prepared.[8]

Some amide derivatives have been reported to form inclusion complex with a wide variety of organic compounds.[9] Optically active amide derivatives are expected to include one enantiomer of a racemic guest selectively. According to this idea, some amide derivatives of tartaric acid (**11-13**) were designed as chiral hosts.[10] As will be described in the following section, these amide hosts were found to be useful for resolution of binaphthol (BNO) (**14**) and related compounds (**15, 16**).

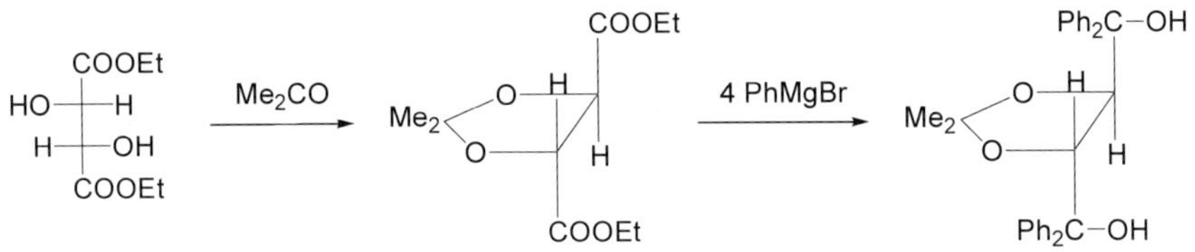

6   **7**   **8**

a: *rac*-form
b: (R, R)-(-)-form
c: (S, S)-(+)-form

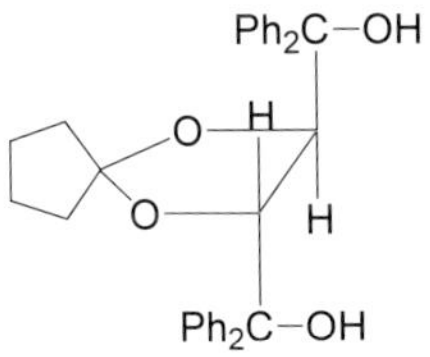

9 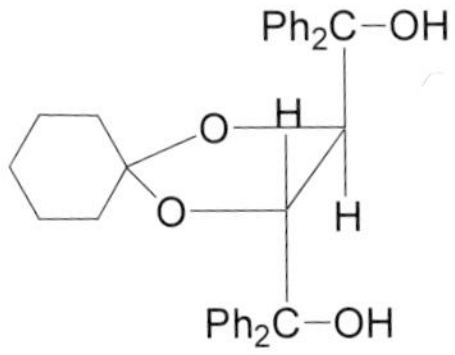   **10**

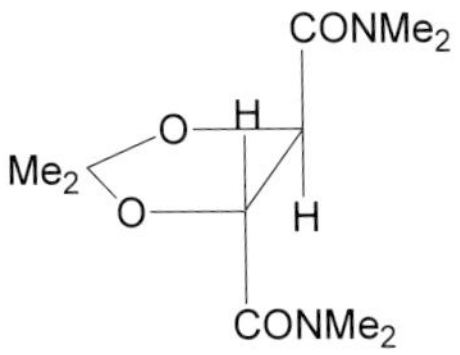

11 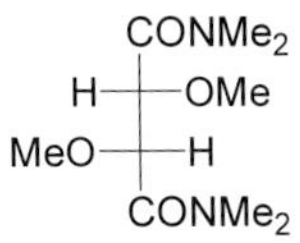   **12** 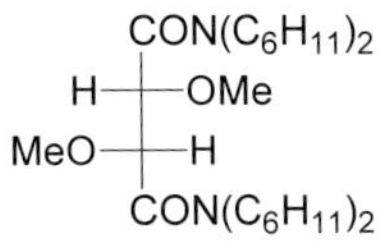   **13**

a: (R, R)-(-)-form
b: (S, S)-(+)-form

2-3.   Binaphthols and Related Compounds

It has been known that phenol derivatives work as a good host compound for various organic guest molecules, since the acidic hydroxyl groups of phenols form a relatively strong hydrogen bond with various organic functional groups.[10] These data strongly suggest that chiral phenol derivatives such as 2,2'-dihydroxy-1,1'-binaphthyl (14) (binaphthol, BNO) and its derivatives (15, 16) work as a good host for optical resolution.

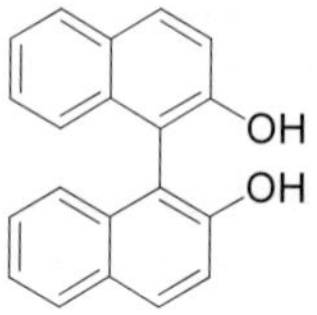

14 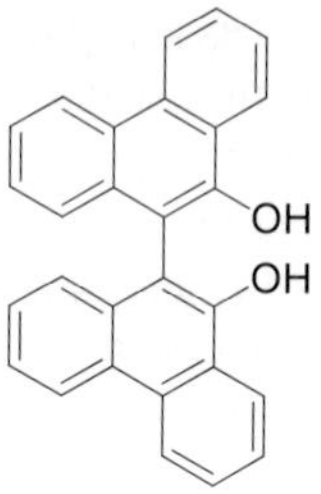   **15** 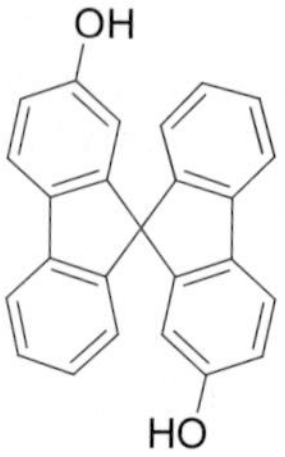   **16**

a: *rac*-form
b: (*R*)-(+)-form
c: (*S*)-(-)-form

Firstly, **14a** was resolved by inclusion complexation with **12a**. For example, when a solution of **12a** (4.06 g, 17.5 mmol) and **14a** (5.0 g, 17.5 mmol) in benzene (5 ml)-hexane (5 ml) was kept at room temperature for 12 h, a 1:1 complex of **12a** and **14b** was formed, which upon recrystallization from benzene gave a pure complex as colorless prisms of mp 149-150 °C (3.70 g, 82%). Column chromatography of the complex on silica gel (benzene) gave **14b** of 100% ee (1.8 g, 72%). From the filtrate left after separation of the 1:1 complex of **12a** and **14b**, **14 c** of 100% ee (1.48 g, 59%) was isolated by inclusion complexation with **12b**.[10]

By similar complexation process of **15a** with **11a** and **11b**, **15c** of 100% ee (74%) and **15b** of 100% ee (80%) were obtained in the yields indicated. Similar treatment of **16a** with **13a** gave **16b** of 100% ee in 90% yield.[10]

Some other optical resolution procedures of *rac*-BNO (**14a**) by complexation with various chiral ammonium salts are summarized in the section 6 of this chapter as an example of the progress on novel enantiomer separation technique.

3.   Optical Resolutions

3-1. General Procedures

A solution of chiral host and racemic guest compounds in an appropriate solvent is kept at room temperature until inclusion complex crystallizes out. In the complexation, it is necessary to use a solvent which does not form inclusion complex with the host compound. The inclusion complex formed is filtered and purified by recrystallization from solvent, if necessary. Host:guest molar ratio is usually a 1:1 or 1:2. The ratio was determined by elemental analysis, NMR spectrum or TG measurement. The purified inclusion complex can be dissociated into the components by an appropriate procedure such as distillation, recrystallization, chromatography, and extraction with base or acid. In case of the inclusion complex of phenol derivative with ammonium salt, it is easily dissociated into the components by dissolving in a mixture of organic solvent and water. The phenol derivative and ammonium salt dissolve in organic solvent and water, respectively. Optical purity of the enantiomer obtained was determined by HPLC on a chiral stationary phase, ^{1}H NMR method using a chiral shift reagent or by comparison of the $[\alpha]_D$ value with that of an authentic sample. All optical purities are shown by the enantiomeric excess (ee) value. All yields of the enantiomers obtained by the resolution are calculated based on the amount of enantiomers containing in the racemic compounds used for the resolution.

Reason for the effective optical resolution by the inclusion complexation with a chiral host has been clarified by X-ray analysis of the complex formed. By the X-ray structural study of the host-guest complex, absolute configuration of the chiral guest resolved has also been elucidated easily, since absolute configuration of the chiral host is known. These X-ray data have been reported in the literature cited together with the detailed experimental procedure of the resolution.

In the case of volatile racemic guest, optical resolution can be carried out by using distillation technique in the presence of a non-volatile chiral host compound. The resolution by distillation is summarized in the section of 5. In the section 5, optical resolution by inclusion crystallization in a suspension medium in hexane or water is also described.

3-2. Hydrocarbons and Halogeno Compounds

Optical resolution of some hydrocarbonds and halogeno compounds by inclusion complexation with the chiral host (**9a**) has been accomplished.[11,12] Preparation of optically active hydrocarbons is not easy and only a few example of the preparation of optically active hydrocarbons have been reported. For example, optically active 3-phenylcyclohexene has been derived from tartaric acid through eight synthetic steps.[11] Although one-step synthesis of optically active 3-methylcyclohexene from 2-cyclo- hexanol by the Grignard reaction using chiral nickel complex as a catalyst has been reported, the enantiomeric purity of the product is low, 15.9%.[11] In this section, much more fruitful results by our inclusion method are shown.

When a solution of *rac*-3-methylcyclohexene (**17a**) (0.58 g, 6.1 mmol) and **9a** (3 g, 6.1 mmol) in ether (15 ml) was kept at room temperature for 12 h, a 2:1 inclusion complex of **9a** and (-)-**17a** (2.5 g, 75%) was obtained as colorless prisms. The crystals were purified by recrystallization from ether to give the inclusion complex (2.4 g, 71%), which upon heating *in vacuo* afforded (-)-**17a** of 75% ee by distillation (0.19 g, 66%). Inclusion complexation with (-)-**17a**, OH absoptions of **9a** (3590 and 3400 cm^{-1}) were shifted to lower frequencies (3400 and 3230 cm^{-1}). Since cyclohexane does not form an inclusion complex with **9a**, hydrogen bonding between π-orbital of **17a** and the OH group of **9a** would be important for the inclusion complex formation. Dissociation energies of the 2:1 complex of **9a** and **17a** were determined to be 45 kJ mol^{-1}. This data show that the stabilization energy of the complex is quite high.

17	**18**	**19**	**20**
a: R = 3-Me		a: R = H	
b: R = 4-Me		b: R = Cl	
c: R = 4-CH=CH$_2$			

By the same inclusion complexation followed by two recrystallizations of the complex formed, 4-methyl- (**17b**) (33% ee, 55%), 4-vinylcyclohexene (**17c**) (28% ee, 73%), bicyclo[4.3]nonane-2,5-diene (**18**) (ee value was not determined, 90%), and 3-chloro- (**19a**, 56% ee, 48%) and 3,4-dichloro-1-butene (**19b**) (ee value was not determined, 42%) were also resolved in the ee values and yields indicated.[11]

Optical resolution of *trans*-1,2-Dichlorocyclohexane (**20**) was also accomplished by complexation with **9a**. When a solution of **9a** (50 g, 0.1 mol) in *rac*-**20** (50 g, 0.33 mol) was kept at room temperature for 12 h, a 2:1 inclusion complex of **9a** and (-)-**20** was obtained as colorless prisms (45.8 g, 80% based on **9a**, mp 108-109 °C). Heating of the complex at 200 C/20 mmHg gave (-)-**20** of 43% ee by distillation (6 g, 77% based on **9a**). The same treatment of the (-)-**20** of 43% ee (6 g) with **9a** (5 g) followed by distillation gave (-)-**20** of 72% ee (0.7 g, 90% based on **9a**). When the (-)-**20** of 72% ee (0.6 g) was treated again with **9a** (0.7 g) as described above, (-)-**20** of 90% ee was obtained as colorless oil (0.06 g, 55%).[12] X-ray analysis of the complex of **9a** with (-)-**20** of 90% ee showed that (-)-**20** molecules exist as diequatorial form in the complex. The absolute configuration of the (-)-**20** was also found to be (*R,R*) by the X-ray study.[12]

3-3.   Amines, Amine *N*-Oxides, Oximes, and Amino Acid Esters

Nitrogen atoms of organic compounds form relatively strong hydrogen bond with OH group of a host compound and amines, amine *N*-oxides, oximes, and esters of amino acids can be resolved efficiently by complexation with a chiral host. Optical resolutions of these compounds are described.

When a solution of **2** (243 g, 1 mol) and *rac*-2-methylpiperazine (**21a**, 100g, 1 mol) in BuOH (50 ml) was kept at room temperature for 12 h, a 2:1 inclusion complex of **2** and (*S*)-(+)-2-methylpiperazine (**21c**) was obtained as colorless prisms, which upon three recrystallizations from BuOH gave pure complex crystals (60 g, 20%). Heating of the crystals *in vacuo* gave **21c** of 100% ee by distillation (9.5 g, 19%).[13] Optical resolution of **21a** can also be accomplished by complexation with the host **3**.

When a solution **3** (242 g, 0.5 mol) and **21a** (100 g ,1 mmol) in MeOH (500 ml) was kept at room temperature for 12 h, a 1:1 complex of **3** and (*R*)-(-)-2-methylpiperazine (**21b**) was obtained as colorless prisms, which upon three recrystallizations from MeOH gave pure crystals (75 g, 26%, mp 86-88 °C). Heating of the crystals *in vacuo* gave **21b** of 100% ee by distillation (12.5 g, 25%). The host compounds left after the distillation can be used again for resolution. Treatments of the filtrate left after the former and the latter resolution experiments with **3** and **2**, respectively, gave optically pure **21b** and **21c**, respectively in the yield around 20%.[13]

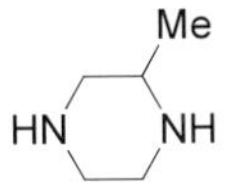

X-ray analysis of the 2:1 complex of **2** and **21c** showed that the two **2** molecules are binding to one **21c** molecule by the formation of two OH---N hydrogen bonds. The data also showed that the combination of **2** of the (*S*)-configuration and **21c** of the (*S*)-configuration is important. This agrees with the fact that **2** does not form complex with **21b** of the (*R*)-configuration.[13]

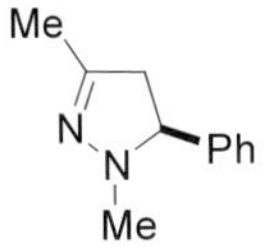

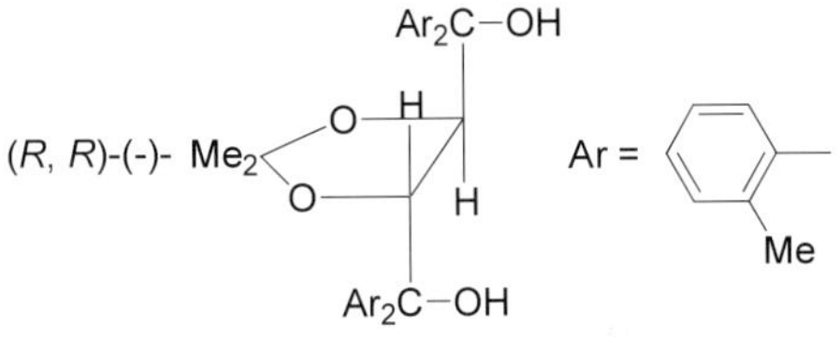

For an optical resolution of 1,3-dimethyl-5-phenyl-∇^2-pyrazoline (**22**), tetra-(*o*-tolyl) derivative of **8a**, (*R*,*R*)-(-)-*trans*-4,4-bis[hydroxydi(*o*-tolyl)methyl]-2,2-di-methyl-1,3-dioxacyclopentane (**23**) was prepared.[7] When a solution of **23** (1.5 g, 2.87 mmol) and *rac*-**22** (1.0 g, 5.75 mmol) in toluene-hexane (1:4, 25 ml) was kept at room temperature for 12 h, a 1:1 inclusion complex of **23** and (*S*)-(-)-**22** was obtained as colorless prisms (0.90 g, mp 128-130 °C), which upon heating *in vacuo* (200 °C/2 mmHg) gave (*S*)-(-)- **22** of 96% ee (0.21 g, 42%).[14] X-ray crystal structure of the complex has also been reported.[14]

The Gabriel synthesis is a classical but useful preparative method for primary amines. Reaction of an alkyl bromide (**24**) with potassium phthalimide (**25**) gives the corresponding N-alkylphthalimide (**26**), which upon treatment with hydrazine followed by KOH affords the primary amine (**27**). When a chiral alkyl halide is used in the Gabriel synthesis, a chiral primary amine is obtained. However, preparation of optically active alkyl halides is not easy. If optical resolution of **26** which has a chiral alkyl group can be done, a new preparative method for optically active amines can be established by a combination of the resolution with the Gabriel synthetic method. Some examples of the combination method are described.

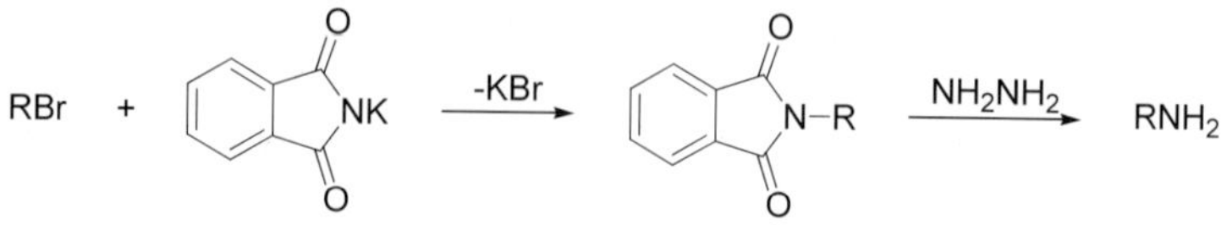

a: R = PhMeCH-
b: R = Me$_2$CHCH$_2$MeCH-
c: R = EtMeCH-
d: R = EtMeCHCH$_2$-

When a solution of **3** and two molar equivalents of *rac*-**26a** in ether-light petroleum was kept at room temperature for 12 h, a crystalline 1:1 inclusion complex of **3** and (+)-**26a** was obtained. Two recrystallizations of the crude crystals from ether-light petroleum gave pure crystals which upon distillation *in vacuo* gave (+)-**26a** of 55% ee. Decomposition of (+)-**26a** with hydrazine gave (+)-**27a** of 55% ee in 40% yield. By the same procedure, (-)-**27b** of 30% ee, (+)-**27c** of 30% ee, and (-)-**27d** (ee was not determined) were obtained.[15]

Reaction of *rac*-1-*tert*-butyl-3-chloroazetidin-2-one (**28**) with **25** gave the *rac*-phthalimide derivative (**29**). Optical resolution of *rac*-**29** was accomplished efficiently by complexation with **15**. When a solution of **15b** and two molar equivalents of *rac*-**29** in benzene-hexane (1:1) was kept at room temperature for 12 h, a crystalline 1:1 inclusion complex of **15b** and (-)-**29** was obtained. After one recrystallization from benzene-hexane, the crystals were chromatographed on silica gel to give pure complex consisting of (-)-**29** of 100% ee in 63% yield. Decomposition of the complex with hydrazine gave optically pure (-)-3-amino-1-*tert*-butylazetidin-2-one (**30**) in 44% yield.[15] Mechanism of the precise chiral recognition between **15b** and (-)-**29** in their 1:1 complex was clarified by X-ray crystal structural analysis.[15]

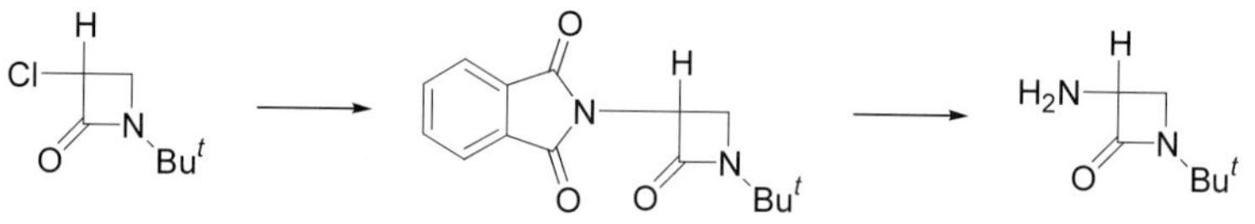

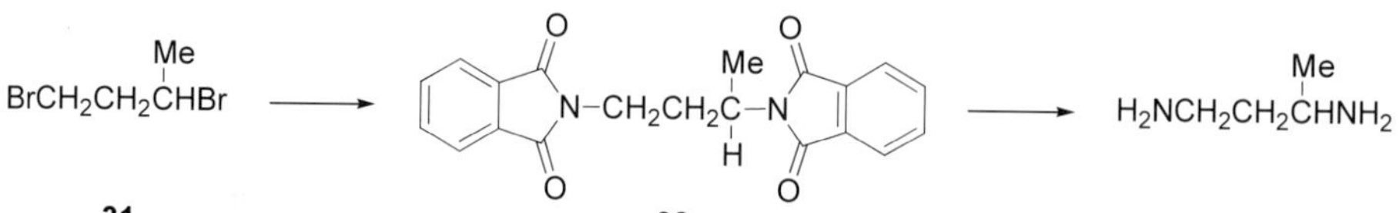

This method can be applied for preparation of an optically active diamine. For example, reaction of *rac*-1,3-dibromobutane (**31**) with **25** gave the *rac*-diphthalimide (**32**). Optical resolution of *rac*-**32** was accomplished efficiently by complexation with **3** to give optically pure **32**. Decomposition of the optically pure **32** gave optically pure (-)-butane-1,3-diamine (**33**) in 50% yield.[15]

Amine *N*-oxides (**34a-e**) were resolved very efficiently by complexation with **14b**. In this case, both enantiomers of **34** were obtained in an optically pure form.[16]
For example, when a solution of **14b** (1.0 g, 3.6 mmol) and *rac*-**34b** (1.2 g, 7.2 mmol) in THF (20 ml)-hexane (10 ml) was kept at room temperature for 5 h, a 1:1 complex of **14b** and (+)-**34b** was obtained as colorless prisms. The crystals were recrystallized from THF-hexane to give pure crystals (0.85 g, 53%, mp 167-169 °C). The complex was separated to its components by column chromatography on silica gel. Firstly, **14a** (0.5 g) was recovered from a fraction eluted by ethyl acetate-benzene (1:4). Secondly, (+)-**34b** of 100% ee (0.29 g, 48%) was obtained from a fraction eluted by MeOH. Evaporation of the filtrate left after separation of the complex between **14b** and (+)-**34b**, gave crude (-)-**34b**. Treatment of the crude (-)-**34b** with **14c** by a similar manner to that described above, followed by column chromatography, yielded finally (-)-**34b** of 100% ee in 40% yield.[16] Compounds **34a** and **34c-e** were also resolved effectively by complexation with **14b**, and the corresponding (+)-enantiomers were obtained in the optical and chemical yields indicated, (+)-**34a** (100% ee, 21%), (+)-**34c** (73% ee, 39%), (+)-**34d** (100% ee, 30%), and (+)-**34e** (100% ee, 68%).[16]

a: R = H; R' = Et
b: R = *m*-Me; R' = Et
c: R = *m*-Me; R' = ^{i}Pr
d: R = *p*-Me; R' = Et
e: R = *p*-Me; R' = ^{i}Pr

34

By X-ray structural study of the complexes of **14b** with (-)-**34b** and of **14c** with (+)-**34b**, mechanism of the chiral recognition has been clarified.[16] Optical purities of all enantiomers obtained by the resolution were determined by ^{1}H NMR measurements in the presence of the new chiral shift reagent **3**.[17]

Optical resolution of two oximes of cyclohexanone derivatives, 4-methyl-1-(hydroxyimino)cyclohexane (**35**) and *cis*-3,5-dimethyl-1-(hydroxyimino)cyclohex- ane (**36**) has been accomplished by an enantioselective complexation with **3**.[18] For example, when a solution of *rac*-**35** and **3** in ether-petroleum ether was kept at room temperature, a 1:1 complex of **3** and (+)-**35** was obtained as colorless needles. Treatment of the complex with alkylamine gave an alkylamine complex of **3** and (+)-**35**. Since the optical purity of the (+)-**35** was not determined directly, its O-benzoyl derivative was prepared and its optical purity was determined to be 79% ee by HPLC method. Therefore, the optical purity of the (+)-**35** obtained by the resolution can be estimated to be higher than 79% ee.[18] By the same treatment of *rac*-**36** with **3**, (+)-**36** of approximately 59% ee was obtained.[18] In order to determined the optical purity of (+)-**35** in its complex with **3** more precisely, Beckmann rearrangement of the oxime in the inclusion compound was carried out. Heating of the inclusion complex of (+)-**35** and **3** with conc H_2SO_4 gave (-)-5-methylcaprolactam of 89% ee (**37**). Therefore, it is certain that the optical purity of the (+)-**35** enantioselectively included in the complex with **3** is higher than 89% ee.[18]

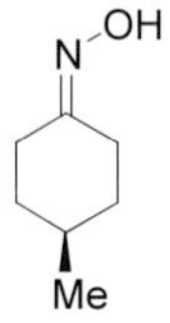

35 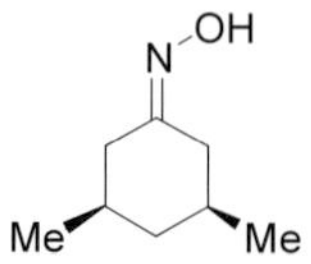   36 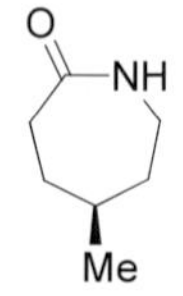   37

Amino acids, especially artificial ones, are interesting targets for optical resolution. Since amino acids themselves are difficultly included with host compounds due to their ionic character, their ester derivatives were resolved.

For example, when a solution of **10a** (2.5 g, 4.9 mmol) and *rac-N*-ethyl-ethoxycarbonylaziridine (**38a**) (1.4 g, 9.8 mmol) in benzene (20 ml)-hexane (20 ml) was kept at room temperature for 5 h, a 1:1 inclusion compound of **10a** and (-)-**38a** was formed as colorless needles (1.9 g, 59%, mp 127-131 °C), which upon distillation in *vacuo* gave (-)-**38a** of 100% ee (0.24 g, 34%).[19] By the same method, **38b-i** were also resolved efficiently (Table 1). Of these, **38b, 38d** and **38i** were resolved more efficiently with the host **9a**. Although the optical purities of the resolved **38c, 38g, 38h**, and **38i** were not determined, these were assumed to be optically pure because theit $[\alpha]_D$ values did not change by repeating the complexation with **10a**. The (+)-**38b** of 64% ee which had been obtained by one complexation gave the optically pure enantiomer by repeating the complexation with **10a**.[19]

Table 1. Optically active **38** obtained by one complex with **9a** or **10a**.

38	Host	Yield (%)	% ee
a	10a	34	100
b	10a	32	---[a]
c	9a	43	64
d	9a	44	100
e	10a	28	100
f	10a	33	100
g	10a	42	---[a]
h	10a	74	---[a]
i	9a	30	---[a]

[a] Purity was not determined.

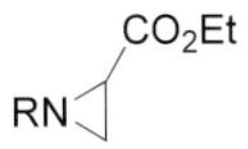

38a: R = Et
38b: R = ^{n}Pr

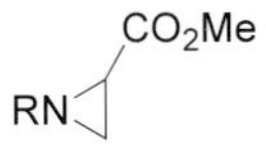

38c: R = Et
38d: R = ^{n}Pr

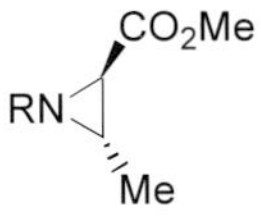

38e: R = ^{n}Pr
38f: R = ^{i}Pr

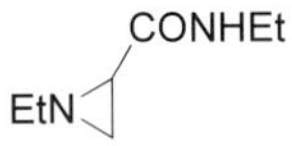

38g

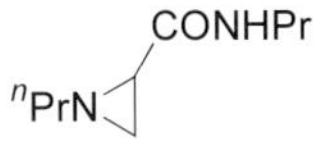

38h

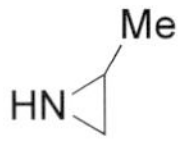

38i

Some real amino acid esters and related compounds were resolved by complexation with **9a** or **10a**. When a solution of **10a** (10.87 g, 21.5 mmol) and rac-methyl 2-aminopropanoate (**39**) (4.43 g, 43.0 mmol) in benzene (9 ml)-light petroleum (9 ml) was kept at room temperature for 4 h, a 2:1 complex of **10a** and (+)-**39** was obtained, after three recrystallizations (1.51 g, 13%, mp 195-197 °C). Heating the complex in vacuo gave (+)-**39** of 100% ee by distillation (0.12 g, 12%).[20] Although rac-methyl 2-aminophenylpropanoate (**40**) was also resolved efficiently by complex- ation with **9a** to give (+)-**40** of 100% ee in 40% yield, optical resolution of rac-ethyl 2-amino-2-phenylethanoate (**41**) gave very poor result (Table 2). Similar resolutions of hydroxycarboxylic acid esters such as methyl 3-hydroxybutanoate (**42**) and its derivatives (**43-45**) were successful (Table 2).[20] In order to know the reason for the efficient chiral recognition ability between the host and guest molecules, X-ray crystal structure of the inclusion complex of **9a** and (+)-**42** was studied.[20]

Table 2. Optical resolution by complexation with **9a** or **10a**.

Guest	Host	Enantiomers obtained		
			Yield (%)	Optical purity (% ee)
39	**10a**	(+)-**39**	12	100
40	**9a**	(+)-**40**	40	100
41	**10a**	(-)-**41**	65	6
42	**9a**	(+)-**42**	44	100
43	**9a**	(+)-**43**	28	100
44	**9a**	(+)-**44**	15	64
45	**10a**	(+)-**45**	45	40

Resolutions of diethyl 2-imidazol-1-ylsuccinates (**46a-e**) by complexation with **8-10** were also successful.[21]

MeCH(NH$_2$)COOMe   PhCH$_2$CH(NH$_2$)COOMe   PhCH(NH$_2$)COOEt   MeCH(OH)CH$_2$COOMe

39   **40**   **41**   **42**

MeCH(OH)CH$_2$COOEt   ClCH$_2$CH(OH)CH$_2$CH$_2$COOMe   MeOOCCH$_2$CH(OH)COOMe

43   **44**   **45**

a: R =   b: R =   c: R =

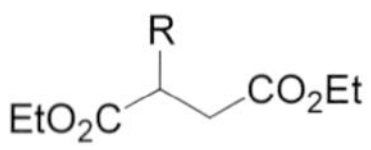

46

d: R =   e: R =

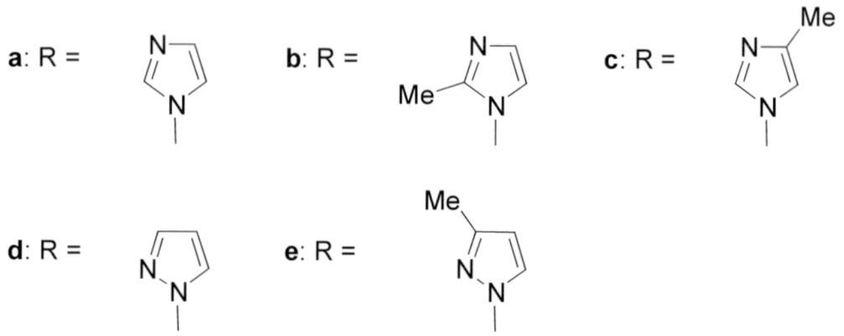

3-4.   Alcohols and Cyanohydrins

Diarylcarbinols (**47**) were easily resolved by complexation with brucine (**4**). For example, when a solution of **4** (1.8 g, 4.56 mmol) and *rac-m*-chlorophenylphenyl- carbinol (**47c**) (1.0 g, 4.57 mmol) in MeOH-hexane (9:1, 2 ml) was kept at room temperature for 2 h, a 1:1 complex of **4** and (+)-**47c** was obtained (1.21 g). Two recrystallizations of the crude crystals from MeOH-hexane (9:1) gave pure crystals (0.8 g, mp 136-137 °C), which upon chromatography on silica gel (AcOEt) gave (+)-**47c** of 99.2% ee in 56.0% yield. Evaporation of the solvent from the filtrate left after separation of the crude inclusion crystals of **4** and (+)-**47c**, followed by chromatography on silica gel and distillation, gave (-)-**47c** of 72.4% ee in 98% yield.[22] The optical purities, and yields of optically active **47a-i** obtained by the same resolution method applied to **47c** are summarized in Table 3.[22]

Table 3. Optical purity and yield of the enantiomer obtained
by complexation with **4**.

47		Enantiomer	
	X	% ee	Yield(%)
a	*m*-Me	92.1	61.3
b	*p*-Me	92.6	2.6
c	*m*-Cl	99.2	56.0
d	*p*-Cl	97.0	30.4
e	*m*-Br	98.0	44.0
f	*p*-Br	100.0	20.0
g	*m*-OMe	93.0	32.0
h	*m*-NO$_2$	99.6	47.8
i	*p*-NO$_2$	85.5	72.0

The secondary alcohols which are substituted with one aryl and one sterically bulky alkyl group, such as **48** and **49**, were also resolved efficiently by the complexation with **4**. For example, when a solution of **4** (29.7 g, 75.4 mmol) and *rac*-**48** (12.3 g, 75.4 mmol) in MeOH (20 ml) was kept at room temperature for 12 h, a 1:1 complex of **4** and (-)-**48** was obtained, after three recrystallizations from MeOH, as colorless prisms (4.20 g, mp 112-115 °C), which upon heating *in vacuo* gave (-)-**48** of 100% ee by distillation (1.22 g, 19.8%). Similarly, *rac*-**49** was resolved to give finally (-)-**49** of 100% ee in 38% yield.[22]

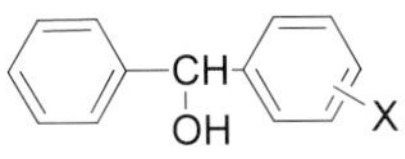

| 47 | 48 | 49 |

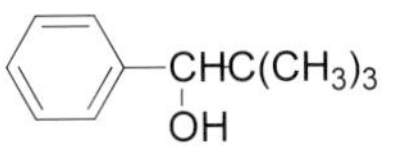

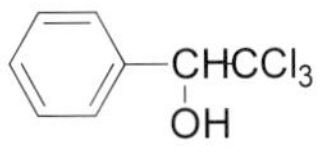

The secondary alcohols which are substituted with one aryl and one sterically less bulky alkyl group were resolved efficiently by complexation with **8a** or **9a**. For example, when a solution of **8a** (10 g, 21.5 mmol) and *rac*-1-phenylethanol (**50**) (2.62 g, 21.5 mmol) in 1:1 toluene-hexane (20 ml) was kept at room temperature for 12 h, a 2:1 complex of **8a** and (-)-**50** was obtained as colorless prisms (9.69 g, 85.7%), which upon heating *in vacuo* gave (-)-**50** of 75.1% ee by distillation (0.95 g, 72.5%). Pure inclusion crystal (7.75 g, 68.4%) obtained by one recrystallization of the crude crystal (9.69 g) was heated *in vacuo* to give (-)-**50** of 98.6% ee (0.75 g, 57.3%).[23] By the same procedure, the secondary alcohols (**51-56**) were also resolved efficiently, although the resolution of **51, 53, 54,** and **56** was accomplished more efficiently by complexation with **9a** instead of **8a** (Table 4).[23]

Table 4. Optical resolution of alcohols (**50 - 56**) by complexation with the host **8a** or **9a**.

| Alcohols | Hosts | | | |
| | 8a | | 9a | |
	Yield (%)	ee (%)	Yield (%)	ee (%)
50	72.5	75.1		
51			79.1	63.6
52	39.7	91.4		
53			79.4	78.8
54			45.3	78.4
55	69.5	97.0		
56			65.0	92.1

PhCH(OH)CH$_3$ H$_3$C—⟨benzene⟩—CH(OH)C$_2$H$_5$ PhCH(OH)C$_2$H$_5$

50   **51**   **52**

OH
53

OH
54

PhCH(OH)C≡CH
55

⟨ring⟩—CH(OH)C≡CH
PhO
56

Me O⟩—⟨—OH
Me O
57a

⟨cyclopentane spiro⟩ O—⟨—OH
O
57b

⟨cyclohexane spiro⟩ O—⟨—OH
O
57c

OH
⟨chain⟩—OH
58

The most interesting application of the resolution method by the complexation with **9** or **10** was accomplished for glycerol acetals, **57a-c**. When a solution of **9a** (3.55 g, 7.24 mmol) and *rac*-**57a** (1.45 g, 10.98 mmol) in 1:1 ether-hexane (20 ml) was kept at room temperature for 12 h, a 2:1 complex of **9a** and (+)-**57a** was obtained as colorless needles (3.57 g, 75%). One recrystallization of the crude crystal gave pure one (2.42 g, 51%), and its heating *in vacuo* gave (+)-**57a** of 100% ee (0.14 g, 30%). Resolutions of **57b** and **57c** by complexation with **9a** and **10a**, respectively, gave finally (+)-**57b** of 100% ee (80%) and (+)-**57c** of 100% ee (20%) in the yields indicated, respectively.[23]

In order to clarify mechanism of the precise chiral recognition between aliphatic secondary alcohols and **9a** or **10a** in their inclusion crystal, X-ray structure of a 1:1 inclusion crystal of **9a** and (+)-1,3-butanediol (**58**) was investigated. Finally, it was found that hydrogen bond between the OH group of **9a** and that on the chiral carbon of **58** plays an important role to arrange both molecules at close positions to be able to recognize the chirality each other.[24]

By combination of two processes of chemical reaction and optical resolution, a novel one-pot preparation method of optically active *sec*-alcohols becomes available. When these processes are accomplished in a water suspension medium, these are really green and sustainable procedure. Some these examples are described.

When a mixture of acetophenone **59a** (1.0 g, 8.3 mmol), NaBH$_4$ (0.94 g, 24.9 mmol), and water (10 ml) was stirred at room temperature for 2 h, *rac*-**60a** was produced. To the water suspension medium of *rac*-**60a** was added powdered **8a** (3.87 g, 8.3 mmol), and the mixture was stirred for 3 h to give a 2:1 inclusion complex of **8a** with (-)-**60a**.[25] Inclusion complex formed was filtered and dried. Heating of the complex *in vacuo* gave (-)-**60a** of 95% ee (0.42 g, 85%). From the filtrate left after separation of the inclusion crystals, (+)-**60a** of 77% ee (0.35 g, 70%) was obtained by extraction with ether. By the same procedure, optically active **60a** and **60c-g** were prepared (Table 5). Solid state and solvent-free organic reactions have been well established[25,26a-c]. Host-guest inclusion complexation in the solid state has also been reported.[26c]

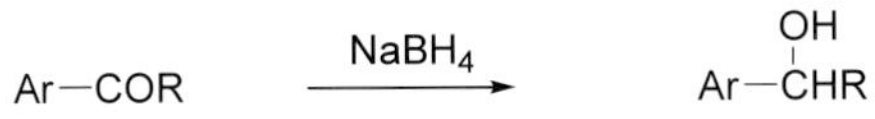

$$\text{Ar-COR} \xrightarrow{\text{NaBH}_4} \text{Ar-CHR}$$
$$\quad\quad\quad\quad\quad\quad\quad\quad\quad\quad\overset{\text{OH}}{|}$$

59 → **60**

a: Ar = Ph; R = Me
b: Ar = Ph; R = Me
c: Ar = 2-pyridyl; R = Me
d: Ar = 3-pyridyl; R = Me
e: Ar = 4-pyridyl; R = Me
f: Ar = 2-furyl; R = Me
g: Ar = 2-thiophenyl; R = Me

Table 5. Result of one-pot preparation method of optically active *sec*-alcohols (**60a-g**) by a combination of reduction of ketone and enantiomeric resolution in a water suspension medium.

Ketone	Host	From complex			From filtrate		
			Yield (%)	% ee		Yield (%)	% ee
60a	8a	(-)-**60a**	85	95	(+)-**60a**	70	77
60b	8a	(-)-**60b**	96	62	(+)-**60b**	50	52
60c	8a	(-)-**60c**	26	76	(+)-**60c**	156	18
60c	9a	(-)-**60c**	44	99	(+)-**60c**	134	40
60c	10a	(-)-**60c**	92	88	(+)-**60c**	76	62
60d	8a	(+)-**60d**	88	>99	(-)-**60d**	86	73
60d	9a	(+)-**60d**	86	96	(-)-**60d**	82	66
60e	9a	(+)-**60e**	80	77	(-)-**60e**	82	36
60f	8a	(-)-**60f**	76	93	(+)-**60f**	96	50
60g	8a	(-)-**60g**	84	86	(+)-**60g**	61	43

The one-pot method is also applicable to the preparation of optically active epoxides and sulfoxides.[25]

In the optical resolution of cyanohydrins, it was first found that brucine (**4**) is a suitable host for the cyanohydrins which substituted with one aromatic group and one bulky alkyl group. In this case, not only a simple enantiomer separation of *rac*-cyanohydrin but also its transformation to one enantiomer occurred and one pure enantiomer was obtained in a yield of more than 100%. For example, when a solution of *rac*-1-cyano-2,2-dimethyl-1-phenylpropanol (**61a**) (1.0 g, 5.3 mmol) and **4** (2.1 g, 5.3 mmol) in MeOH (2 ml) was kept in a capped flask for 12 h, a 1:1 brucine complex of (+)-**61a** (2.08 g, 134%, mp 112-114 °C) separated out as colorless prisms. Decomposition of the complex with dil HCl gave (+)-**61a** of 97% ee (0.67 g, 134%). From the filtrate, *rac*-**61a** (0.33 g, 33%) was obtained.[27a] The yield of (+)-**61a** of more than 100% shows a transformation of (-)-**61a** to (+)-**61a** through racemization during the complexation due to the base (**4**)-catalyzed equilibrium shown in (eq. 1). The yield of (+)-**61a** increased up to 200% by leaving the MeOH solution to evaporate gradually during the complexation. For example, when a solution of *rac*-**61a** (1.0 g) and **4** (2.1 g) in MeOH (2 ml) was kept in an uncapped flask at room temperature for 1, 3, 6, 12, and 24 h, amount of solvent decreased to 1,9, 1.8, 1.6, 1.3, and 0.6 ml, respectively, and the (+)-**61a** was obtained in the optical and chemical yields indicated after the complexation for 1 (55% ee, 40%), 3 (80% ee, 80%), 6 (95% ee, 110%), 12 (96, 160%), and 24 h (97% ee, 200%).[26] When the inclusion complexation of the (+)-**61a** of 97% ee with **4** is repeated once again, 100% optically pure enantiomer was obtained.[27a]

a: R^1 = H; R^2 = *t*Bu

b: R^1 = *p*-Cl; R^2 = *t*Bu

c: R^1 = *p*-Me; R^2 = *t*Bu

d: R^1 = *p*-OH; R^2 = *t*Bu

e: R^1 = *m*-OH; R^2 = *t*Bu

f: R^1 = H; R^2 = CCl$_3$

g: R^1 = *m*-PhO; R^2 = H

h: R^1 = H; R^2 = *n*Bu

i: R^1 = H; R^2 = *i*Pr

j: R^1 = H; R^2 = *n*Pr

k: R^1 = H; R^2 = Et

l: R^1 = H; R^2 = Me

m: R^1 = H; R^2 = H

a: R^1 = *t*Bu; R^2 = Me

b: R^1 = *i*Pr; R^2 = Me

c: R^1 = Et; R^2 = Me

d: R^1 = *s*Bu; R^2 = H

e: R^1 = *i*Pr; R^2 = H

f: R^1 = ClCH$_2$; R^2 = H

g: R^1 = Me; R^2 = H

By the same procedure, optically pure (+)-**61b** (177%), (-)-**61c** (94%), (-)-**61d** (132%), (-)-**61e** (110%), and (+)-**61f** (84%) were obtained in the yields indicated.[27a]

Very interestingly, all cyanohydrins (**61h-m**) which are substituted with one phenyl group and one less bulky alkyl group or hydrogen atom do not form inclusion complex with **4**. However, the cyanohydrins which are substituted with two alkyl groups (**62a-c**) or with one alkyl group and one hydrogen atom (**62d-f**) formed complex with **4** and were resolved.[27b]

The chiral hosts **8a** and **9a** were found to be useful for the resolution of cyano-hydrins which cannot be resolved with **4**. For examples, **61g** and **61m** were resolved by complexation with **9a** and **8a**, respectively, to give (-)-**61g** of 72.5% ee (70%) and (+)-**61m** of 100% ee (47.6%), respectively, in the yield indicated. The most simple chiral cyanohydrin derived from acetaldehyde (**62g**) was resolved by complexation with **8a** and optically pure (+)-**62g** was obtained in 52.6% yield.[23]

3-5   Epoxides and Oxaziridines

Optically pure 2,3-epoxycyclohexanones (**63-65**) were first prepared by resolu- tion of their *rac*-derivatives by complexation with **3**,[28] although optically active 2,3-epoxycyclohexanone of 20% ee has been prepared in 1980 by an enantioselective epoxidation of cyclohexen-2-one.[28] When a solution of **3** (5.1 g, 10.6 mmol) and *rac*-**64** (5.94 g, 42.4 mmol) in 1:1 ether-light petroleum (20 ml) was kept at room temperature for 6 h, a 1:1 complex of **3** and (-)-**64** was obtained as colorless prisms, which upon Kugelrohr distillation *in vacuo* gave (-)-**64** of 90% ee. Two recrystallizations of the 1:1 complex of **3** and (-)-**64** of 90% ee (4.68 g) from ether-light petroleum (50 ml) gave pure complex (2.74 g, 34%, mp 117-118 °C), which upon Kugelrohr distillation *in vacuo* gave (-)-**64** of 100 ee (0.9 g, 30%).[28] By the same procedure, (+)-**63** of 100% ee (18%) and (+)-**65** of 100% ee (35%) were obtained in the yields indicated.[28]

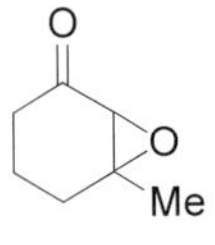

63 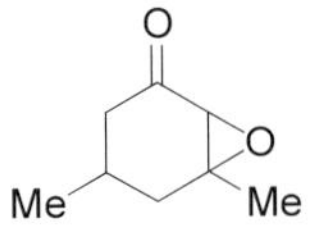   **64** 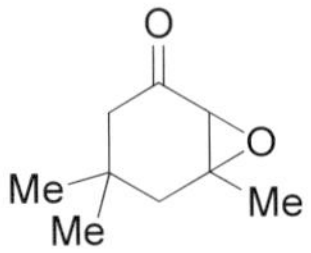   **65**

Preparation of optically active β-ionone epoxide by a solid state kinetic resolution in the presence of the chiral host **10a** is also possible. When a mixture of **10a**, β-ionone (**66**) and *m*-chloroperbenzoic acid (MCPBA) is ground by mortar and pestle in the solid state, (+)-**67** of 88% ee was obtained.[29] Mechanism of the kinetic resolution is shown below. Of course, all processes proceed in the solid state. Firstly, oxidation of **66** with MCPBA gives *rac*-β-ionone epoxide (**67**). Secondly, enantioselective inclusion of (+)-**67** with **10a** occurs. Thirdly, uncomplexed (-)-**67** is oxidized to give the Baeyer- Villiger oxidation product (-)-**68** of 72% ee. This is the first example of the resolution by an enantioselective inclusion complexation in the solid state.

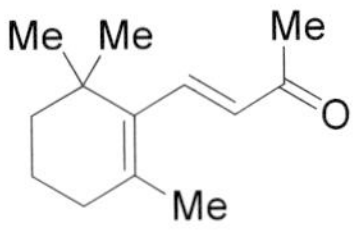

66 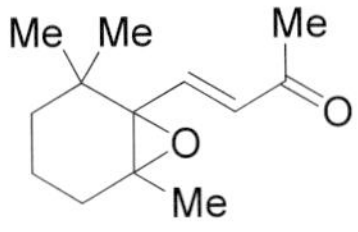   **67** 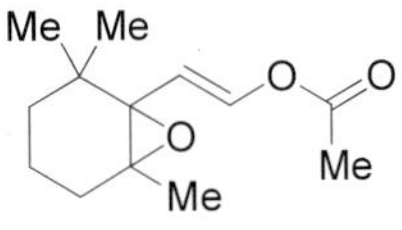   **68**

Oxaziridines (**69a-d**) were also resolved efficiently by complexation with **10a**, and (+)-**69a** (99% ee, 69%), (+)-**69b** (98% ee, 68%), (+)-**69c** (90% ee, 56%), and (+)-**69d** (90% ee, 59%) were obtained in the optical and chemical yields indicated.[30] Resolution of aziridine derivatives is described in the section 3-3.

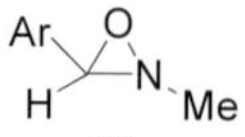

a: Ar = C_6H_5
b: Ar = p-MeC_6H_4
c: Ar = p-ClC$_6$H$_4$

69

d. Ar = [pyridyl structure]

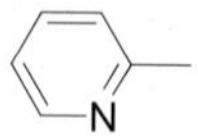

3-6   Ketones, Esters, Lactones, and Lactams

Since optically active carbonyl compounds are important as chiral synthons of various fine chemicals, drugs, and bioactive materials, optical resolutions of carbonyl derivatives are very important and interesting problem. In 1983, we have reported the resolution of methylcycloalkanones with **3** as the first successful example of the resolution by complexation with a chiral host compound.[4] For example, when a solution of **3** (19.2 g, 39.8 mmol) and *rac*-3-methylcyclhexanone (**70**) (17.8 g, 159 mmol) in 1:1 ether-petroleum ether (100 ml) was kept at room temperature for 6 h, a 1:2 complex of **3** and (+)-**70** was obtained, after two recrystallizations, as colorless prisms (11.6 g, 41%), which upon distillation *in vacuo* gave (+)-**70** of 66% ee (3.5 g, 39.3%).[4] By repeating the complexation of **3** and the (+)-**70** of 66% ee, pure complex was obtained (4.1 g, 15%), which upon heating *in vacuo* gave (+)-**70** of 100% ee (1.16 g, 13%).[32] By the same procedure, *rac*-3-methylcylopentanone (**71**) and 5-methyl-γ-butyrolactone (**72**) were resolved to give finally optically pure (-)-**71** (6%) and (-)-**72** (5%) in the yields indicated.[4]

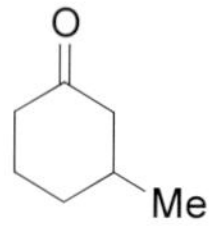

70 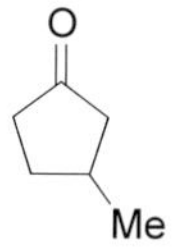   **71** 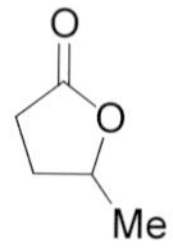   **72**

73

a: R = H
b: R = COCH$_3$
c: R = COC$_2$H$_5$
d: R = COC$_3$H$_7$

e: R = [tetrahydropyranyl structure]

(*S, S*)-(-) Ph–C≡R≡C–Ph with Cl, OH substituents

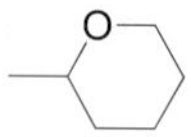

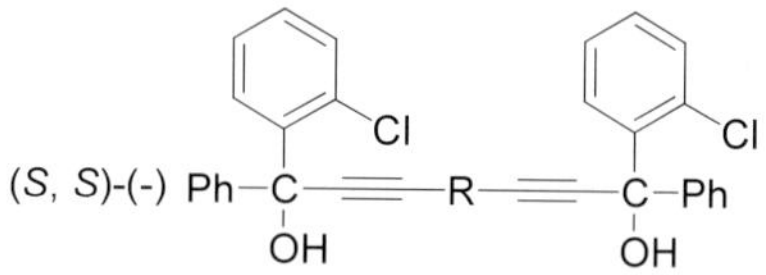

74

a: R = [p-phenylene structure]

b: R = [m-phenylene structure]

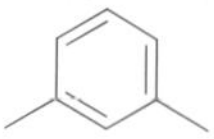

Although the optically active 4-hydroxycyclo-2-pentenone (**73a**) is a very important starting material of synthesis of prostaglandins, it is not easy to obtain optically pure **73a** efficiently. The biphenanthrol host **15** was found to be a good host for resolution of ester derivatives (**73b-d**) and tetrahydropyranyl ether (**73e**) of **73a**. For example, when a solution of **15b** (0.8 g, 2.07 mmol) and *rac*-**73b** (0.58 g, 4.14 mmol) in MeOH (5 ml) was kept at room temperature for 12 h, a 1:1 complex of **15b** and (-)-**73b** (0.88 g) was obtained as colorless prisms. Recrystallization of the complex from MeOH gave pure complex (0.51 g, 47%, mp 154-156 °C), which upon heating *in vacuo* gave (-)-**73b** of 100% ee (0.13 g, 45%).[32] From the filtrate left after the separation of a crude 1:1 complex of **15b** and (-)-**73b**, (+)-**73b** of 100% ee (0.16 g, 55%) was obtained by treatment with **15c**. By the same method, optically pure (-)-**73c** (28%), (-)-**73d** (51%) and (-)-**73e** (51%) were obtained in the yields indicated.[32] Unfortunately, **63a** was not efficiently resolved with **15**, although it forms a 1:1 complex with **15**. Furthermore, it is difficult to hydrolyze **73b-d** into **73a** without racemization. However, since **73e** is easily hydrolyzed without racemization under mild conditions, resolution of **73e** has an advantage. Hydorlysis of optically pure (-)-**73e** with dil HCl gave optically pure (-)-**73a** in almost quantitative yield.[32]

As described above, direct resolution of **73a** is difficult. However, it was found that simple benzene derivatives of **3**, (*S,S*)-(-)-1,4-bis[3-(*o*-chlorophenyl)-3-hydroxy- 3-phenyl-1-propynyl]benzene (**74a**) and (*S,S*)-(-)-1,3-bis[3-(*o*-chlorophenyl)-3-hydro- xy-3-phenyl-1-propynyl]benzene (**74b**) are useful for the resolution of **73a**. **74a** and **74b** were prepared by a coupling reaction of **2** with *p*- and *m*-dibromobenzene, respectively.[32] When a solution of **74a** (5.0 g, 8.94 mmol) and *rac*-**73a** (1.75 g, 17.9 mmol) in EtOH (10 ml) was kept at room temperature for 12 h, a 1:1:1 inclusion crystal of **74a**, (-)-**73a**, and EtOH was obtained as colorless prisms, after one recrystallization from EtOH, 3.63 g (57.7%, mp 70-75 °C). Heating of the crystal *in vacuo* gave (-)-**73a** of 100% ee (0.48 g, 54.9%).[34] When the inclusion complexation was carried out in toluene, a 1:2 inclusion crystal of **74a** and (+)-**73a** was obtained. After three recrystallizations from toluene, the inclusion crystal was heated *in vacuo* to give (+)-**73a** of 76.6% ee in 9.2% yield.[34] By the same resolution method by the complexation with **74a**, **75** was resolved to give optically pure (+)-**75** in 47.3% yield.[34] Bicyclic ketones **76** and **77** which are also important materials for prostaglandin synthesis were resolved by inclusion complexation with **74b** to give optically pure (-)-**76** and (+)-**77** in 25.9 and 29% yields, respectively.[34]

Although the chiral host **74** is useful for resolution of cyclic ketones **73a**, **75**, **76**, and **77**, it is not useful for the resolution of lactones. However, the host **3** is useful for resolution of lactones **78**, **79**, **80**, **81**, and **82** which are important key compounds for synthesis of various physiologically active materials including prostaglandins. For example, when a solution of **3** (10.0 g, 20.7 mmol) and *rac*-**78** (10.3 g, 82.8 mmol) in 2:1 ether-light petroleum (45 ml) was kept at room temperature for 6 h, a 1:2 complex of **3** and (+)-**78** (15.0 g, 99%) was obtained, which upon distillation gave (+)-**78** of 8.5% ee (5.0 g, 97%). Ten recrystallizations of the 1:2 complex of **3** and (+)-**78** of 8.5% ee (15 g) from ether (40 ml) gave pure complex as colorless needles (1.67 g, 11%, mp 118-120 °C), which upon distillation *in vacuo* gave (+)-**78** of 100% ee (0.52 g, 10%).[34] Optical resolution of **79**, **80**, **81**, **82**, and **83b** by the complexation with **3** gave optically pure (+)-**79** (10%), (+)-**80** (13%), (+)-**81** (32%), (-)-**82** (20%) and (+)-**83b** (24%) in the yields indicated.[34]

75   **76**   **77**   **78**

79   **80**   **81**   **82**   **83**

a: R = H
b: R = Me
c: R = Et

84   **85**   **86**

For resolution of some bicyclic ketones such as **83a-c**, **84**, **85**, and **86**, the chiral host **8** is very effective. For example, when a solution of **8a** (2.33 g, 5 mmol) and *rac*-**86** (1.79 g, 10 mmol) in benzene-hexane (4:1) (25 ml) was kept at room temperature for 12 h, a 1:1 complex of **8a** and (-)-**86** was obtained, after two recrystallizations from benzene, as colorless prisms (2.30 g, 71%, mp 186-189 °C), which upon heating *in vacuo* gave (-)-**86** of 100% ee (0.63 g, 70%). The filtrate left after separation of the 1:1 complex of **8a** and (-)-**86** was evaporated to give crude (+)-**86** of 79% ee (0.9 g). When the crude (+)-**86** (0.9 g) was treated with **8b** (2.36 g) as above, (+)-**86** of 100% ee was obtained (0.77 g, 86%).[7] The same treatments of **83a-c**, **84**, and **85** with **8a** gave optically pure (-)-**83a** (41%), (-)-**83b** (62%), (-)-**83c** (43%), (+)-**84** (80%), and (-)-**85** (58%), respectively, in the yields indicated.[7] Mechanism of these precise chiral recognitions between **8** and bicyclic ketones have been clarified by X-ray study of two inclusion complexes of **8a** with (-)-**83b** and with (-)-**86**.[35]

In the optical resolution of bicyclo[2.2.1]heptanones (**87a**, **88-90**), bicyclo[2.2.2]- octanones (**91-94**) and bicyclo[3.2.1]octanone (**95**) by complexation with various chiral host compounds, some best host-guest combinations were found. Resolutions of **88**, **89**, **90**, and **92** were accomplished efficiently by complexation with **3** to give (+)-**88** (100% ee, 33%), (-)-**89** (100% ee, 16%), (+)-**90** (100% ee, 60%), and (-)-**92** (100% ee, 41%), respectively, in the optical and chemical yields indicated.[36] However, resolutions of **93** and **95** were accomplished efficiently by complexation with **8a** to give optically pure (-)-**93** and (-)-**95** in 56 and 48% yields, respectively. On the other hand, resolution of **94** can be accomplished only by complexation with **15c** to give finally (-)-**94** of 100% ee in 31% yield.[36] Mechanism of these chiral recognition in the inclusion complex crystal has been studied by X-ray analysis.[36] Nevertheless, none of **3**, **8a** and **15c** is applicable to the resolution of **87a** and **91**. Although resolution of **87a** by complexation with brucine (**4**) gave (1*R*,4*R*)-(+)-**87a** of 27% ee in 40% yield,[37] all attempts of the resolution of **91** were failed. The big difference between **91** and **92** in the efficiency of the resolution is very interesting. Interestingly, however, an amide derivative of **87a**, 2-azabicyclo-[2.2.1]- hept-3-one (**87b**) was resolved by complexation with **4** and (1*R*,5*S*)-(-)-**87b** of 92% ee was obtained in 13% yield.[37] X–Ray crystal structure of the complex of **4** with (1*R*,5*S*)-(-)-**87b** was studied.[37]

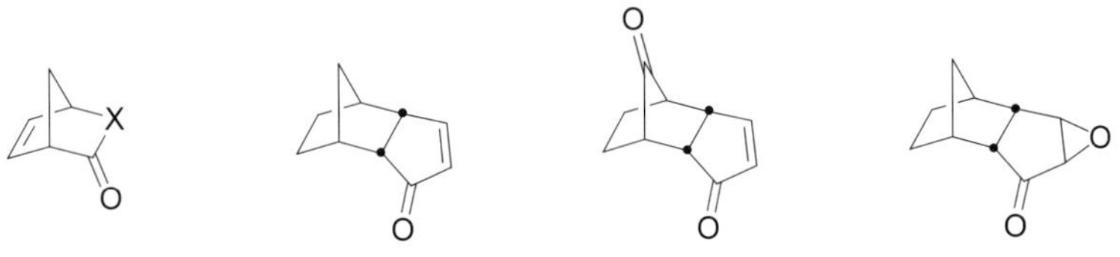

87

a: X = CH$_2$

b: X = NH

88

89

90

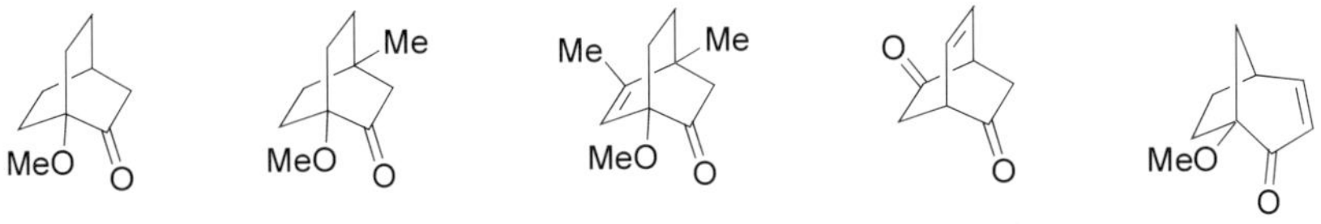

91

92

93

94

95

Some simple lactone derivatives were easily resolved by complexation with the chiral hosts (**8-10**) derived from tartaric acid. For example, when a solution of **10a** (2.0 g, 3.95 mmol) and *rac*-5-ethoxyfuran-2(5*H*)-one (**96b**) (1.01 g, 7.89 mmol) in 1:1 toluene-benzene (20 ml) was kept at room temperature for 24 h, a 1:1 inclusion complex was obtained as colorless needles (1.74 g) which upon heating *in vacuo* (220 °C/20 mmHg) gave (+)-**96b** of 88% ee (0.16 g, 32%).[38] The host-guest inclusion complexation can also be carried out in a water suspension medium. When a suspension of finely powdered **10a** (1.5 g, 1.98 mmol) and *rac*-**96b** (0.51 g, 3.98 mmol) in water (10 ml) containing hexadecyltrimethylammonium bromide (0.05 g) as a surfactant was stirred at room temperature for 6 h, a 1:1 inclusion complex of **10a** and (+)-**96b** was formed as crystals, which upon heating *in vacuo* (220 °C/20 mmHg) gave (+)-**96b** of 98% ee (0.06 g, 24%). When the recrystallization method is used for a complexation of **8a** and *rac*-**96b**, (-)-**96b** of 100% ee was obtained in 10% yield.[38] Finally, **96a**, **96c-e**, and **97** were efficiently resolved by complexation with **8**, **9**, or **10** in an organic solvent or in a water suspension medium to give (-)-**96a** (92% ee, 41%), (-)-**96c** (94% ee, 25%), (+)-**96d** (90% ee, 46%), (+)-**96e** (92% ee, 68%), and (-)-**97** (96% ee, 50%), respectively, in the optical and chemical yields indicated.[38]

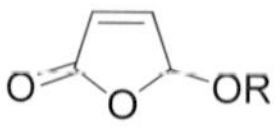

$$\text{96}$$

a: R = Me
b: R = Et
c: R = nPr
d: R = iPr
e: R = C$_6$H$_{11}$

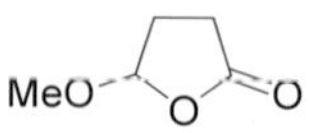

$$\text{97}$$

Optical resolutions of bicyclic acid anhydride (**98**), lactones (**99, 100**), and carboximides (**101, 102**) had also been investigated. For example, when a solution of **8a** (5.0 g, 21 mmol) and *rac-cis*-3-methylcyclohex-4-ene-1,2-dicarboxylic acid anhydride (**98**) (3.6 g, 42 mmol) in 1:1 ether-hexane (60 ml) was kept at room temperature for 12 h, a 1:1 inclusion complex of **8a** and (1R,2R)-(+)-**98** was obtained, after two recrystallizations, as colorless prisms (3.6 g, 53%, mp 118-121 C), which upon heating *in vacuo* (170-200 C/5 mmHg) gave (1R,2R)-(+)-**98** of 100% ee by distillation (0.93 g, 52%).[39] By the same procedures, **99-102** were resolved efficiently (Table 6).[39]

Table 6. Optical resolution of **98-102** by complexation with **8-10**.

Guest	Host	Product	
		Optical purity (% ee)	Yield (%)
98	8a	(+)-98 100	52
99a	9a	(+)-99a 43	59
99b	9a	(+)-99b 100	56
100a	10a	(-)-100a 67	63
100b	10a	(+)-100b 100	28
101a	8a	(+)-101a 33	90
101a	9a	(-)-101a 100	16
101b	10a	(+)-101b 100	50
101c	10a	(+)-101c 100	38
102a	10a	(+)-102a 100	30
102b	8a	(+)-102b 100	26
102b	9a	(+)-102b 100	63
102c	9a	(+)-102c 98	19
102d	9a	(+)-102d 98	40

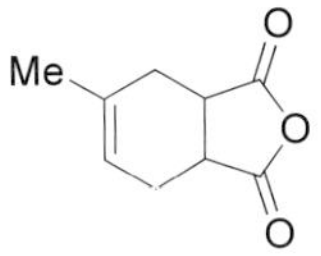

98 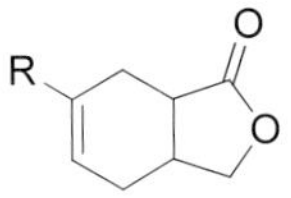   **99** 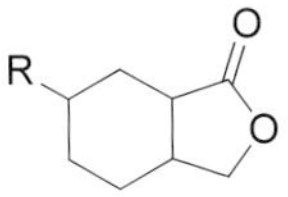   **100**

a: R = H
b: R = Me

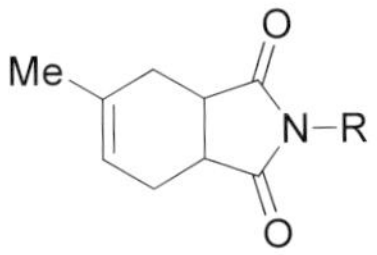

101   **102**

a: R = Me   a: R = Me
b: R = Et   b: R = Et
c: R = Ph   c: R = nPh
                 d: R = iPr

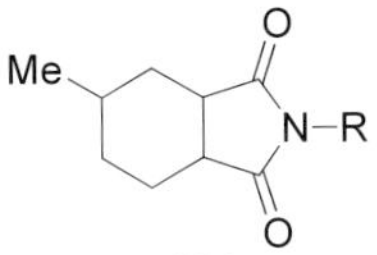

Pantolactone, dihydro-3-hydroxy-4*H*-dimethyl-2(3*H*)-furanone (**103**) which is an important starting material of the synthesis of pantothenic acid, was also easily resolved by complexation with **10a**. When a solution of **10a** (5.5 g, 9.93 mmol) and *rac*-**103** (2.6 g, 20 mmol) in 1:1 benzene-hexane (20 ml) was kept at room temperature for 1 h, a 1:1 complex of **10a** and (*S*)-(-)-**103** was obtained, after two recrystallizations from 1:1 benzene-hexane, as colorless needles (2.05 g), which upon heating *in vacuo* gave (*S*)-(-)-**103** of 99% ee (0.39 g, 30%).[40] In order to clarify the mechanism of the precise chiral recognition between **10a** and (*S*)-(-)-**103**, their inclusion complex crystal was studied by X-ray analysis[40] and by AFM technique.[41]

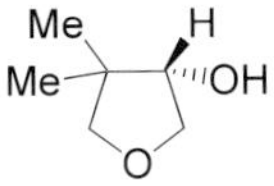

103

Monoterpenes have been used as chiral building blocks for the total synthesis of various biologically active natural products. However, some naturally occurring monoterpenes such as α- and β-terpene exist as a mixture of both enantiomers with variable enantiomeric excess. Therefore, it is difficult to obtain optically pure monoterpenes and their derivatives from natural sources. Optically impure terpenes such as verbenone (**104b**) and apoverbenone (**104a**) were purified by inclusion complexation with **23** or **8a**. For example, when a solution of **23** (22 g, 42 mmol) and (1*S*,5*S*)-(-)-verbenone of 78% ee (**104b**) (7.5 g, 50 mmol) in hexane (125 ml) was kept at room temperature for 12 h, a 1:1 inclusion complex of **23** and (1*S*,5*S*)-(-)-**104b** was obtained, after three recrystallizations from hexane, as colorless prisms (13.5 g, 48%, mp 111-112 C). The complex (9.1 g) was treated with MeOH (10 ml) to give MeOH complex crystal of **23**. From the filtrate left after separation of the MeOH complex of **23**, (1*S*,5*S*)-(-)-**104b** of 99% ee (2.01 g) was obtained.[42] By similar treatment of **104a** of 91% ee with **8a** as above, (1*R*,5*R*)-(+)-**104a** of 98% ee was obtained.[42]

104

a: R = H
b: R = Me

Optically active glycidic esters (**105**) are useful synthon for various biologically active substances. Chiral hosts derived from tartaric acid were found to be effective for the resolution of **105**. When a solution of **9a** (4.29 g, 8.71 mmol) and *rac*-ethyl 2,2-diethylglycidate (**105g**) (1.5 g, 8.71 mmol) in ether (7.5 ml)-hexane (2 ml) was kept at room temperature for 12 h, a 2:1 inclusion complex of **9a** and (+)-**105g** as colorless prisms (3.18 g, 62%), which upon heating at 170 °C/2 mmHg gave (+)-**105g** of 100% ee (0.47 g, 62%). From the ether-hexane solution left after separation of the 2:1 inclusion crystal of **9a** and (+)-**105g**, (-)-**105g** of 45% ee was isolated (1.03 g, 137%).[43] By the same procedure, **105a-f** and **105h-j** were resolved by complexation with **9a** or **10a** (Table 7). Although efficiencies for the resolution of **105b** and **105c** are not good, those of **105e-h** are excellent. In the latter case, inclusion complexes initially obtained by complexation experiment are not necessary to purify further by recrystallization. Efficiencies for the resolution of **105d, 105i** and **105j** are moderate (Table 7).[43] Inclusion complexation can also be carried out by a suspension method in hexane or water. For example, when a suspension of powdered **9a** (1.72 g, 3.48 mmol) and oily **105g** (0.6 g, 3.48 mmol) in water (4.3 ml) containing hexadecyltrimethylammonium bromide (17.2 mg) as a surfactant was stirred at room temperature for 24 h, a 2:1 inclusion complex of **9a** and (+)-**105g** was obtained as colorless powder (1.83 g, 91%), which upon heating at 170 °C/2 mmHg gave (+)-**105g** of 100% ee (0.27 g, 91%). From the aqueous layer left after the separation of the 2:1 inclusion complex of **9a** and (+)-**105g**, (-)-**105g** of 85% ee was obtained by distillation (0.33 g, 109%). The same inclusion complexation of **10a** with **105f** and with **105g** in a water suspension gave optically pure (+)-**105f** (78%) and (+)-**105g** (95%), respectively, in the yields indicated.[43] Suspension method in hexane is also available.[43]

105

a: $R^1 = R^2 = R^3 = R^4 = Me$
b: $R^1 = R^2 = R^3 = Me; R^4 = Et$
c: $R^1 = R^2 = R^4 = Me; R^3 = Et$
d: $R^1 = R^2 = Me; R^3 = R^4 = Et$
e: $R^1 = R^2 = R^4 = Me; R^3 = H$
f: $R^1 = R^2 = Me; R^3 = H; R^4 = Et$
g: $R^1 = R^2 = R^4 = Et; R^3 = H$
h: $R^1R^2 = (CH_2)_5; R^3 = H; R^4 = Me$
i: $R^1R^2 = (CH_2)_5; R^3 = H; R^4 = Et$
j: $R^1 = R^3 = R^4 = Me; R^2 = H$

Table 7. Resolution of **105a-j** through inclusion complexation with **9a** or **10a**.

HOST	Guest	Product		Yield (%)
			Optical purity (% ee)	
9a	105a	(+)-**105a**	100	9
9a	105b	(+)-**105b**	10	37
10a	105b	(+)-**105b**	18	38
9a	105c	(-)-**105c**	10	72
9a	105d	(+)-**105d**	66	69
10a	105d	(+)-**105d**	47	40
9a	105e	(+)-**105e**	100	63
10a	105f	(+)-**105f**	96	32
9a	105g	(+)-**105g**	100	62
10a	105g	(+)-**105g**	100	23
9a	105h	(-)-**105h**	100	31
10a	105h	(-)-**105h**	100	51
10a	105i	(-)-**105i**	54	63
9a	105j	(-)-**105j**	93	42

For resolution of simple ester derivatives, biphenanthryl host **15** was found to be useful. For example, when a suspension of powdered **15c** (1.0 g, 2.59 mmol) and oily *rac*-ethyl 4-chloro-3-hydroxybutyrate (**111**) (0.79 g, 5.18 mmol) in hexane (2 ml) was kept at room temperature for 12 h, a 1:1 complex of **15c** and (-)-**111** was obtained as colorless prisms (0.81 g, 58%, mp 94-97 °C), which upon heating *in vacuo* gave (-)-**111** of 95.3% ee by distillation (0.23 g, 57%).[32] Evaporation of the filtrate left after the separation of the above complex gave (+)-**111** of 40.2% ee (0.53 g) which upon treatment with **15b** as above resulted in (+)-**111** of 97.8% ee (0.27 g, 68%). When the complexation of (-)-**111** of 95.3% ee with **15c** and of (+)-**111** of 97.8% ee with **15b** is repeated again, (-)-**111** and (+)-**111** of 100% ee were obtained.[32]

By the same one complexation with **15c**, **107-109, 112** and **113** were resolved to give (-)-**107** (69.4% ee, 75%), (-)-**108** (100% ee, 86%), (+)-**109** (92.7% ee, 86%), (+)-**112** (60% ee, 81%), and (-)-**113** (58.0% ee, 85%), respectively, in the optical and chemical yields indicated. Interestingly, however, **106** and **110** did not form inclusion complex with **15**.[32] although *rac*-methyl 2-chloropropionate (**113**) can be resolved by complexation with **15c**.[44] In order to clarify the mechanism of the precise chiral recognition between **15c** and optically active esters, X-ray crystal   structures of the

1:1 complexes of **15c** with (*S*)-(-)-**107** and with (*S*)-(-)-**111** were studied.[44]

The chiral host **15c** was found to be effective for resolution of β-lactams. For example, when a solution of **15c** (1.0 g, 2.59 mmol) and *rac*-β-lactam (**114**) (0.6 g, 5.82 mmol) in benzene (10 ml) was kept at room temperature for 12 h, a 1:1 inclusion complex of **15c** and (+)-**114** was obtained, after one recrystallization from benzene, as colorless prisms (1.02 g, mp 187-190 °C). Chromatography of the complex on silica gel using THF as a solvent gave (+)-**114** of 100% ee (0.23 g, 77%).[45] Similar treatment of *rac*-**115** with **15c** gave optically pure (+)-**105** in 63% yield. X-ray crystal structure of a 1:1 complex of **15c** and (+)-**115** has been studied.[45]

MeCH(OH)COOMe	MeCHClCOOMe	MeCH(OPh)COOMe	MeCH(OH)CH$_2$COOEt
106	**107**	**108**	**109**

MeCHClCH$_2$COOMe	ClCH$_2$CH(OH)CH$_2$COOMe	MeOOCCH(OH)CH$_2$COOMe	MeCH(NH$_2$)COOEt
110	**111**	**112**	**113**

114   **115**

3-7   Sulfoxides, Sulfoximines, Selenoxides, Arsineoxides, Phosphinates and Phosphine Oxides

For optical resolution of sulfoxides, binaphthol host **14** was found to be very effective. For example, when a solution of **14b** (3.0 g, 10.5 mmol) and *rac*-methyl *m*-methylphenyll sulfoxide (**116c**) (3.23 g, 21.0 mmol) in benzene-hexane (1:1, 20 ml) was kept at room temperature for 12 h, a 1:1 complex of **14b** and (+)-**116c** was obtained, after one recrystallization from benzene, as colorless prisms (3.57 g, 77%, mp 152-154 °C), which upon chromatography on silica gel gave (+)-**116c** of 100% ee (77%) and recovered **14b** (77%) in the yields indicated.[46] The mothor liquor left from the initial complexation experiment was evaporated to dryness and the residue was chromatographed to give (-)-**116c** of 62% ee (1.30 g). When a solution of the (-)-**116c** of 62% ee (1.30 g) and **14c** (2.41 g, 8.44 mmol) in benzene (10 ml) was kept at room temperature for 12 h, a 1:1 complex of **14c** and (-)-**116c** was obtained, after one recrystallization from benzene, as colorless prisms (1.79 g, 39%, mp 152-154 °C), which upon chromatography gave (-)-**116c** of 100% ee (0.62 g, 38%) and recovered **14c** (1.16 g, 48%).[46] By the same procedure, **116d** was also resolved easily by **14** to give (+)- and (-)-**116d** of 100% ee in good yields. However, **116e** was poorly resolved to give approximately 5% ee enantiomer, and **116b** and **116e** did not form inclusion complex with **14**.

116

a: R^1 = H; R^2 = Me

b: R^1 = o-Me; R^2 = Me

c: R^1 = m-Me; R^2 = Me

d: R^1 = m-Me; R^2 = Et

e: R^1 = p-Me; R^2 = Me

117

a: R = nBu

b: R = iBu

c: R = sBu

d: R = nPr

e: R = iPr

f: R = Et

Some dialkyl sulfoxides (**117**) were also resolved by complexation with **14**. n-Butyl methyl sulfoxide (**117a**) and methyl n-propyl sulfoxide (**117d**) were easily resolved with **14** to give optically pure (+)- and (-)-enantiomers of **117a** and **117d**, respectively in good yields. However, resolution of **117b** and **117f** was not effective and approximately 25% ee enantiomers of **117b** and **117f** were obtained by one complexation with **14b** or **14c**. However, **117c** and **117e** did not form complex with **14**.[46]

A reverse resolution of **14a** by complexation with an optically active sulfoxide is also available. For example, when a solution of (-)-**116c** of 100% ee (1.40 g, 15 mmol) and **14a** (2.60 g, 9.10 mmol) in benzene-hexane (1:2, 15 ml) was kept at room temperature for 12 h, a 1:1 complex of (-)-**116c** and **14c** was obtained, after one recrystallization from benzene, as colorless prisms (2.0 g, 100%), which upon chromatography gave **14c** of 100% ee (1.08 g, 83%) and (-)-**116c** of 100% ee (0.52 g, 74%). From mother liquor, **14b** of 100% ee was obtained in 76% yield.[46]

Mechanism of the mutual chiral recognition between **14b** and (+)-**116c** was studied by X-ray analysis of their inclusion crystal. By this study, the absolute configuration of the (+)-**116c** was determined to be the (R).[47]

Optically active vinyl sulfoxide was prepared by a combination of resolution and elimination reaction. Firstly, inclusion complexation of rac-2-chloroethyl m-tolyl- sulfoxide (**118**) and **14b** in benzene gave, after two recrystallizations from benzene, a 1:1 complex of **14b** and (+)-**118** of 100% ee in 72% yield. Secondly, treatment of the complex with 10% NaOH gave optically pure (+)-m-tolyl vinyl sulfoxide (**119**) by HCl elimination as colorless oil. Rapid polymerization of the (+)-**119** proceeded by treatment with BuLi or BuMgBr at -78 °C to give optically active polymer (**120**). Oxidation of **120** with H_2O_2 gave optically active polysulfone (**121**).[48]

Although **14b** is not effective for the resolution of **116a**, the host **10a** is very effective for the resolution of not only the **116a** but also benzyl methyl sulfoxide (**122**) and alkyl phenylsulfinates (**123**). By complexation with **10a**, (*R*)-(+)-**116a** (100% ee, 56%), (-)-**122** (100% ee, 15%), (+)-**123a** (69% ee, 18%), (+)-**123b** (77% ee, 20%), and (-)-**123c** (56% ee, 27%) were obtained in the optical and chemical yields indicated.[49]

122

123

a: R^1 = H; R^2 = Me
b: R^1 = *p*-Me; R^2 = Me
c: R^1 = *p*-Me; R^2 = Et

Optically active sulfoximines are useful synthons for various biological substances and usually prepared by stereoselective imination reaction of optically active sulfoxide. Direct optical resolution of sulfoximine is more useful but no any successful method is reported. For the resolution of some sulfoximines (**124a-g**), chiral host **3** and **14** were found to be very effective.[50] For example, when a solution of **14b** (6.7 g, 23 mmol) and *rac*-methy *m*-tolylsulfoximine (**124d**) (7.9 g, 46 mmol) in benzene (60 ml) was kept at room temperature for 12 h, a 1:1 complex of **14b** and (+)-**124d** of 90% ee was obtained as colorless prisms (8.5 g, 80%). The crude crystals were purified by two recrystallizations from benzene to give the complex of optically pure (+)-**124d** (7.9 g, 74%). The complex was decomposed by treating with 3% NaOH, and water insoluble part was taken up in benzene. From the dried benzene solution, (+)-**124d** of 100% ee was obtained (2.8 g, 70%). The benzene solution left after the separation of the crude 1:1 complex of **14b** and (+)-**124d** was treated with **14c** (3 g) to give finally (-)-**124d** of 100% ee (3.2 g, 81%). In both complexation, optically pure **14b** and **14c** were recovered by acidification of the NaOH solution.[50] By the same method, **124b** and **124e** were resolved with **14b** and gave optically pure (-)-**124b** and (-)-**124e** in 37 and 50% yields, respectively. However, resolution of **124a** was not effective and (-)-**124a** of 35% ee was obtained in 45% yield by repeating five times recrystallization of its complex with **14b** from benzene. **124c, 124f,** and **124g** did not form complex with **14b**. These results show that the efficiency of the resolution is best when the alkyl group is methyl or ethyl and the aryl group is *m*-tolyl. Since this tendency is similar to that in the case of sulfoxide, the efficiency of the resolution of **124** is probably dependent on packing of **14** and **124** molecules in crystalline lattice of their complex as has been reported for the complex of **14b** and (+)-**116c**.[46]

124

a: R^1 = H; R^2 = Me
b: R^1 = H, R^2 = Et
c: R^1 = o-Me; R^2 = Me
d: R^1 = m-Me; R^2 = Me
e: R^1 = m-Me; R^2 = Et
f: R^1 = m-Me; R^2 = nPr
g: R^1 = p-Me; R^2 = Me

125

a: R^1 = Me; R^2 = iC_5H_{11}
b: R^1 = Et, R^2 = nC_6H_{13}

Although **14b** did not form complex with dialkyl sulfoximines (**125**), **3** formed complex with some of them, and some were resolved efficiently by the complexation. For example, when a solution of **3** (3.1 g, 7.5 mmol) and *rac*-methyl *iso*-pentyl- sulfoxyimine (**125a**) (1.0 g, 7.5 mmol) in ether (20 ml) was kept at room temperature for 12 h, a 1:1 complex of **3** and (-)-**125a** of 17% ee was obtained (2.6 g, 126%). The crude complex was purified by four recrystallizations from ether to give the complex of (-)-**125a** of 100% ee (0.92 g, 90%). Column chromatography of the complex on silica gel gave (-)-**125a** of 100% ee (0.4 g, 80%). By the same method, **125b** was resolved to give finally (-)-**125b** of 100% ee in 88% yield.[50]

Despite various attempts, no optically pure selenoxide has been obtained so far, since racemization easily occurs *via* the hydrate (eq. 2). It was found that **3** and **14** can be used for the resolution of selenoxides (**126**). Owing to the racemization of optically active **126** through the equilibrium shown in eq. 2, it might be possible to isolate one enantiomer as an inclusion complex in yields of >100% in the presence of a chiral host. When a solution of **14b** (1.0 mmol) and *rac*-ethyl m-tolylselenoxide (**126f**) (1.0 mmol) in acetone (10 ml)-hexane (10 ml) was kept at room temperature for 4 h, a 1:1 complex of **14b** and (-)-**126f** of almost 100% optically purity (0.31 g, 123%, mp 85 °C) was obtained. By the same procedure, **126a-e** and **126g** were also easily resolved to give inclusion complexes with **14b** of almost optically pure (+)-**126a** (72%), (+)-**126b** (25%), (+)-**126c** (120%), (-)-**126d** (64%), (+)-**126e** (99%), and (-)-**126g** (20%) in the yields indicated.[51] Optical purities of **126a-g** were determined for their complexes with **14b** by the NMR method using the chiral shift reagent, Eu(hfc)$_3$. Optically active free **126** can be isolated from their complex easily by column chromatography on Al$_2$O$_3$. However half lives of the optically active **126c** (6.5 min), **126d** (19.5 min), **126f** (9.5 min), **126g** (8.5 min) are rather short in MeOH at 19 °C as indicated. Half life of optically active **126d** (222 min) is relatively long in CHCl$_3$ at 19 °C.[51]

$$A-\overset{O}{\underset{\cdot\cdot}{\overset{\|}{Se}}}-B \quad \underset{\longleftarrow}{\xrightarrow{H_2O}} \quad A-\overset{OH}{\underset{OH}{\overset{|}{Se}}}-B \quad \underset{\longleftarrow}{\xrightarrow{-H_2O}} \quad A-\overset{\cdot\cdot}{\underset{OH}{\overset{|}{Se}}}-B \quad (eq.\ 2)$$

126

a: R^1 = H; R^2 = Me   e: R^1 = m-Me; R^2 = Me
b: R^1 = H; R^2 = Et   f: R^1 = m-Me; R^2 = Et
c: R^1 = o-Me; R^2 = Me   g: R^1 = p-Me; R^2 = Et
d: R^1 = o-Me; R^2 = Et

127

a: R^1 = $CH_3(CH_2)_3$-; R^2 = Et
b: R^1 = $CH_3(CH_2)_4$-; R^2 = Me
c: R^1 = $CH_3(CH_2)_4$-; R^2 = Et

Some dialkyl selenoxides (**127a-c**) were also resolved efficiently by complexation with **3** or **14b** to give optically almost pure (+)-**127a** (154%), (-)-**127b** (79%), and (-)- **127c** (74%) in the yields indicated.[51]

Optical resolution of selenoxides by complexation is more efficient than that of sulfoxides. Although efficiency of the resolution for o- and p-tolyl-substituted sulfo- xides is not good, the efficiency for selenoxides with the same substituent is good. In order to clarify the mechanism of the efficient chiral recognition, X-ray crystal structure of a 1:1 complex of **14b** and (-)-**126f** was studied.[52]

$$(+)-\ R^1-\overset{O}{\underset{R^3}{\overset{\|}{As}}}\diagdown R^2 \quad \underset{\longleftarrow}{\xrightarrow{H_2O}} \quad R^1-\overset{OH}{\underset{OH}{\overset{|}{As}}}\diagup R^2_{R^3} \quad \underset{\longleftarrow}{\xrightarrow{-H_2O}} \quad (-)-\ R^1-\overset{O}{\underset{R^3}{\overset{\|}{As}}}\diagdown R^2 \quad (eq.\ 3)$$

128

a: R^1 = H; R^2 = iPr
b: R^1 = o-Me; R^2 = iPr
c: R^1 = m-Me; R^2 = Et
d: R^1 = m-Me; R^2 = iPr
e: R^1 = p-Me; R^2 = Et
f: R^1 = p-Me; R^2 = iPr

Resolution of arsine oxides is also very difficult, because optically active arsine oxide rapidly recemizes in the presence of a small amount of water through the equilibrium shown in (eq. 3). By inclusion complexation with **14c**, **128a-f** were resolved quite efficiently to give (+)-**128a** (65%ee 52%), (+)-**128b** (64% ee, 64%), (+)-**128c** (42% ee, 91%), (+)-**128d** (64% ee, 50%), (+)-**128e** (40% ee, 57%), and (+)-**128f** (65% ee, 38%) as 1:1 inclusion complexes with **14c** in the optical and chemical yields indicated. Optical purities of the (+)-**128a-f** in their complexes with **14c** were determined by ^{1}H NMR spectral measurements of the complexes in CDCl$_3$ in the presence of the chiral shift reagent **3**.[53] It has been reported that the chiral hosts **3**,[17] **8a**,[54] and **14b**[17,55] work as a good chiral shift reagent.

Resolution of phosphinates and phosphine oxides is also difficult. These, however, were easily resolved by complexation with **14b** or **14c**. For example, when a solution of **14b** (3.5 g, 12 mmol) and rac-**129c** (4.5 g, 24 mmol) in benzene (50 ml) was kept at room temperature overnight, a 1:1 complex of **14c** and (+)-**129c** was obtained, after two recrystallizations from benzene, as colorless plates (1.95 g, 34%, mp 140-142 °C), which upon column chromatography on silica gel (benzene) gave (+)-**129c** of 100% ee (0.70 g, 31%). The filtrate left after the separation of the crude complex of **14c** and (+)-**129c** was evaporated to dryness, and the residue was chromatographed on silica gel (benzene) to give crude (-)-**129c** (2.7 g). The crude (-)-**129c** and **14b** (4.1 g) were dissolved in benzene (30 ml) and the solution was kept at room temperature overnight to give a 1:1 complex of **14b** and (-)-**129c**, after two recrystallizations from benzene, as colorless plates (3.1 g, 53.9%, mp 140-142 °C), which upon column chromatography on silica gel (benzene) gave finally (-)-**129c** of 100% ee (1.12 g, 50%).[56]

129

a: R = H
b: R = o-Me
c: R = m-Me
d: R = p-Me

130

a: R = H
b: R = m-Me

By the same method, *rac*-**129a-b** and *rac*-**129d** were efficiently resolved with **14c** to give optically pure (+)-**129a** (12%), (+)-**129b** (47%), and (+)-**129d** (32%), respectively in the yields indicated. From the filtrate left after the separation of the complexes of (+)-**129b** and (+)-**129d** with **14c**, optically pure (-)-**129b** (14%) and (-)-**129d** (14%) were obtained in the yields indicated by treatment with **14b**.[56]

By the same method, *rac*-phosphine oxides **130a** and **130b** were also resolved efficiently by complexation with **14c** to give optically pure (-)-**130a** and (-)-**130b** in 60 and 33% yields, respectively. Treatment with **14b** of the filtrate left after separation of the complexes of (-)-enantiomers, optically active (+)-**130a** (30%) and (+)-**130b** (16%) were obtained.[56] X-ray crystal structures of the 1:1 complexes of **14b** with (+)-**129a** and of **14c** with (-)-**129a** were analyzed.[56]

4. Optical Resolution by Inclusion Complexation with an Achiral Guest Compound

Resolution of some hosts can be accomplished by inclusion complexation with an achiral guest compound.

4-1. Resolution of 7-Bromo-1,4,8-triphenyl-2,3-benzo[3.3.0]octa-2,4,7-trien-6-one

The title compound (**131**)[57] has been found to include a wide variety of solvent molecules as guests and form crystalline inclusion crystals.[58] In the inclusion crystallizations, racemates or congromerates of **131** were formed depending on the choice of solvent. In the latter case, the inclusion crystals consisting of one enantiomer of **131** were formed preferentially and the optical resolution of **131** could be performed.

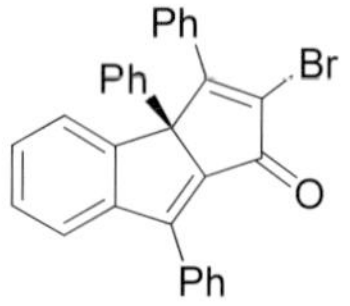

131

Recrystallization of *rac*-**131** from the solvents shown in Table 8 gave a 1:1 inclusion crystal of the *rac*-**131** with the solvent as yellow crystals (Table 8).

Table 8. Formation of 1:1 inclusion crystals of *rac* compound of **131** with solvent (guest).

Solvent (guest)	mp (°C)	Solvent (guest)	mp (°C)
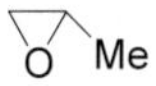 (epoxide) Me	112-115	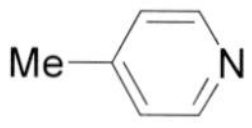 Me—(pyridine)N	100-102
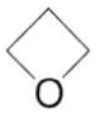 (oxetane)	nd[a]	PhNH$_2$	59-61
 (furan)	nd[a]	 N—N (DABCO)	145-149
 (cyclohexene oxide)	109-112	 Me$_2$N—⟨⟩—NMe$_2$	177-179
CHCl$_3$	101-105	 MeO—⟨⟩—OMe	159-165
CHBr$_3$	nd[a]	PhCl	88-90
CH$_2$Cl$_2$	127-131	 Cl—⟨⟩—Cl	132-138

[a] not distinct.

On the other hand, recrystallization of the *rac*-**131** from the solvents shown in Table 9 gave 1:1 inclusion crystals of the optically active **131** with the solvent as yellow crystals (Table 9). For example, recrystallization of the *rac*-**131** (mp 185-186 °C) from THF gave a 1:1 inclusion complex of the optically active **131** and THF, each single crystal consisting of optically pure (+)- or (-)-**131**. By seeding with one crystal of optically pure **131** during the recrystallization of *rac*-**131**, a large quantity of the optically pure **131** could be obtained.[58] For example, to a solution of the *rac*-**131** (10 g) in THF (50 ml) was added one piece of the crystal of the (+)-**131**-THF complex and the mixture was kept at room temperature for 12 h to give the (+)-**131**-THF complex, after one recrystallization from THF, 0.81 g (100% ee, 14%, mp112-114 °C). Distillation of THF from the complex *in vacuo* gave (+)-**131** of 100% ee (0.7 g, 14%, mp 234-236 °C). To the filtrate left after filtration of the crude (+)-**131**-THF complex, one piece of the crystal of the (-)-**131**-THF complex was added and the solution was kept at room temperature   for 12 h to give the (-)-**131**-THF complex, after two recrystallizations from THF, 1.1 g (100% ee, 19%). Distillation of THF from the complex *in vacuo* gave (-)-**131** of 100% ee (0.96 g, 19%). Optical purity of the (+)- and (-)-**131** was determined by HPLC using a column containing an optically active solid phase.[58]

Table 9. Formation of 1:1 inclusion crystals of *rac* mixture of **131** with solvent (guest).

Solvent (guest)	mp (°C)	Solvent (guest)	mp (°C)
 (tetrahydrofuran)	112-114	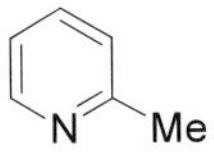 (2-methylpyridine, N–Me)	135-139
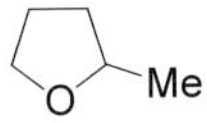 (2-methyltetrahydrofuran, O–Me) **132**	nd[a]	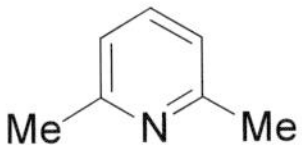 (2,6-dimethylpyridine, Me–N–Me)	98-102
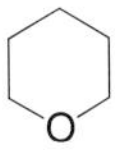 (tetrahydropyran)	120-125	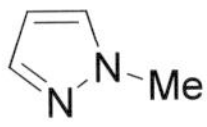 (1-methylpyrazole, N–N–Me)	119-123
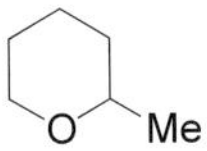 (2-methyltetrahydropyran, O–Me) **133**	nd[a]	MeI	nd[a]
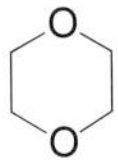 (1,4-dioxane)	122-128	EtBr	nd[a]
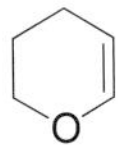 (3,4-dihydro-2H-pyran)	nd[a]	EtI	nd[a]
(pyridine)	nd[a]	CCl$_4$	127-130
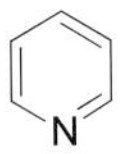		CBr$_4$	129-134

[a] not distinct.

It is interesting to note that the solvents which form complexes with *rac*-**131** do not form complexes with optically active **131** and *vice versa*. For example, the powdered (+)-**131** obtained by the evaporation of THF from the (+)-**130**-THF complex turns to guest-free crystals by recrystallization from solvents indicated in Table 8. The big difference of the role between the solvents shown in Tables 8 and 9 is interesting. X-ray crystal structural analysis of a 1:1 complex of (*S*)-(-)-**131** and 1,2-dichloroethane showed that the latter is accommodated in the cavity of the inclusion complex as a form of nearly eclipsed chiral rotamer, but no any significant interaction between the host and guest molecules is present.[59]

By using optically active **131** for the complexation, some guest compounds could be resolved. For example, when a solution of (-)-**131** (1.3 g) in *rac*-2-methyl-tetrahydrofuran (**132**) (5 g) was kept at room temperature   for 12 h, a 1:1 inclusion complex of (-)-**131** and (-)-**132** was formed as yellow prisms (1.32 g), which upon distillation gave (R)-(-)-**132** of 32% ee (0.1 g, 4%). By the same method, 2-methylpyran (**133**) was resolved by (-)-**131** to give (R)-(-)-**133** of 30% ee in 8% yield. Although the efficiency of the resolution is not high, chiral recognition between **131** and **132** or **133**, which have no binding groups such as hydroxyl for hydrogen bonding, is interesting.[58]

Further interesting enantioselective complexation between **131** and **133** has been observed. When a single crystal of (+)-**131**-(+)-**133** complex was used to seed at the recrystallization of rac-**131** (2.0 g) from rac-**133** (20 g), inclusion crystals of (+)-**131** of 99.0% ee and (+)-**133** of 29.4% ee were obtained (0.22 g, 18.6%).[58]

4-2. Resolution of rac-Guests by Complexation with the Achiral Host, 2,3,6,7,10,11- Hexahydroxytriphenylene

It was found that the title achiral host molecules (**134**) are arranged in a chiral form in their inclusion complex with a guest. By using this phenomenum, some rac-guests were resolved by complexation with achiral **134**.

134   **135**

Firstly, it was found that **134** forms inclusion complexes with various solvents such as n-PrOH (1:3), i-PrOH (1:2), n-BuOH (1:3), cyclopentanol (1:3), cyclohexanol (1:2), acetone (1:2), acetonylacetone (2:3), cyclopentanone (1:3), cyclopentenone (1:3), 2-methylcyclopentanone (1:3), 3-methylcyclopentanone (1:3), cyclohexenone (1:3), cyclohexanone (1:3), 2-methylcyclohexanone (1:3), 3-methylcyclohexanone (1:3), γ-butyrolactone (1:3), THF (1:2), 1,4-dioxane (1:2), and DMF (1:2) in the host:guest ratios indicated. Among these inclusion complexes, complexes with i-PrOH, cyclopentanone, cyclopentenone and cyclohexenone were found to be chiral by CD spectral measurements in the solid state.[60] In order to clarify the reason for the generation of chirality in these complexes, X-ray structure of a 1:3 complex of **134** with cyclopentenone was studied. The X-ray analysis showed that **134** molecules form a chiral helical column through hydrogen bond formation, and cyclopentenone molecules are binding to the OH groups of **134** in the chiral column.[60] By using the complexation of the chiral helices of **134** with guest, optical resolution of rac-guest was carried out. For example, recrystallization of **134** from 2-methylcyclopentanone (**135**) gave their 1:3 inclusion complex crystals. Heating of one piece of the crystal which shows a (+)-Cotton effect in the region of 300 nm in vacuo gave (+)-**135** of 34% ee by distillation. Heating of the other piece of crystal which shows a (-)-Cotton effect in the region of 300 nm in vacuo gave

(-)-**135** of 37% ee.[60]

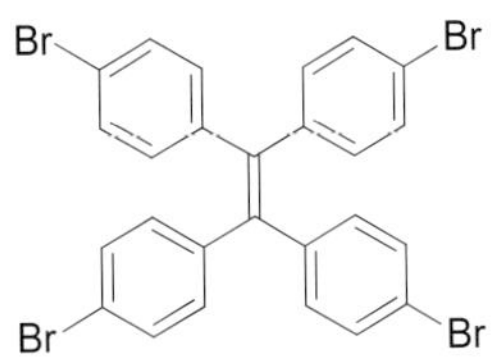

136	*P*-**136**	*M*-**136**

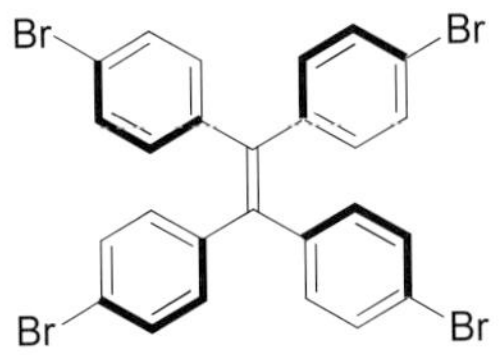

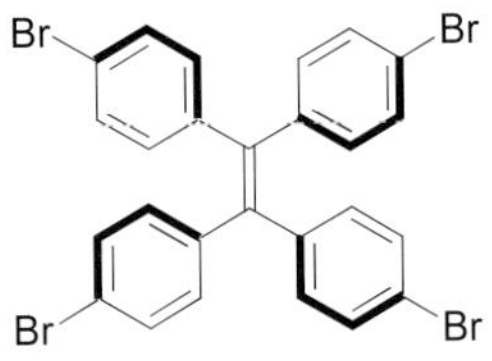

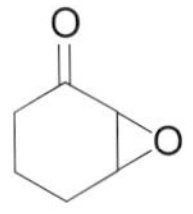

a: X = H
b: X = *o*-Me
c: X = *p*-Me

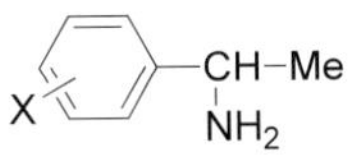

137	**138**	**139**

Chiral arrangement of achiral host, tetra(*p*-bromophenyl)ethylene (**136**) in an inclusion complex with guest is also interesting. In the inclusion complexes with acetone (1:2), 1,4-dioxane (1:1), benzene (1:1), *p*-xylene (1:1), cyclohexanone (1:1), THF (1:2), toluene (1:1), and β-picoline (1:1) obtained in the ratios indicated by recrystallization of **136** from these liquid guest, **136** molecules in the complexes with the former four guests are arranged in a chiral form. The chirality of **136** is generated by twisting of the *p*-bromophenyl groups as shown in *P*-**136** and *M*-**136**. In the complxes with the latter four guests, any chirality generated.[61] Interestingly, however, when **136** was exposed to the following guest vapor at room temperature for 24 h, a single crystal of pure **136** was gradually changed into the chiral polycrystalline inclusion complexes with THF (1:2), 1-4-dioxane (1:1), benzene (1:1), *p*-xylene (1:1), and β-picoline (1:1) of the host-guest ratios indicated. The racemic-to-chiral transformation through gas-solid reaction is also interesting. X-Ray structural study of these inclusion complexes of *P*-**136** and *M*-**136** has been done.[61]

5. Optical Resolution by Inclusion Crystallization in Suspension Media and by Fractional Distillation

It was found that efficient inclusion crystallization can be accomplished simply by mixing powdered crystalline host and a hydrophobic guest compound in hexane or water. By using the inclusion crystallization in suspension media, very efficient optical resolution method was established. Furthermore, when inclusion complexation between chiral host and *rac*-guest in the solid state is combined with a distillation procedure, optical resolution can easily be accomplished by the fractional distillation procedure.

For example, when a suspension of powdered **8a** (1.0 g, 2.14 mmol) and *rac*-1-phenylethanol (**50**) (0.262 g, 2.14 mmol) in hexane (10 ml) was stirred at room temperature for 6 h, a 2:1 inclusion crystal of **8a** and (-)-**50** (1.12 g) was obtained. Heating of the filtered inclusion crystal *in vacuo* gave (-)-**50** of 95% ee

(0.112 g, 85%). By a similar procedure, various alcohols and epoxides were resolved efficiently (Table 10). In some cases, the host **9a, 10a,** and **3** are more effective than **8a,** as shown in Table 10.[62] The enantioselective inclusion crystallization was also found to proceed efficiently in aqueous suspension. For example, when a suspension of powdered **8a** (1.0 g, 2.14 mmol) and *rac-***50** (0.26 g, 2.14 mmol) in water (10 ml) containing *N*-hexadecyl- trimethylammonium bromide (0.1 g) as a surfactant was stirred at room temperature for 24 h, a 2:1 inclusion crystal of **8a** and (-)-**50** was obtained. Heating the filtered inclusion crystal *in vacuo* gave (-)-**50** of 98% ee (0.11 g, 85%). This result clearly shows that the enantioselective inclusion crystallization occurs efficiently in the solid state. By a similar procedure, some alcohols, hydroxycarboxylic acid esters, epoxycyclohexanones, and epoxides were also resolved efficiently (Table 10). In the case of **42, 109,** and **137,** however, no inclusion crystallization occurred in water suspension.[63]

Table 10. Results of optical resolution using the inclusion crystalization by suspension method.

Host	Guest	Medium	Product		
				Yield (%)	% ee
8a	50	Hexane	(-)-**50**	85	95
8a	50	H$_2$O	(-)-**50**	85	98
8a	52	Hexane	(-)-**52**	75	100
8a	52	H$_2$O	(-)-**52**	75	98
8a	55	Hexane	(+)-**55**	89	92
8a	55	H$_2$O	(+)-**55**	76	100
9a	42	Hexane	(+)-**42**	80	80
9a	109	Hexane	(+)-**109**	93	78
9a	63	Hexane	(-)-**63**	82	100
8a	63	H$_2$O	(-)-**63**	73	100
9a	137	Hexane	(+)-**137**	78	75
3	65	Hexane	(+)-**65**	57	98
3	65	H$_2$O	(+)-**65**	85	97
9a	105g	Hexane	(+)-**105g**	75	100
9a	105g	H$_2$O	(+)-**105g**	89	100
10a	105f	Hexane	(+)-**105f**	78	100
10a	105f	H$_2$O	(+)-**105f**	80	100
9a	105e	Hexane	(+)-**105e**	59	70
9a	105e	H$_2$O	(+)-**105e**	52	86
10a	138	Hexane	(+)-**138**	76	75
10a	138	H$_2$O	(+)-**138**	74	47

Efficiency of the resolution by the suspension method is sometimes higher than that by the recrystallization method. For example, **64** was resolved by inclusion crystallization with **3** in ether followed by two recrystallizations of the inclusion crystal, and (-)-**64** of 100% ee has been obtained in 35% yield from the purified inclusion crystal by distillation.

By using **10a** and **10b** host compounds in the suspension method, both enantiomers of guest compound can be obtained easily in optically pure state. For example, keeping a suspension of **10a** (1.0 g, 1.97 mmol) and *rac*-**65** (0.5 g, 3.96 mmol) in hexane (10 ml) for 12 h gave an inclusion crystal (1.1 g) which upon distillation *in vacuo* gave (-)-**65** of 100% ee (0.16 g, 65%). From the filtrate left after separation of the inclusion crystal, (+)-**65** of 48% ee (0.29 g, 114%) was obtained by distillation. The crude (+)-**65** was treated with **10b** in hexane (8.5 ml) as above to give finally (+)-**65** of 100% ee (0.16 g, 62%).[62]

When the resolution by the suspension method is combined with distillation, both enantiomers can be separated easily by fractional distillation. For example, after a suspension of **8a** (1.0 g, 2.14 mmol) and *rac*-**50** (0.26 g, 2.14 mmol) in hexane (1 ml) was kept at room temperature for 1 h, the mixture was distilled *in vacuo* to give initially uncomplexed (+)-**50** of 59% ee (0.16 g, 125%) at lower boiling temperature. The residue was then distilled at elevated temperature and (-)-**50** of 97% ee (0.09 g, 69%) was obtained at higher boiling temperature. By the same combination method, *rac*-**50** was separated into enantiomers by fractional distillation in the presence of **3**, and (-)-**50** of 68% ee (121%) and (+)-**50** of 95% ee (63%) were obtained at lower and higher boiling temperature, respectively. When the resolution by the combination methods is repeated for the partially resolved samples, optically pure enantiomers can be obtained easily. Since the combination method is very simple and effective, this might be useful for large scale resolution in industries.

The resolution method of enantiomers by distillation has been established as a practically usefull technique by accumulation of many successful examples.[63]

The heating of a mixture of **10a** and two molar equivalents of *rac*-(*p*-tolyl)ethylamine (**139c**) in the Kugelrohr apparatus at 70 °C/2 mmHg gave (+)-**139c** of 98% ee in 102% yield by distillation, and the further heating of the residue at 150 °C/2 mmHg gave (+)-**139c** of 100% ee in 98% yield. By the same procedure, epoxides, alcohols, hydroxycarboxylic acid esters, amino alcohols, and amines were resolved efficiently (Table 11).[63] In Table 11, only the enantiomers which form an inclusion complex with the chiral host used and then distill at a relatively high temperature are listed.

Table 11. Enantiomer resolution by fractional distillation in the present of a chiral host

Host	rac-Guest	Distillation times repeated	Chiral product	
			Yield (%)	% ee
9a	137	2	60	94
9a	141	1	87	3
9a	63	2	60	93
10a	63	2	55	98
3	65	3	31	95
10a	105f	1	58	92
9a	105g	1	74	92
10a	105g	1	78	90
10a	138	4	10	90
8a	50	2	40	96
8a	52	2	47	92
8a	55	3	17	96
140	142	2	39	90
9a	143a	3	30	94
9a	143b	3	39	92
9a	143c	1	44	0
9a	144	2	33	100
10a	144	2	37	100
140	145	1	95	25
8a	139a	1	3	62
10a	139b	3	18	95
10a	139c	1	98	100
8a	146	1	98	42

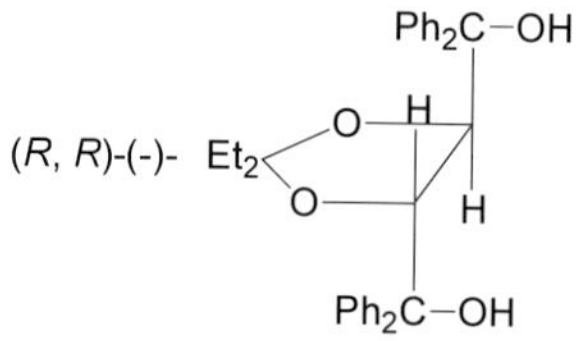

(R, R)-(-)- Et_2

Ph_2C-OH

Ph_2C-OH

140

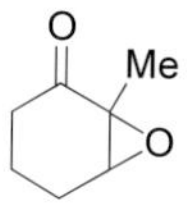

141

$MeCHCH_2OH$
OH

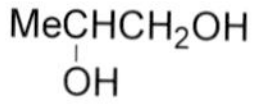

142

$$XCH_2CHCH_2COOR$$
$$OH$$

143

a: R = Me; X = H
b: R = Et; X = H
c: R = Me; X = Cl

$$MeCHCH_2OH$$
$$NH_2$$

144

$$MeCHCH_2NH_2$$
$$OH$$

145

146

In some cases, enantiomers of high optical purity cannot be obtained by one distillation procedure. In such case, almot optically pure samples can be obtained by repeating the distillation procedure several times. In a special case, almost optically pure (+)- and (-)-enantiomers are obtained by the resolution method using both the (+)- and (-)-host compounds. For example, heating a mixture of **10a** (1.6 g, 3.2 mmol) and *rac*-**63** (0.8 g, 6.3 mmol) at 70 °C/2 mmHg gave (+)-**63** of 58% ee (0.42 g, 105%) by distillation. Further heating of the mixture at 150 °C/2 mmHg gave (-)-**63** of 84% ee (0.36 g, 93%). By repeating the resolution of (-)-**63** of 84% ee in the presence of **10a**, (-)-**63** of 98% ee was obtained in 55% yield (Table 11). On the other hand, heating of a mixture of (+)-**63** of 58% ee (0.42 g) and **10b** at 70 °C/2 mmHg gave (-)-**63** of 31% ee (0.2 g, 51%). Further heating of the mixture at 150 °C/2 mmHg gave (+)-**63** of 98% ee (0.27 g, 72%).[63]

6. Resolution of 2,2'-Dihydroxy-1,1'-binaphthyl and Related Compounds

Resolution method of a racemic compound can be improved. As an example of the improvement, various resolutions of the title compounds are described.

Resolution of **14a** by complexation with **12** is described in the section 2-3.[10] However, much more efficient resolution of **14a** by complexation with *N*-alkyl cinchonidinium halides (**147**) was accomplished.[64] For example, when a solution of *N*-benzylcinchonidinium chloride (**147a**) (0.74 g, 1.76 mmol) and **14a** (1 g, 3.5 mmol) in MeOH (20 ml) was kept at room temperature for 6 h, a 1:1 inclusion complex of **147a** and **14b** was obtained as colorless prisms (0.89 g, 72%, mp 240-244 °C). The 1:1 complex was decomposed with dil HCl. Subsequent extraction with AcOEt, followed by evaporation of the solvent gave, after one recrystallization from MeOH, **14b** of 100% ee (0.3 g, 60%). This method is also effective for the separation of the enantiomers of the 6,6'-dibromo derivative (**148a**). In this case, when a solution of **147a** (0.95 g, 2.26 mmol) and **148a** (2.0 g, 4.5 mmol) in MeOH (2 ml)-AcOEt (10 ml) was kept at room temperature for 6 h, a 1:1 inclusion complex of **147a** and **148c** was obtained as color- less prisms (1.7 g, 89%, mp 229-232 °C). Decomposition of the complex with dil HCl gave **148c** of 99% ee (0.82 g, 82%). From the filtrate left after the separation of the inclusion complex, **148b** of 79% ee (1.1 g, 110%) was isolated by treatment with dil HCl.

Interestingly, however, **147a** was not suitable for the separation of the enantio-mers of **15**, because **147a** did not form an inclusion complex with **15**. In contrast this separation was accomplished by the use of *N*-butylcinchonidinium bromide (**147b**). For example, when a solution of   **147b** (0.14 g, 0.33 mmol)   and **15a** (0.25 g, 0.65

mmol) in MeCN (10 ml) was kept at room temperature for 24 h, a 1:1 inclusion complex of **147b** and **15b** was obtained as colorless prisms (0.22 g, 83%, mp 168-169 °C). Decomposition of the complex with dil HCl gave **15b** of 100% ee (0.1 g, 80%). From the filtrate left after the separation of the inclusion complex, **15c** of 58% ee (0.15 g, 80%) was isolated by treatment with dil HCl.[64]

(8S, 9R)-(-)-

147

a: R = PhCH$_2$; X = Cl
b: R = nBu; X = Br

148

a: *rac*-form
b: (*R*)-(+)-form
c: (*S*)-(-)-form

X-Ray crystal structure analysis of the complexes of **147a** with **14b** and of **147b** with **15b** showed that the complexes are mainly constructed *via* formation of an intermolecular hydrogen bond between X$^-$ of **147** and the OH group of **14** or **15**. It was also found that an intramolecular hydrogen bond between the OH group of **147** and X$^-$ plays an important role in forming the complx.[65] This finding suggests that a chiral compound which has R$_3$N$^+$ X$^-$ and OH groups would be a good host for resolution of **14a**, **15a**, and **148a**. According to this idea, chiral β-hydroxy trimethylammonium bromide hosts **149a-d** were prepared from the corresponding β-amino acids by transformation of their NH$_2$ and COOH groups into Me$_3$N$^+$Br$^-$ and CH$_2$OH, respectively, *via* the three step reaction shown in (eq. 4).[66]

When a solution of (+)-**149c** (0.5 g, 2.08 mmol) and **14a** (1.19 g, 4.16 mmol) in EtOH (6 ml) was kept at room temperature for 24 h, a 1:1 inclusion complex of (+)-**149c** and **14b** was obtained, after one recrystallization from ether, as colorless prisms. The pure crystals were dissolved in an ether-water mixture. From the ether solution, **14b** of 100% ee was obtained (0.42 g, 70%). From aqueous solution, **149c** was recovered unchanged (0.43 g, 86%). By the same complexation experiments of **14a** with (-)-**149a**, (+)-**149b**, and (+)-**149d**, **14c** (100% ee, 68%), **14b** (100% ee, 36%), and **14c** (100% ee, 48%) were obtained, respectively, in the optical and chemical yields indicated.[66]

a: R = Me
b: R = iPr
c: R = iBu
d: R = sBu

It was also found that **147** is useful for the resolution of 2,2'- (**150-152**) and

4,4'-dihydroxybiphenyl derivatives (**153**).[67] When a solution of **147a** (1.74 g, 4.13 mmol) and *rac*-4,4',6,6'-tetramethylbiphenyl (**150a**) (2 g, 8.26 mmol) in EtOH (10 ml) was kept at room temperature for 12 h, a 1:1 inclusion complex of **147a** and **150b** was formed. Two recrystallizations of the complex from EtOH gave the pure complex (1.1 g, mp 138-140 °C), which upon mixing with AcOEt (20 ml)-water (20 ml) decomposed into the components. From the AcOEt solution,, **150b** of >99.9% ee was obtained as colorless needles (0.53 g, mp 187-189 °C). For the resolution of *rac*-4,4'- dihydroxy-2,2',3,3',6,6'-hexamethylbiphenyl (**153a**), however, **9a** is very effective. When a solution of **9a** (4.56 g, 9.27 mmol) and **153a** (5 g, 18.5 mmol) in dibutyl ether (20 ml)-hexane (10 ml) was kept at room temperature for 12 h, a 1:1 inclusion complex of (+)-**153b** and **9a** was obtained (3.9 g), which upon recrystallization from dibutyl ether-hexane (2:1) gave the pure complex (2.43 g, 34%, mp 135-137 °C). The pure complex was dissolved in 10% NaOH. From the aqueous NaOH solution, **153b** of >99.9% ee was obtained by acidification with dil HCl followed by recrystallization as colorless prisms (0.76 g, 30%, mp 167-168 °C).[67]

It was also found that **151** can easily be resolved by complexation with optically active *trans*-1,2-diaminocyclohexane (**154**).[68] During the resolution study, we also found the very interesting result that the chiral recognition ability between **151** and **154** in the inclusion complex depends on their molar ratio. Hence, although **154** formed a 1:1 complex with racemic **151a**, **154** formed a 2:1 complex with optically active **151c**. Resolution of **151a** was accomplished only in the latter case. When a solution of **154c** (0.76 g, 6.66 mmol) and **151a** (2.0 g, 6.68 mmol) in toluene (0.3 ml) was kept at room temperature for 2 h, a 1:1 complex of **154** and **151a** was formed as colorless needles (2.43 g, 88%, mp 140-141 °C). On the other hand, when a solution of **154** (1.85 g, 16.2 mmol) and **151a** (2.5 g, 8.1 mmol) in toluene (3 ml) was kept at room temperature for 1 h, a 2:1 complex of **154** and **151b** was formed, after recrystallization from toluene, as colorless needles (1.45 g, 67%, mp 116-117 °C). The crystals were dissolved in a mixture of AcOEt (20 ml)-dilute HCl (20 ml), and the AcOEt layer was evaporated to give **141b** of 100% ee as colorless needles (0.78 g, 62%, mp 233-235 °C).[68]

2,2'-Dihydroxy-3,3',4,4',6,6'-hexamethyl-1,1'-biphenyl (**152**), however, formed only a 1:1 inclusion complex with **154** in which precise chiral recognition is present. When a solution of **154** (0.42 g, 3.68 mmol) and **152a** (2.0 g, 7.4 mmol) in toluene (0.3 ml) was kept for 2 h, a 1:1 complex of **154** and **152b** was formed, after one recrystallization from toluene, as colorless needles (0.8 g, mp 117-125 °C). Treatment of the complex by the procedure applied for the resolution of **151a** gave **152b** of 100% ee as colorless needles (0.39 g, 39%, mp 127 °C).[68]

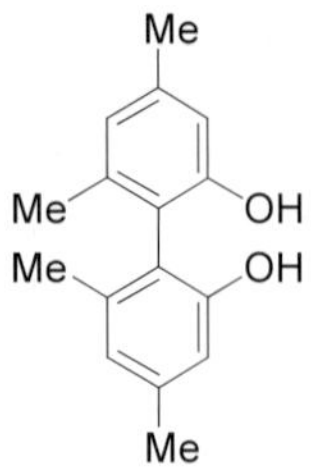

150 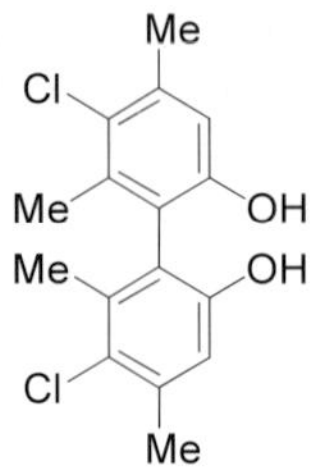   **151** 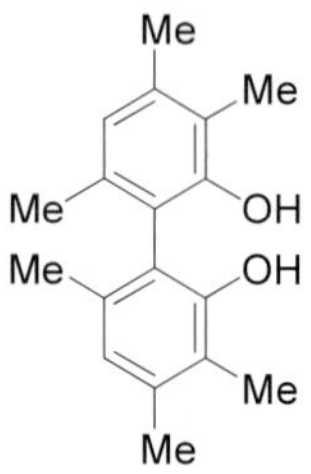   **152** 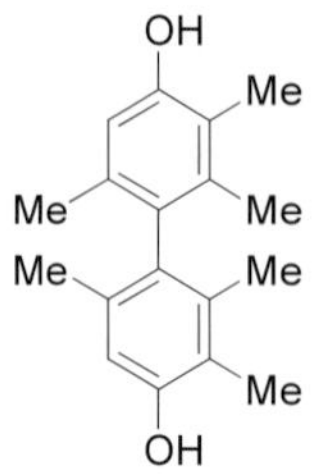   **153**

a: *rac*-form
b: (+)-form
c: (−)-form

Much more efficient resolution of **14a** was accomplished by inclusion complexation with (*S*)-(−)-*N*-(3-chloro-2-hydroxypropyl)-*N*,*N*,*N*-trimethyl-ammonium chloride (**155**).[69] By this complexation method, both eantiomers **14b** and **14c** were obtained in good optical and chemical yields. When a solution of **14a** (100 g, 0.35 mol) and **155** (39.5 g, 0.21 mol) in EtOH (800 ml) was kept at room temperature overnight, a 1:1 inclusion complex of **14c** and **155** precipitated as colorless crystals (60.3 g, 73%) . Treatment of the complex (50.3 g) with a 1:1 ether-water mixture (1 l) resulted in   dissociation into **14c** and **155**, which dissolved in ether and water, respectively. From the ether solution, **14c** of 99.5% ee (34.5 g, 69%) was obtained, whereas from aqueous solution **155** of 99% ee (22.7 g, 57%) was recovered unchanged. The EtOH solution left after separation of the complex of **14c** and **155** by filtration is subsequently dried. The remaining residue after treatment by a 1:1 ether-water mixture (1 l) gave an ether solution rich in **14b**. From the aqueous solution, **155** of 99% ee (15.8 g, 40%) was recovered unchanged. When the ether solution was kept at room temperature overnight, **14a** crystallized out as a 1:2 complex with ether, whereas **14b** of 99% ee (31.5 g, 63%) was obtained from the remaining ether solution. By this experiment, **14b** of 99% ee and **14c** of 99.5% ee were obtained in 63 and 69% yields, respectively, and **155** of 99% ee was recovered unchanged totally in 97% yield.[69]

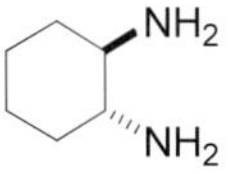

(*S*)-(−)- $Me_3\overset{+}{N}CH_2\underset{CH_2Cl}{CHOH}$ Cl^-

154 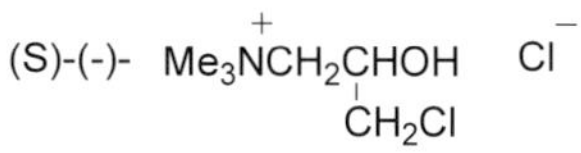   **155**

In this resolution of **14a** with **155**, the following two findings are skillfully applied. One is that only **14a** forms a 1:2 complex with ether and crystallized out, and optically pure **14b** or **14c** which does not form complex with ether can be obtained from the ether solution. The other one is that complex of **14b** or **14c** with **155** is easily dissociated into the components by treatment with ether-water mixture to give an ether solution of **14b** or **14c** and an aqueous solution of **155**.[69]

7.   Conclusions

Although some kinds of optically active compounds can be prepared by an asymmetric synthesis using a chiral catalyst, this method is not applicable for preparation of all kinds of compounds. Furthermore, optical yields of the product are not always very high. On the contrary, optical resolution method by inclusion complexation with a chiral host is applicable to various kinds of guest compounds as described in this chapter. When optically pure product cannot be obtained by one resolution procedure, perfect resolution can be accomplished by repeating the process, although asymmetric synthetic process cannot be repeated. Especially, optical resolutions by inclusion complexation with a chiral host in a water suspension medium and by fractional distillation in the presence of a chiral host are valuable as green and sustainable processes.

8. References

1) Toda, F.; and Akagi, K. (1968) Molecular Complexes of Acetylene Alcohols with n- and π-Bonds, *Tetrahedron Lett.*, 3695-3698.
2) Toda, F., D. L. Ward., and Hart, H. (1981) Wheel-and-Axile Design as a Source of Host-Guest Compounds. The Crystal Structure of the 2:1 Acetone Tetraphenyl-2,4-hexadiyne-1,6-diol Complex, *Tetrahedron Lett.*, **22**, 3865-3868.
3) Toda, F. (1987) Isolation and Optical Resolution of Materials Utilizing Inclusion Crystallization, *Top. Curr. Chem.*, **140**, 43-69.
4) Toda, F., Tanaka, K., Omata, T., Nakamura, K., and Oshima, T. (1983) Optical Resolution of 3-Methylcyclohexanones and 5-Methyl- -butyrolactone by Complexation with Optically Active 1,6-Bis(*o*-halophenyl)-1,6-diphenylhexa-2,4-diyne-1,6-diol, *J. Am. Chem. Soc.*, **105**, 5151.
5) Toda, F., and Tanaka, K. (1981) A New Optical Resolution Method of Tertiary Acetylenic Alcohol Utilizing Complexation with Brucine, *Tetrahedron Lett.*, **22**, 4669-4672.
6) a) Toda, F., Tanaka, K., Ueda, H., and Oshima, T. (1983) Chiral Recognition in Complexes of Tertiary Acetylenic Alcohols and Sparteine; Mutual Optical Resolution by Complex Formation, *J. Chem. Soc., Chem. Commun.*, 743-744. b) Toda, F., Tanaka, K., Ueda, H., and Oshima, T. (1985) Complex of Tertiary Acetylenic Alcohol and Brucine or Sparteine. X-ray Crystal Structural Study and Application to Optical Resolution, *Israel J. Chem.*, **25**, 338-345. c) Toda, F., and Tanaka, K. (1986) Optical Resolution of Tertiary Acetylenic Alcohols and Secondary Alcohols by Complexation with Achiral Amines, *Chem. Lett.*, 1905-1908. d) Yabuki, T., Yasui, M., Tezuka, H., Harada, S., Kasai, N., Tanaka, K., and Toda, F. (1988) The Crystal and Molecular Structures of 1:2 Complexes of *N,N'*-Dimethylpiperazine with 1-(*o*-Chlorophenyl)- 1-phenyl-2-propyn-1-ol, *Chem. Lett.*, 749-752.
7) Toda, F., and Tanaka, K. (1988) Design of a New Chiral Host Compound, trans-4,5-Bis(hydroxydiphenylmethyl)-2,2-dimethyl-1,3-dioxacyclopentane. An Effective Optical Resolution of Bicyclic Enones Through Host-Guest Complex Formation, *Tetrahedron Lett.*, **29**, 551-554.
8) Toda, F., Sato, A., Tanaka, K., and Mak, T. C. W. (1989) Optical Resolution of Pantolactone by Inclusion Crystallization with (*R,R*)-(-)-*trans*-2,3-Bis(diphenyl-hydroxymethyl)-1,4-dioxaspiro[4.5]decane and Crystal Structure of the Resulting 1:1 Complex, *Chem. Lett.*, 873-876.
9) a) Toda, F., Tanaka, K., Tagami, Y., and Mak, T. C. W. (1985) Inclusion Properties of *N,N,N,N*-Tetraisopropyloxamide and Crystal Structure of Its 1:1 Complex with 1-Methylnaphthalene, *Chem. Lett.*, 195-378. b) Toda, F., Tanaka, K., and Mak, T. C. W. (1986) Molecular Complexes of *N,N,N,N*-Tetraisopropyloxamide. Crystal Structure of Its 1:1

Adduct with *m*-Cresol, and Its 1:1 and 1:2 Adducts with *p*-Cresol, *Bull. Chem. Soc. Jpn.*, **59**, 1189-1194. c) Toda, F., Tagami, Y., and Mak, T. C. W. (1986) Separation of Isomeric Alcohols by Selective Complexation with *N,N,N,N*-Tetracyclohexylfumar- amide and *N,N,N,N*-Tetraisopropylfumaramide, and Structures of Two Resulting Crystalline Adducts, *Chem. Lett.*, 1909-1912. d) Toda, F., Tagami, Y., Hao, Q., Fan, H.-F., and Mak, T. C. W. (1986) Crystal Structure of a 1:1:1 Molecular Complex of (*R*)-(+)-2,2'-Dihydroxy-1,1'-dinaphthyl, *N,N,N,N*-Tetra(2-butyl)terephthalamide, and Methanol, *Chem. Lett.*, 1913-1916. e) Toda, F., Kai, A., Yagi, M., and Mak, T. C. W. (1987) Inclusion Properties of the New Host Compound 1,2,4,5-Tetra- (morpholinocarbonyl)benzene and X-Ray Crystal Structure of Its 1:4 Complex with Water, *Chem. Lett.*, 1025-1028. f) Toda, F., Kai, A., Tagami, Y., and Mak, T. C. W. (1987) A Novel Host Compound with High Inclusion Ability, *N,N,N,N*-Tetracyclo- hexyl-2,2'-biphenyldicarboxamide, and Crystal Structure of Its 1:1 Phenol Complex, *Chem. Lett.*, 1393-1396.

10) a) Toda, F., and Tanaka, K. (1988) Efficient Optical Resolution of 2,2'-Dihydroxy-1,1'-binaphthyl and Related Compounds by Complexation with Novel Chiral Host Compounds Derived from Tartaric Acid, *J. Org. Chem.*, **53**, 3607-3609. b) Toda, F., Tanaka, K., Nassimbeni, L. R., and Niven, M. (1988) Efficient Optical Resolution of 2,2'-Dihydroxy-1,1'-binaphthyl and 10,10'-Di-hydroxy-9,9'-biphenan- thryl by Complex Formation with Novel Chiral Host Compounds Derived from Tartaric Acid, *Chem. Lett.*, 1371-1374.

11) Miyamoto, H., Sakamoto, M., Yoshioka K., Takaoka, R., and Toda, F. (2000) Resolution of Hydrocarbons by Inclusion Complexation with a Chiral Host Compound, *Tetrahedron Asymm.*, **11**, 3045-3048 and references cited therein.

12) Kato, M., Tanaka, K., and Toda, F. (2001) Stereoselective Inclusion and Structure of Equaroeial-*trans*-1,2-dichlorocyclohexane, *Supramol. Chem.*, **13**, 175-180.

13) Toda, F., Tanaka, K., and Kido, M. (1988) Optical Resolution of 2-Methylpiperazine by Complex Formation with Optically Active 1-Phenyl-1-(*o*-chlorophenyl)- prop-2-yn-1-ol and 1,6-Diphenyl-1,6-(*o*-chloro-
phenyl)hexa-2,4-diyne-1,6-diol, *Chem. Lett.*, 513-516.

14) Toda, F., Tanaka, K., Infantes, L., Forces-Forces, C., Claramunt, R. M., and Elguero, J. (1995), Optical Resolution of 1,3-Dimethyl-5-phenyl-∇^2-pyrazoline by Diastereo- isomeric Complex Formation with an Optically Active Host Compound: X-Ray and Molecular Structure of the Complex, *J. Chem. Soc., Chem. Commun.*, 1453-1454.

15) Toda, F., Soda, S., and Goldberg, I. (1993) Preparation of Optically Active Amines by a Combination of Gabriel Synthesis and Optical Resolution. X-Ray Crystal Structure of the Adduct Between (-)-10,10'-Dihydroxy-9,9'-biphenanthryl and *N*-(1-*tert*-Butyl-2-oxoazetidin-3-yl)phthalimide, *J. Chem. Soc., Perkin Trans.* 1, 2357-2361.

16) Toda, F., Mori, K., Stein, Z., and Goldberg, I. (1989) Optical Resolution of Amine N-Oxides by Diastereoisomeric Complex Formation with Optically Active Host Compound, *Tetrahedron Lett.*, **30**, 1841-1844.

17) Toda, F., Mori, K., Okada, J., Node, M., Itoh, A., Oomine, K., and Fuji, K. (1988) New Chiral Shift Reagents, Optically Active 2,2'-Dihydroxy-1,1'-binaphthyl and 1,6-Di(*o*-chlorophenyl)-1,6-diphenylhexa-2,4-diyne-1,6-diol, *Chem.Lett.*, 131-134.

18) Toda, F., and Akai, H. (1990) Isolation of Optically Active Oximes, *J. Org. Chem.*, **55**, 4973-4974.

19) Mori, K., and Toda, F. (1990) Efficient Optical Resolution of Aziridines by Complexation with Optically Active Host Compounds, *Tetrahedron Asymm.*, **1**, 281-282.

20) Toda, F., Sano, A., Nassimbeni, L. R., and Niven, M. L. (1991) Optical Resolution of Amino Acid and Hydroxycarboxylic Acid Esters by Complexation with Optically Active Host Compounds: a Crystallographic Result, *J. Chem. Soc., Perkin Trans.* 2, 1971-1975.

21) Zaderenko, P., Lopez, P., Ballesteros, P., Takumi, H., and Toda, F. (1995) Resolution of 2-Azol-1-ylsucccinic Esters by Enantioselective Inclusion Methodology, *Tetrahedron Asymm.*,

6, 381-384.

22) Toda, F., Tanaka, K., and Koshiro, K. (1991) A New Preparative Method of Optically Active Diarylcarbinols, *Tetrahedron Asymm.*, **2**, 873-874.

23) Toda, F., Matsuda, S., and Tanaka, K. (1991) Efficient Resolution of Secondary Alcohols, Cyanohydrins, and Glycerol Acetates by Complexation with the Host   Derived from Tartaric Acid, *Tetrahedron Asymm.*, **2**, 983-986.

24) Nishikawa, K., Matsumoto, A., Tsukada, H., Shiro M., and Toda, F. (1997), Inclusion Complex   of   (*S*)-1,3-Butanediol   with   (*S,S*)-(+)-*trans*-2,3-Bis(hydroxyl-diphenylmethyl)-1,4-dioxaspito[4.4]nonane, *Acta Crystallograph. Sec. C*, 351-353.

25) Miyamoto, H., Yasaka, S., Takaoka, R., Tanaka, K., and Toda, F. (2001) One-pot Preparation of Optically Active *sec*-Alcohols, Epoxides, and Sulfoxides by a Combination of Synthesis and Enantiomeric Resolution with Optically Active Hosts in a Water Suspension Medium, *Eanantiomer,* **6**, 51-55.

26) a) Toda, F., Tanaka, K., and Sekikawa, A., (1987) Host-Guest Complex Formation by a Solid-Solid Reaction, *J. Chem. Soc., Chem. Commun.*, 279-280. b) Toda, F. (1995) Solid State Organic Chemistry; Efficient Reactions, Remarkable Yields, and Stereo- selectivity, *Acc. Chem. Res.*, **12**, 480-486. c) Tanaka, K. and Toda, F. (1987) Solvent- Free Organic Synthesis, *Chem. Rev.*, **100**, 1025-1074.   27) a) Toda, F., and Tanaka, K. (1983) Conversion of Racemic Cyanohydrin into One Optically Acrtive Iosmer in the Presence of Brucine, *Chem. Lett.*,661-664. b) Tanaka, K. and Toda, F. (1987) Chiral Recognition and Optical Resolution of Cyanohydrin by Complexation with Brucine, *Nippon Kagaku Kaishi,* 456-459.

28) Tanaka, K., and Toda, F. (1983) Preparation of Optically Pure 2,3-Epoxycyclo- hexanones, *J. Chem. Soc., Chem. Commun.*, 1513-1514.

29) Toda, F., Mori, K., Matsuura, Y., and Akai, H. (1990) Solid State Kinetic Resolution of b-Ionone Epoxide and Dialkyl Sulphoxides in the Presence of Optically Active Host Compounds. The First Enantioselective Host-Guest Inclusion Complexation in the Solid State, *J. Chem. Soc., Chem. Commun.* 1591-1593.

30) Toda, F., and Ochi, M. (1996) Enantioselective Oxidation of Imines in Inclusion Complexes with a Chiral Host Compound and Preparation of Optically Pure Oxaziridines by Enantiomer Resolution, *Enantiomer,* **1**, 85-88.

31) Toda, F., and Tanaka, K. (1988) A New Chiral Host Compound 10,10'-Dihydroxy-9,9'-biphenanthryl. Optical Resolution of Propionic Acid Derivatives, Butyric Acid Derivatives, and 4-Hydroxycyclopent-2-en-1-one Derivatives by Complexation, *Tetrahedron Lett.*, **29**, 1807-1810.

32) Tanaka, K., Kakinoki, O., and Toda, F. (1992) Regio- and Enantioselective Photodimerisation of Cyclohex-2-enone as an Inclusion Complex with a New Optically Active Host, (-)-1,4-Bis[3-(*o*-chlorophenyl)-3-hydroxy-3-phenyl- 1-propynyl]benzene: Preparation of the Optically Pure (-)-*syn-trans*-Dimer of Cyclohex-2-enone, *J. Chem. Soc., Perkin Trans.* 1, 307.

33) Tanaka, K., Kakinoki, O., and Toda, F. (1992) Efficient Resolution of Some Key Compounds of Prostaglandin Synthesis, *Tetrahedron Asymm.*, **3**, 517-520.

34) Toda, F., and Tanaka, K. (1985) Optical Resolution of Key Compounds of Prostaglandin Synthesis and Related Compounds, *Chem. Lett.*, 885-888.

35) Nassimbeni, L. R., Niven, M. L., Tanaka, K., and Toda, F. (1991) On the Optical Resolution of Bicyclic Enones Through Host-Guest Complex Formation: The Crystallographic Result, *J. Crystallograph. and Spectroscop. Res.*, **21**, 451-457.

36) Toda, F., Tanaka, K., Marks, D., and Goldberg, I. (1991) Optical Resolution of Bicyclo[2.2.1]heptanone, Bicyclo[2.2.2]octanone, and Bicyclo[3.2.1]octanone Derivatives by Inclusion Complexation with Optically Active Host Compounds, *J. Org. Chem.*, **56**, 7332-7335.

37) Tanaka, K., Kato, M., and Toda, F. (2001) Optical Resolution of 2-Azabicyclo-

[2.2.1]hept-5-en-3-one by Inclusion Complexation with Brucine, *Heterocycles*, **54**, 405-410.

38) Toda, F., Tanaka, K., Leung, C. W., Meetsma, A., and Feringa, B. L. (1994) Preparation of Optically Active 5-Alkoxyfuran-2(5*H*)-ones and 5-Methoxydihydro- furan-2(3*H*)-one by Chiral Inclusion Complexation, *J. Chem. Soc., Chem. Commun.*, 2371-2372.

39) Toda, F., Miyamoto, H., and Ohta, H. (1994) Efficient Optical Resolution of *cis*-4-Methylcyclohex-4-ene-1,2-dicarboxylic Anhydride, *cis*-4-Methylcyclohex-4-ene-1,2-dicarboximide, and Their Derivatives by Complexaion with Optically Active Host Compounds Derived from Tartaric Acid, *J. Chem. Soc., Perkin Trans* 1. 1601-1604.

40) Toda, F., Sato. A., Tanaka, K., and Mak, T. C. W. (1989) Optical Resolution of Pantolactone by Inclusion Crystallization with (*R*,*R*)-(-)-*trans*-2,3-Bis(diphenyl-hydroxylmethyl)-1,4-dioxaspiro[4.5]decane, and Crystal Structure of the Resulting 1:1 Complex, *Chem. Lett.*, 873-876.

41) Kaupp, G., Schmeyers, Y., Toda, F., Takumi, H., and Koshima, H. (1996) Mechanism of Solid-Solid Resolution of Pantolactone, *J. Phys. Org. Chem.*, **9**, 795-88.

42) Toda, F., Tanaka, K., Watanabe, M., Abe, T., and Harada, N. (1995) Enantiomer Resolution by Crystallization with Chiral Host: Application to Monoterpenes, Verbenone and Apoverbenone, *Tetrahedron Asymm.*, **6**, 1495-1498.

43) Toda, F., Takumi, H., and Tanaka, K. (1995) A Simple Preparative Method for Optically Active Glycidic Esters, *Tetrahedron Asymm.*, **6**, 1059-1062.

44) Lee, G.-H., Wang, Y., Tanaka, K., and Toda, F. (1988) The Crystal Structure of 1:1 Complexes of (*S*)-(-)-10,10'-Dihydroxy-9,9'-biphenanthryl with (*S*)-(-)-Methyl 2-Chloropropionate and (*S*)-(-)-Methyl 3-Chloro-2-hydroxybutyrate, *Chem. Lett.*, 781-784.

45) Toda, F., Tanaka, K., Yagi, M., Stein, Z., and Goldberg, I. (1990) New Preparative and Optical Resolution Method for b-Lactams, *J. Chem. Soc., Perkin Trans.* 1, 1215-1216.

46) Toda, F., Tanaka, K., and Nagamatsu, S. (1984) Mutual Optical Resolution of 2,2'-Dihydroxy-1,1'-binaphthyl and Alkyl Aryl or Dialkyl Sulfoxides by Complex Formation, *Tetrahedron Lett.*, **25**, 4929-4932.

47) Toda, F., Tanaka, K., and Mak, T. C. W. (1984) Mutual Optical Resolution of Bis-b-naphthol and Sulfoxides. Absolute Configuration and Crystal Structure of a 1:1 Molecular Complex, *Chem. Lett.*, 2085-2088.

48) Toda, F., and Mori, K. (1986) Induced Asymmetric Polymerisation of Optically Active Vinyl Sulphoxide, *J. Chem. Soc., Chem. Commun.*, 1059-1060.

49) Toda, F., Tanaka, K., and Okuda, T. (1995) Optical Resolution of Methyl Phenyl and Benzyl Methyl Sulfoxides and Alkyl Phenylsulfinates by Complexation with Chiral Host Compounds Derived from Tartaric Acid, *J. Chem. Soc., Chem. Commun.*, 639-640.

50) Mori, K., and Toda, F. (1988) Optical Resolution of Sulfoximines by Complex Formation with Optically Active 2,2'-Dihydroxy-1,1'-binaphthyl or 1,6-Di(*o*-chlorophenyl)-1,6-diphenylhexa-2,4-diyne-1,6-diol, *Chem. Lett.*, 1997-2000.

51) Toda, F., and Mori. K. (1986) Optical Resolution of Selenoxides by Complexation with Optically Active 2,2'-Dihydroxy-1,1'-binaphthyl or 1,6-Di(*o*-chlorophenyl)-1,6-diphenylhexa-2,4-diyne-1,6-diol, *J. Chem. Soc., Chem. Commun.*, 1357-1359.

52) Fujiwara, T., Tanaka, N., Ooshita, R., Hino, R., Mori, K., and Toda, F. (1990) Crystal and Molecular Structure of the Crystalline Host-Guest Complex between (*R*)-(+)-2,2'-Dihydroxy-1,1'-binaphthyl and (*S*)-(-)-(Ethyl *m*-Tolyl Selenoxide), *Bull. Chem. Soc. Jpn.*, **63**, 249-251.

53) Mori, K., and Toda, F. (1990) Optical Resolution of Arsine Oxides by Complexation with Optically Active [1,1'-Binaphthyl]-2,2'-diol, *Bull. Chem. Soc. Jpn.*, **63**, 2127-2128.

54) Tanaka, K., Ootani, M., and Toda, F. (1992) Optically Active *trans*-Bis-(hydroxydiphenylmethyl)-2,2-dimethyl-1,3-dioxacyclopentane and Its Derivatives as Chiral Shift Reagents for the Determination of Enantiomeric Purity and Absolute Configuration, *Tetrahedron Asymm.*, **3**, 709-712.

55) Toda, F., Mori, K., and Okada, J. (1988) New   Chiral   Shift Reagents, Optically Active

2,2'-Dihydroxy-1,1'-binaphthyl and 1,6-Bis(o-chlorophenyl)-1,6-diphenyl-hexa-2,4-diyne-1,6-diol, *Chem. Lett.*, 509-512..

56) Toda, F., Mori, K., Stein, Z., and Goldberg, I. (1988) Optical Resolution of Phosphinates and Phosphine Oxides by Complex Formation with Optically Active 2,2'-Dihydroxy-1,1'-binaphthyl and Crystallographic Study of Two Diastereomeric Complexes with $(CH_3)(C_6H_5)(OCH_3)PO$, *J. Org. Chem.*, **53**, 308-312.

57) Toda, F., Sasaoka, M., Toda, Y., Iida, K., Hino, T., Nishiyama, Y., Ueda, H., and Oshima, T. (1983) A Rubrene Problem. Reaction of 3,3,4,4-Tetrabromo-1,2-bis-(diphenylmethylene)cyclobutane and Dichloroacetic Acid, *Bull. Chem. Soc. Jpn.*, **56**, 3314-3318.

58) Toda, F., and Tanaka, K. (1990) Guest (Solvent)-Dependent Enantioselective Crystallization of 7-Bromo-1,4,8-triphenyl-2,3-benzo[3.3.0]octa-2,4,7-trien-6-one as the Inclusion Complex, *Tetrahedron Asymm.*, **1**, 359-362.

59) Toda, F., Tanaka, K., and Kuroda, R. (1997) Isolation of a Nearly Eclipsed Chiral Rotamer of 1,2-Dichloroethane as an Inclusion Crystal with a Chiral Host Compound, *Chem. Commun.*, 1227-1228.

60) Toda, F., Tanaka, K., Matsumoto, T., Nakai, T., Miyahara, I., and Hirotsu, K. (2000) A New Host 2,3,6,7,10,11-Hexahydroxytriphenylene Which Forms Chiral Inclusion Crystalline Lattice: X-ray Structural Study of the Chiral Crystalline Lattice, *J. Phys. Org. Chem.*, **13**, 39-45.

61) Tanaka, K., Fujimoto, D., Oeser, T., Irngartinger, H., and Toda, F. (2000) Chiral Inclusion Crystallization of Tetra(p-bromophenyl)ethylene by Exposure to the Vapor of Achiral Guest Molecules: A Novel Racemic-to-Chiral Transformation Through Gas-Solid Reaction, *Chem. Commun.*, 413-414.

62) Toda, F., and Tohi, Y. (1993) Novel Optical Resolution Methods by Inclusion Crystallisation in Suspension Media and by Fractional Distillation, *J. Chem. Soc., Chem. Commun.*, 1238-1240.

63) a) Toda, F., and Takumi, H. (1996) Separation of Enantiomers by Fractional Distillation in the Presence of a Chiral Host Compound, *Enantiomer*, **1**, 29-33. b) Kaupp, G. (1994) Resolution of Racemates by Distillation with Inclusion Compounds, *Angew. Chem. Int. Ed. Engl.*, **33**, 728-729.

64) Tanaka, K., Okada, T., and Toda, F. (1993) Separation of the Enantiomers of 2,2'-Dihydroxy-1,1'-binaphthyl and 10,10'-Dihydroxy-9,9'-biphenanthryl by Complex-ation with N-Alkylcinchonidinium Halides, *Angew. Chem. Int. Ed. Engl.*, **32**, 1147-1148.

65) Toda, F., Tanaka, K., Stein, Z., and Goldberg, I. (1994) Optical Resolution of Binaphthol and Biphenanthryl Diols by Inclusion Crystallization with N-Alkyl- cinchonidinium Halides. Structural Characterization of the Resolved Materials, *J. Org. Chem.*, **59**, 5748-5751.

66) Toda, F., and Tanaka, K. (1997) New Chiral Ammonium Salt Hosts Derived from Amino Acids: Very Efficient Optical Resolution of 2,2'-Dihydroxy-1,1'-binaphthyl by Complexation with These Host Compounds, *Chem. Commun.*, 1087-1088.

67) Tanaka, K., Moriyama, A., and Toda, F. (1996) New Preparative Method for Optically Active 2,2'- and 4,4'-Dihydroxybiphenyl Derivatives. A New Chiral Host Compound 4,4'-Dihydroxy-2,2',3,3',6,6'-hexamethylbiphenyl, *J. Chem. Soc., Perkin Trans.* 1, 603-604.

68) Tanaka, K., Moriyama, A., and Toda, F. (1997) Novel Chiral Recognition in Host-Guest Inclusion Complexes Depending on Their Molar Ratios: Efficient Resolution of 2,2'-Dihydroxy-1,1'-binaphthyl Derivatives and CD Spectral Study of Inclusion Complex Crystals, *J. Org. Chem.*, **62**, 1192-1193.

69) Toda, F., Yoshizawa, K., Hyoda, S., Toyota, S., Chatziefthimiou. S., and Mavridis, I. M. (2004) Efficient Resolution of 2,2'-Dihydroxy-1,1'-binaphthyl by Inclusion Complexation with Chiral N-(3-Chloro-2-hydroxypropyl)-N,N,N-tri-methylammonium Chloride, *Org. Biomol. Chem.*, **2**, 449-451.

CRYSTALLINE DIPEPTIDES:
THEIR MOLECULAR RECOGNITION TO BE USEFUL TO ENANTIOMER SEPARATION

Katsuyuki Ogura and Motohiro Akazome
Department of Materials Technology, Faculty of Engineering, Chiba University
1-33 Yayoicho, Inageku, Chiba 263-8522, Japan

1. Introduction

Molecular recognition is a very important process in this world. Enzyme, antibodies, membranes, and biological receptors use the phenomenon of molecular recognition. The "lock-and-key" model that was proposed by Fischer is commonly useful for understanding the phenomenon of the molecular recognition. The complementarily of a lock and a key is essential to this model, in which the lock is the molecular receptor and the key is the substrate that is recognized by the lock to give a defined receptor-substrate complex. This chemistry brought about a new direction for science called "supramolecular chemistry", the chemistry beyond the molecule, where noncovalent bonds and spatial fit between molecular individuals form specific host-guest complexes. Furthermore, organic chemists also have been developing novel synthetic systems for recognizing various substrates. Among them, the chemistry of inclusion crystals creates one of the most useful and informative fields of molecular recognition. Hitherto, the phenomena that host molecules form crystals with appropriate guest molecules have been investigated.[1] Inclusion crystals of urea and bile acid were investigated at the early stage of this field. Besides these hosts, many kinds of host molecules such as "wheel-and-axle"-type diols,[2] cholic acid,[3] tartaric acid derivatives,[4] binaphthols,[5] and resorcinol derivatives[6] have been investigated.

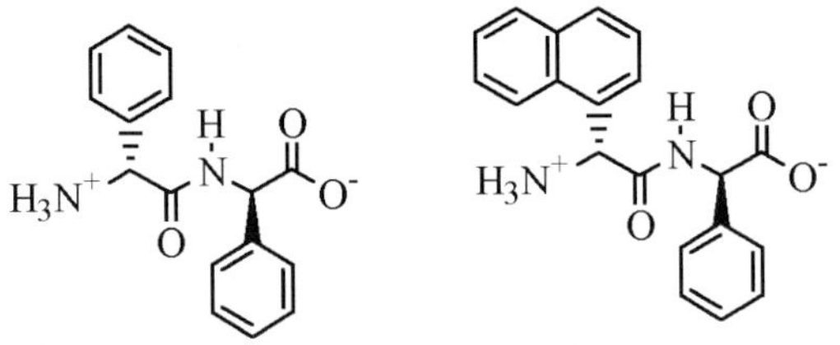

Figure 1. Dipeptide hosts

About ten years ago,[7] we found that a simple dipeptide, (R)-phenylglycyl-(R)-phenylglycine (**1**), forms an inclusion crystal with a sheet

49

F. Toda (ed.), Enantiomer Separation, 49-72.

structure. This finding is much interesting from a viewpoint of asymmetric recognition of a chiral guest by a simple dipeptide. Here, we would like to survey our investigations that are concerned with the molecular recognition of the dipeptide (**1**) and its related molecule, (*R*)-naphthylglycyl-(*R*)-phenylglycine (**2**), as well as their application to the enantiomeric separation.

2.   Sheet Structures in Crystalline Dipeptides

Before our investigation on the inclusion phenomena of the dipeptide (**1**) and its related compounds, it had been reported that water, methanol, dimethyl sulfoxide etc. were included by crystalline dipeptides that consist of naturally occurring α-amino acids such as alanine, valine, and leucine, but their phenomena have not

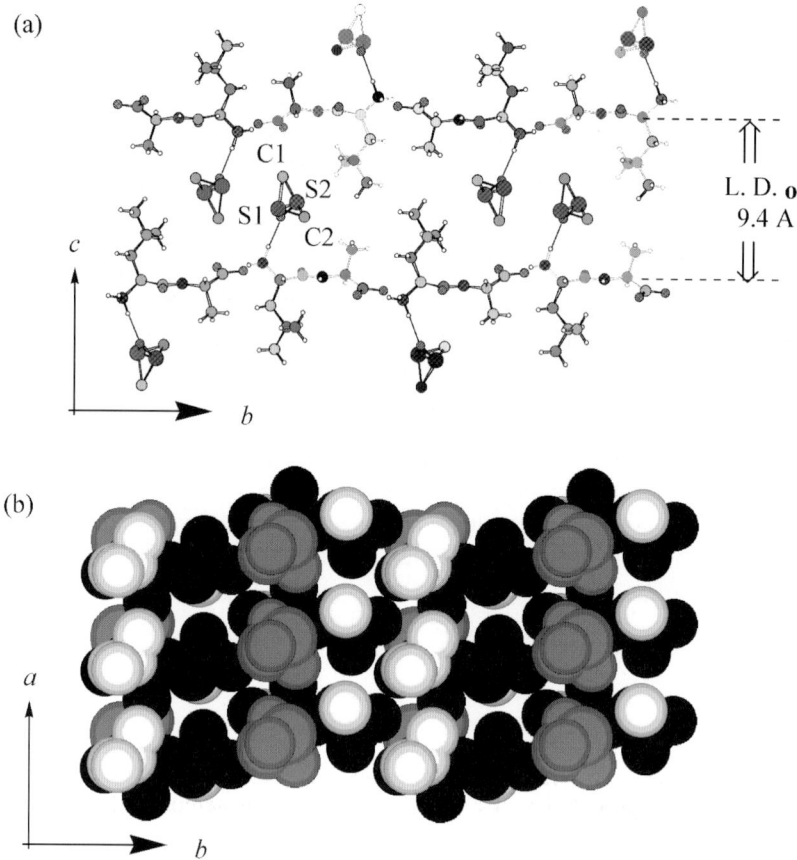

Figure 2. The structure of DMSO-included Crystal of L-leucyl-L-alanine. (a) A b-c plane. The term "L.D." means the layer distance given by PXRD. (b) The sruecture (a-b plane) along with the included DMSO sepresented by CPK model.

been investigated systematically.[8] We reexamined the crystal structure of the inclusion compound derived from L-leucyl-L-alanin and DMSO. The results are shown in Figure 2. As secribed in the litherature,[8] the sulfur atom of DMSO occupired two sites (6:4) to have a trigonal-bipyramidal geometry.

Our attention was paid to the fact that the dipeptide molecues construct a layer structure in a crystalline state. We thought that, if the fexible alky group of these dipeptides are replaced by an aryl group, the resulting sheet structure would have pillars. Namely, the rigid aryl groups can be regarded as pillars that separate the sheets from each other to create the stable cavity for a guest, as shown in Figure 3.

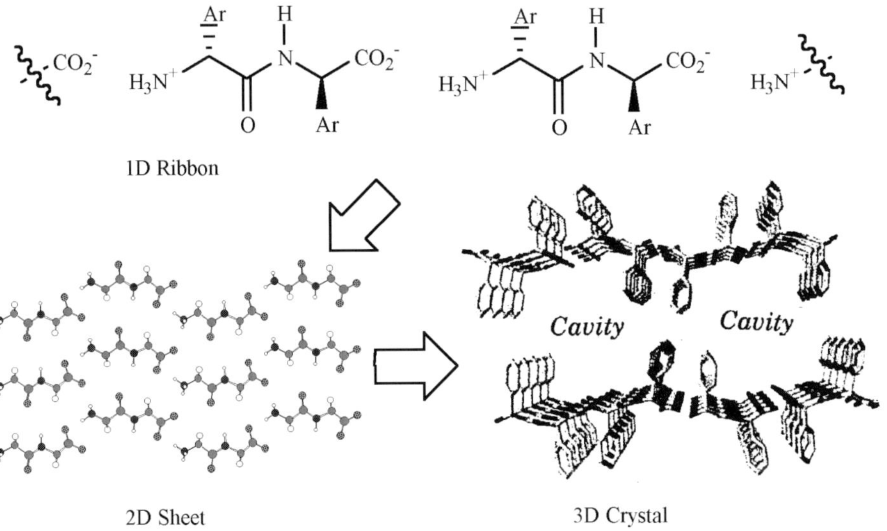

Figure 3. Assembly of dipeptides to form 3-dimensional structure.

There are only a few reports on the organic host molecules that form a precise two-dimensional network (sheet structure) followed by the inclusion of a guest molecule.[9] Ward and his co-workers developed pillared two-dimensional hydrogen-bonded networks comprising guanidinium ions and disulfonate ions, in which the disulfonate ions act as pillars that connect opposing hydrogen-bonded sheets with adjustable porosity.[10]

The inclusion compounds of **1** and **2** were obtained by crystallization from an appropriate solvent system containing a guest (vide infra). (R)-Phenylglycyl-(R)-phenylglycine (**1**) molecules self-assemble to form a layer structure and (S)-isopropyl phenyl sulfoxide molecules are stereospecifically included in the void between the layers (Figure 4).[7] A sheet structure was also observed in THF-included crystals of (R)-(1-naphthyl)glycyl-(R)-phenylglycine (**2**), which were formed by crystallization of **2** from a mixed solvent of tetrahydrofuran (THF) and methanol.[11] The sheet structure (Figure 5) is analogous to that of the

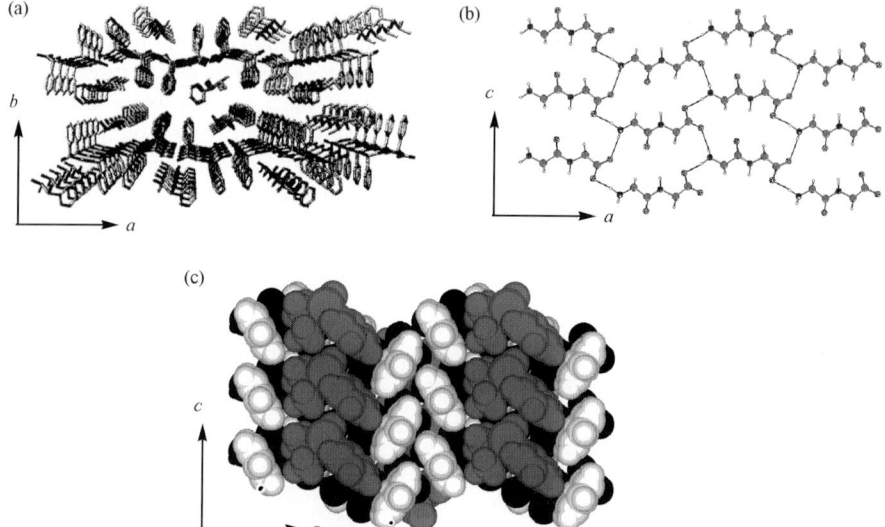

Figure 4. Inclusion compound of (*S*)-isopropyl phenyl sulfoxide by **1**: (a) perspective view of crystals; (b) sheet structure of **1**; (c) CPK model. For clarity, phenyl groups of **1** and guest are colored white and gray, respectively.

above sulfoxide-included compound of **1**. THF molecules were linked to $^+NH_3$ of dipeptide *via* hydrogen bonding and accommodated in the void between the walls of the naphthyl and phenyl groups. Diethyl ether-included compound of **2** showed the similar sheet structure.

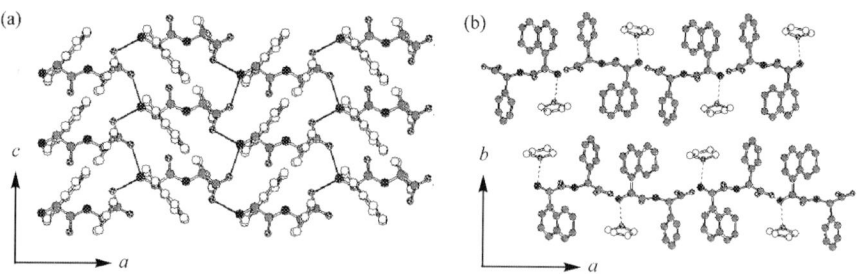

Figure 5. Parallel β-sheet-like structure of THF-included compound of **2**. To improve clarity, all hydrogen atoms are omitted. (a) a-c face. THF is omitted, and the naphthyl and phenyl groups are colored white. (b) a-b face. The carbons of THF are colored white

When allylic alcohols were used as a guest molecule instead of the ethers (THF and diethyl ether), the arrangement of the dipeptide molecules was changed to an antiparallel β-sheet-like structure.[19] By crystallization from methanol-methallyl alcohol, we obtained the single crystal of an inclusion compound, which consisted of **2**, methanol, and methallyl alcohol. The X-ray structure shows an antiparallel mode where two kinds of dipeptide ribbon structures are bound with one another by multiple hydrogen bonding (Figure 6).   Two the ionic pair (COO⁻•••⁺NH₃) and four hydrogen bonding between two peptide ribbons are observed.   Thus, four additional hydrogen bonds play an important role in the antiparallel mode.   This is in sharp contrast with the parallel mode.   Methanol and methallyl alcohol exist as guest molecules in pocket-type cavities surrounded by naphthyl and phenyl groups that stand perpendicularly to the sheet.   In addition, the inclusion compound of 2-propen-1-ol (allyl alcohol) was also obtained and confirmed by the single-crystal X-ray crystallographic analysis to exhibit the same antipallarel mode.

(a)   (b)

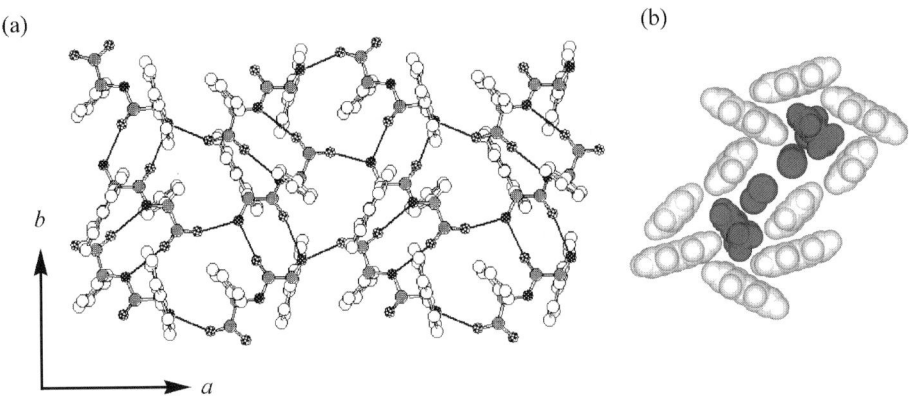

Figure 6.   Antiparallel β-sheet-like structure of **2** included methallyl alcohol and methanol.   (a) a-b face. To improve clarity, guest molecules are omitted, and the naphthyl and phenyl groups are colored white. (b) Guests (gray) in the pocket-type void surrounded by naphthyl and phenyl groups (white)

Here, we wish to mention the mechanism for alterating the sheet structure that is parallel or antiparallel according to the kind of the guest.   The dipeptide (**1** or **2**) can be regarded as a linear rod that has a carboxyl group at one end and an amino group at another end.   The intermolecular salt formation between the amino group and the carboxyl group of the dipeptide molecules results in the formation of a one-dimensional ribbon structure (Figure 7).   In the case that the guest is a sulfoxide or an ether, it has a Lewis base site.   Therefore, the guest is bound to the ammonium hydrogen of the ribbon structure.   The third proton of the ammonium group contributes to the formation of the sheet structure that is parallel-like.   When the guest is an alcohol, the hydroxy proton behaves as the BrØnsted acid to form a hydrogen-bond with the carboxylate site of the ribbon structure.   The ammonium proton does not link to the guest, but it is bound to the amido carbonyl oxygen of the neighboring ribbon to form an eight-membered hydrogen-bond cycle and, as a result, form anti-parallel sheet structure.   Thus, (*R*)-(1-naphthyl)glycyl-(*R*)-phenylglycine (**2**) was shown to have the ability to optionally form "parallel or antiparallel" β-sheet-like structures depending on the included molecules, ethers or allylic alcohols.   This is regarded as a prototype that the guest molecules control β-sheet structures.

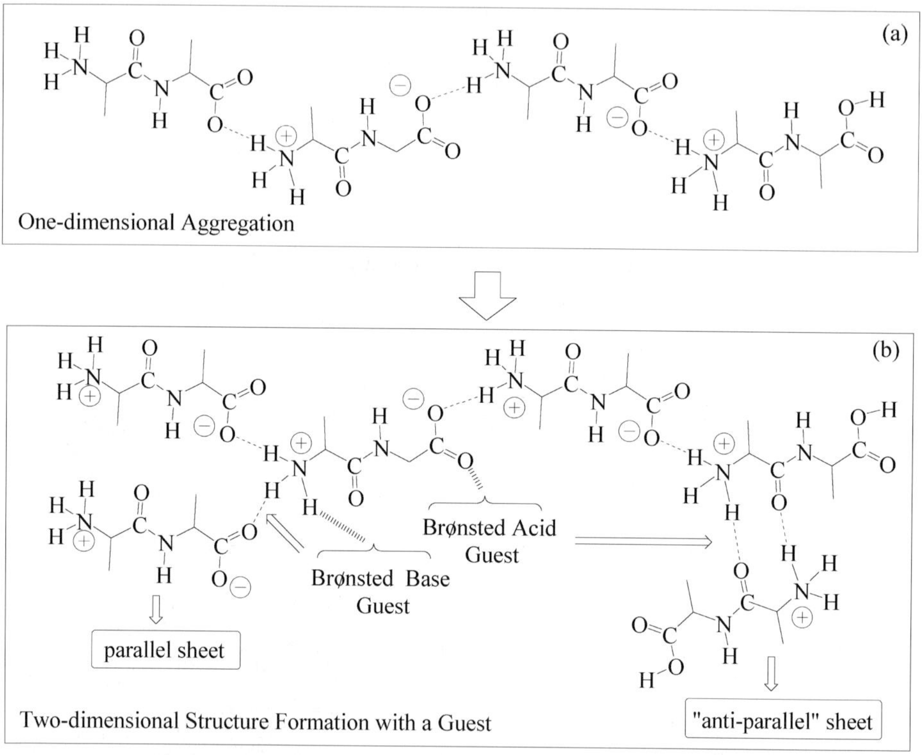

Figure 7. Formation of the sheet structure in inclusion crystals with a Lewis base guest or a Lewis acid guest. (a) Formation of one-dimensional ribbon by head-to-tail arrangement (salt formation) of the dipeptide molecules. (b) Two-dimensional arrangement of the ribbons under the influence of the guest.

3. Enatiomeric Separation of α-Hydroxy Esters by Crystalline Dipeptides.

As mentioned above, the guest choices the sheet struycture, "parallel" or "antiparallel" β-sheet-like structure formed in the inclusion crystals of the dipeptide (**2**). Interestingly, the pocket-type cavity surrounded by naphthyl and phenyl groups appears on the "antiparallel" mode. Since the pocket-type cavity in the antiparallel mode of **2** is chiral, we investigated its asymmetric recognition for chiral α-hydroxy esters,[13] which have two functional groups to bring about synergistic hydrogen bonding.

By crystallization of a mixture of the dipeptide (**2**) and racemic methyl lactate from methanol, asymmetric recognition occurred to give an inclusion compound that contains the (*S*)-form of methyl lactate in 89% ee.[14] X-ray crystallographic study elucidated that, in the inclusion compound, the dipeptide molecules were arranged in antiparallel β-sheet-like structure to accommodate the α-hydroxy ester in the pocket-type cavity.

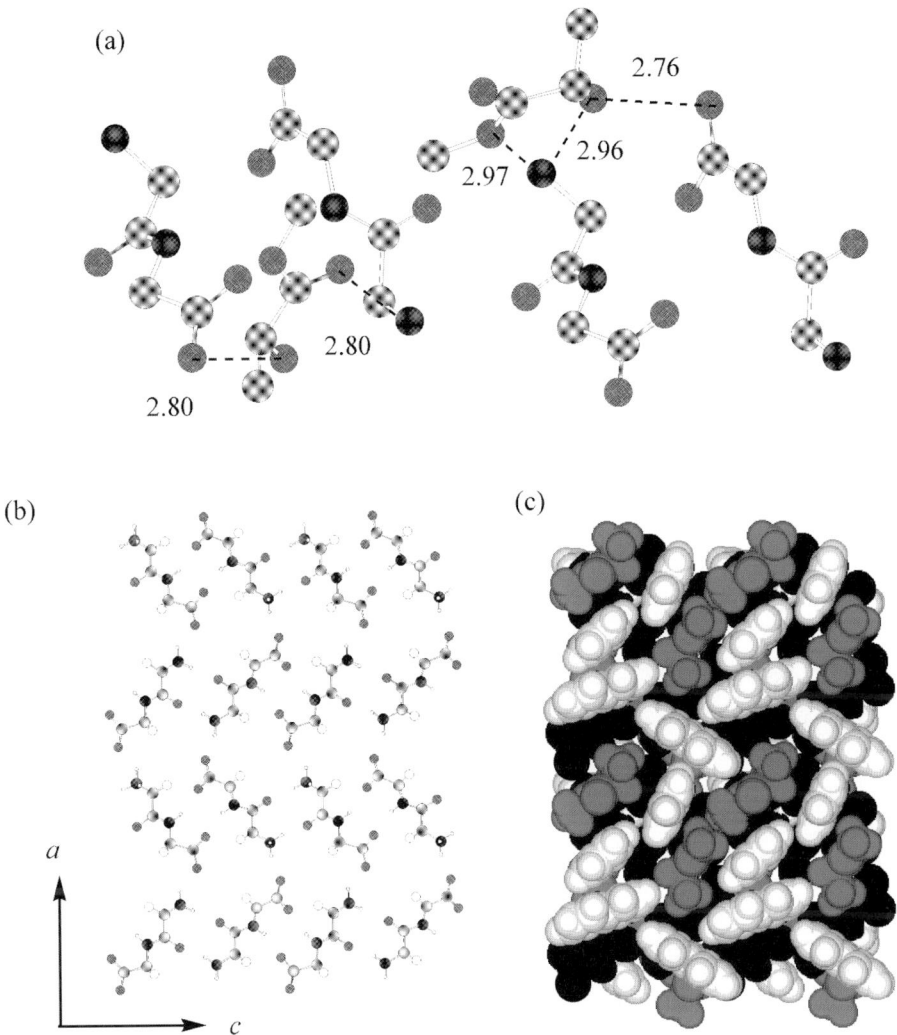

Figure 8. Inclusion compound of (*S*)-methyl lactate by **2**. (a) Host-guest hydrogen bonding (Å).
(b) Sheet structure of dipeptide backbone. (c) Packing (CPK model) of the inclusion compound.

Furthermore, we examined how other α-hydroxy esters are included into the crystals of
2. The results are summarized in Table 1, informing that a small ester part (R^2=methyl
or ethyl) in the α-hydroxy esters is necessary to the formation of their inclusion crystals
and their (S)-enantiomers are included with a high selectivity.

TABLE 1.   Enantiomeric Recognition of α-Hydroxy Esters by the Dipeptide (**2**).

entry	α-hydroxy ester		guest/host[a]	% ee	config
	R^1	R^2			
1	Me	Me	1.00	98	S
2	Me	Et	1.00	97	S
3	Me	n-Pr	not included		
4	Me	i-Pr	not included		
5	Et	Me	1.00 (0.00, 0.28)	87	S
6	-C(CH$_3$)$_2$CH$_2$-		0.50 (1.00, 0.36)	92	S
7	n-Pr	Me	0.53 (1.09, 0.00)	89	S
8	n-Bu	Me	0.50 (1.05, 0.00)	>99	S
9	i-Bu	Me	0.52 (1.10, 0.00)	87	S

[a] Determined by ^{1}H NMR and elemental analysis. The ratios of methanol and water, respectively, in the inclusion compound are shown in the parenthesis.

Furthermore, it should be noted that pocket-type cavities are always formed in these inclusion crystals.   These cavities are similar to each other, as shown in Figure 9.

Thus, the arrangement of the host (**2**) molecules in the inclusion compound is governed by the size or shape of α-hydroxy esters (guest) to form optional sheet structures with adjustable cavities for highly enantiomeric inclusion.

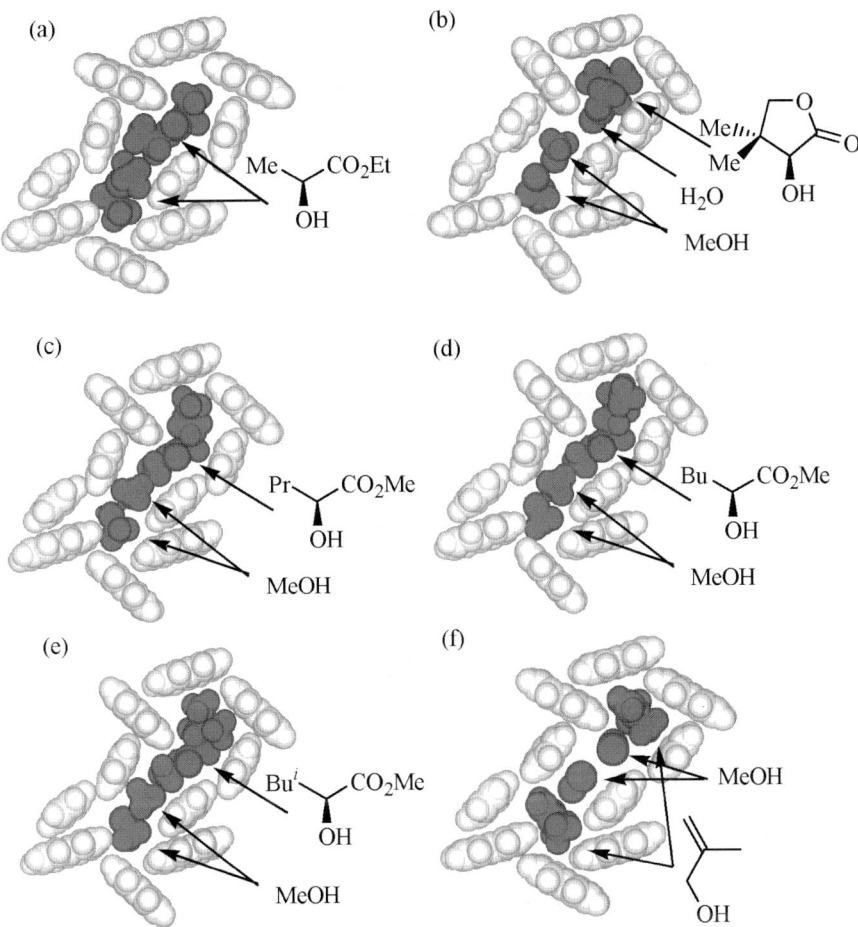

Figure 9. Cavities of the inclusion compounds. (a) **2** • (*S*)-ethyl lactate (1:1), (b) **2** • .(*S*)-dihydro-3-hydroxy-4,4-dimethyl-2(3H)-furanone • MeOH • H₂O (2:1:2:1), (c) **2** • (*S*)-methyl α-hydroxypantanoate • MeOH (2:1:2), (d) **2** • (*S*)-methyl α-hydroxyhexanoate•MeOH (2:1:2), (e) **2** • (*S*)-methyl α-hydroxy-γ-methylpantanoate • MeOH, (f) **2** • methallyl alcohol• MeOH (1:1:1)

4.   Inclusion of Ethers and PEGs

In Section 2, it was mentioned that (*R*)-(1-naphthyl)glycyl-(*R*)-phenylglycine (**2**) forms inclusion compounds with ethers (tetrahydrofuran and diethyl ether). X-Ray crystallographic study showed that the dipeptide molecules are arranged into a "parallel" β-sheet-like structure. Interestingly, (*R*)-phenylglycyl-(*R*)-phenylglycine (**1**) also forms an inclusion compound with THF or ether, which has the similar parallel β-sheet-like structure, but it is so unstable that the included ether molecules escape from the crystal lattice at room temperature to give the original crystals of **1**.

We also found the inclusion of 1,2-dimethoxyethane and its derivatives (guests) between the layers in crystalline dipeptide (**1** or **2**). [15]

TABLE 2. Inclusion of ethers by crystallization of dipeptides

Entry	Guest	1		2	
		Ef.[a]/%	L. D.[b]/Å	Ef.[a]/%	L. D.[b]/Å
1	‵O⌒⌒O‵	100	11.8	100	12.1
2	(OMe, OMe)	98	13.5	100	13.7
3	(OMe, OEt)	83	13.4	78	13.8
4	(OMe, O^{i}Pr)	42	13.4	not included	
5	(OEt, OEt)	not included		not included	
6	(OMe, OMe allyl)	94	15.2	90	15.2
7	(OMe, OMe indane)	78	15.7	88	15.0

[a] Ef. means the ratio (mol%) of the guest based on the dipeptide in the inclusion compound. [b] L.D. means the layer distance measured by PXRD.

Packing views of the crystal structures for the DME- and o-dimethoxybenzene-included compounds of **2** are illustrated in Figure 10. The inclusion compounds have the same layer structure, where the dipeptide molecules were arranged into the parallel β-sheet-like structure. The neighboring naphthyl and phenyl rings stacked each other with the aid of edge-to-face interaction to form a wall on the layer.[16,17] The benzene rings of the o-dimethxoybenzene molecules effectively stacked between the wall of the naphthyl and phenyl groups (Figure 10b). As seen from the layer distances (L. D.) in Table 2 that increase with the increasing molecular length, the guest molecules act as pillars that support the layers.

In the DME-included compound of **2**, DME molecules are regularly arranged in the one-dimensional channel. If the terminal methyl groups of DME are intermolecularly linked, the arranged DME molecules would be regarded as the dimethyl ether of poly(ethylene glycol)s [PEG]   Hence, we investigated whether

(a)   (b)

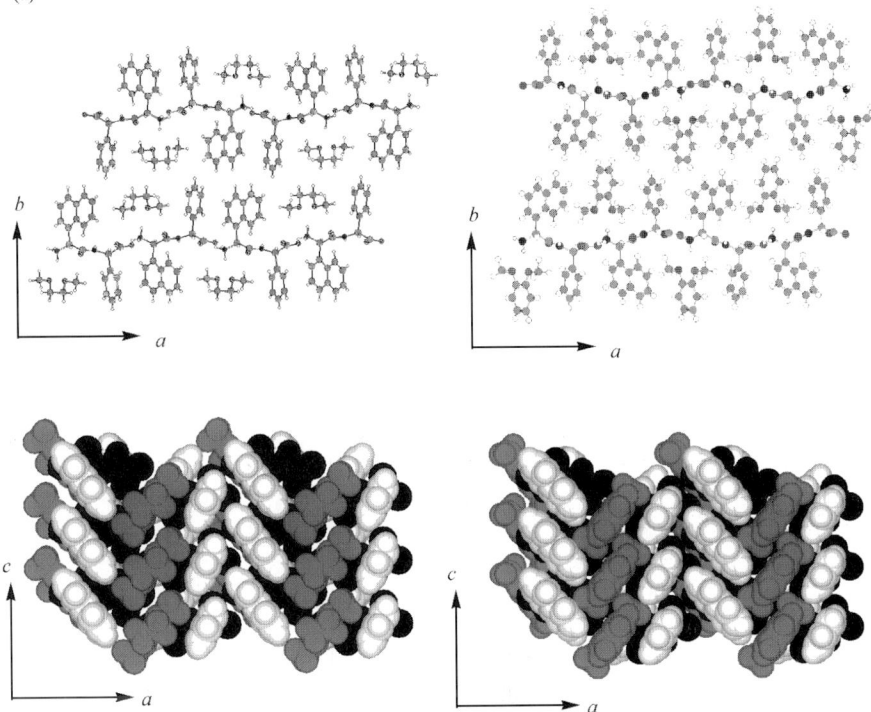

Figure 10. Inclusion compounds of 1,2-dimethoxyethane (a) and *o*-dimethoxybenzene (b) by **2**. Layer sturcture and packing (CPK model) of these inclusion compounds.   For clarity, phenyl groups of **2** and guest are colored white and gray, respectively

the molecules of **2** form an inclusion compound of an oligo(ethylene glycol) dimethyl ether [CH$_3$O-(CH$_2$CH$_2$O)$_n$-CH$_3$; PEG DM(n)]. [18]   The illustration of Table 3 shows that the calculated molar ratios can be estimated according to an equation (n+1)/2, in which n is the number of the ethylenoxy unit.

Ten kinds of PEG DM(n = 1-7, 9, 12, and 20) were selected as the guest molecule. Fortunately, the inclusion compounds were obtained from **2** and all of the selected PEG DM(n) by cocrystallization from methanol.   As summarized in Table 3, PEG DM(n) molecules were efficiently included in the crystal lattice of **2**.   As expected, the observed ratios were proportional to the length of the polymer chain and comparable to the calculated ratios.   In their PXRD patterns, all of these inclusion compounds showed a strong diffraction peak at 12.0 - 12.3 Å.   These values are the same as the layer distance of the DME-included compound of **2**.   These facts suggested that oligo(ethylene glycol) dimethyl ethers are accommodated in the channel constructed by the molecules of **2**.   When hepta(ethylene glycol) dimethyl ether [PEG DM(7)] was used as the guest, we obtained a single crystal suitable for X-ray crystallographic analysis at 173 K. By regarding a series of the guest molecules in the channel as the repeating of a -C-C-O-C-C-O- moiety, their conformation was resolved and refined isotropically.

TABLE 3. Inclusion of poly(ethylene glycol) dimethyl ethers by the dipeptide (**2**)

Guest	Calcd. H/G[a]	Obs. H/G[b]	L.D.[c]
n =1	1.00	1.00	12.1
n =2	1.50	1.56	12.2
n =3	2.00	1.92	12.2
n =4	2.50	2.50	12.2
n =5	3.00	3.10	12.0
n =6	3.50	3.15	12.3
n =7	4.00	3.80	12.1
n =9	5.00	5.00	12.3
n =12	6.50	6.40	12.1
n =20	10.50	10.50	12.1

[a] Calculated values of the host/guest ratio. [b] The host/guest ratio determined by NMR. [c] The layer distance measured by PXRD.

A herringbone motif of naphthyl and phenyl groups constructs the wall of the one-dimensional channel and the PEG DM(7) molecules are accommodated in the channel via hydrogen bonding (O···N distance; 2.98 Å). The conformation of PEG DM(7) to be $G^+TG^-G^+TG^-$, where T, G^+, and G^- denote *trans*, right-handed *gauche*, and left-handed *gauche* conformations, respectively (Figure 11b). A similar conformation of a PEG chain was observed in a PEG-HgCl$_2$ as shown in Figure 11c.[19] Thus, the included PEGs adopt a conformation that meets not only the spatial restriction of the inclusion channel, but also the requirement of the above-mentioned hydrogen bonding. It should be noted that the ethylenoxy unit seems to be essential for the inclusion in the one-dimensional channel because tris(trimethylene glycol) dimethyl ether [CH$_3$O(CH$_2$CH$_2$CH$_2$O)$_3$CH$_3$] was not included at all.

In the inclusion compounds of such alcoholic guests with the dipeptide (**2**), the one-dimensional channel was not constructed, but a zero-dimensional pocket-type cavity was formed.[14] Therefore, we wondered whether poly(ethylene glycol)s (PEG) are included in the crystal lattice of **2**.   In fact, tri(ethylene glycol), tetra(ethylene glycol), and tri(ethylene glycol) monomethyl ether did not form the

(a)

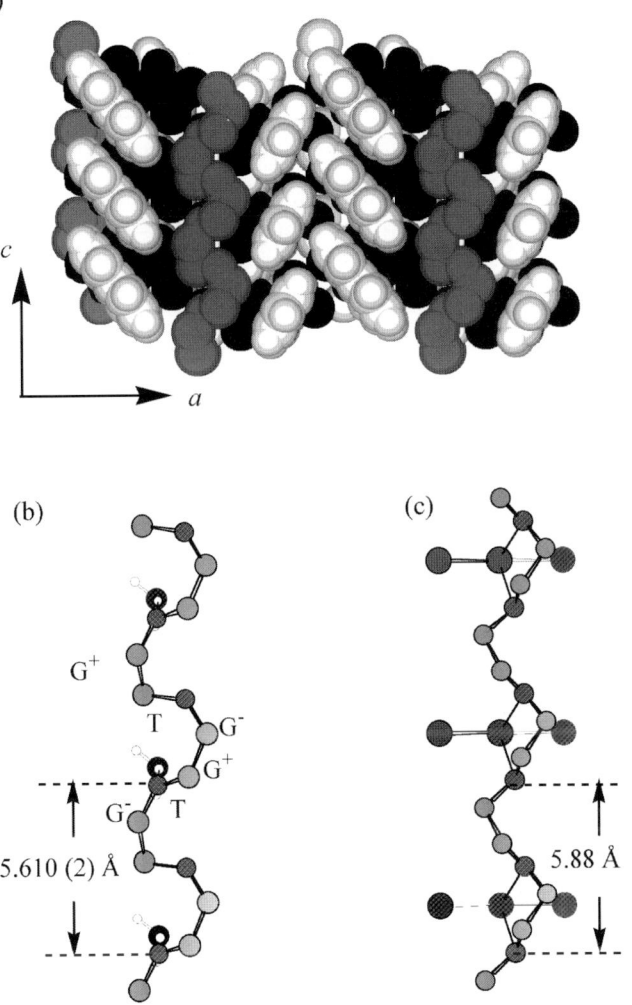

Figure 11. (a) Inclusion compound of MeO(CH$_2$CH$_2$O)$_7$Me by **2**. The guest molecules was resolved as a   PEG.   (b) Conformation of PEG in the channel.   (c) PEG-HgCl$_2$ complex

corresponding inclusion crystals.   However, PEGs with the molecular weight of 400 - 20,000 succeeded in the formation of the corresponding inclusion compounds.   The ratios of **2** : PEG are in good accordance with the values calculated by the equation (n+1)/2 (*vide post*).   The layer distances of these inclusion compounds, which were assigned by their PXRD analysis, are in the range of 12.0 - 12.1 Å, suggesting that the present inclusion compounds have the layer structure similar to that of the DME inclusion compound.

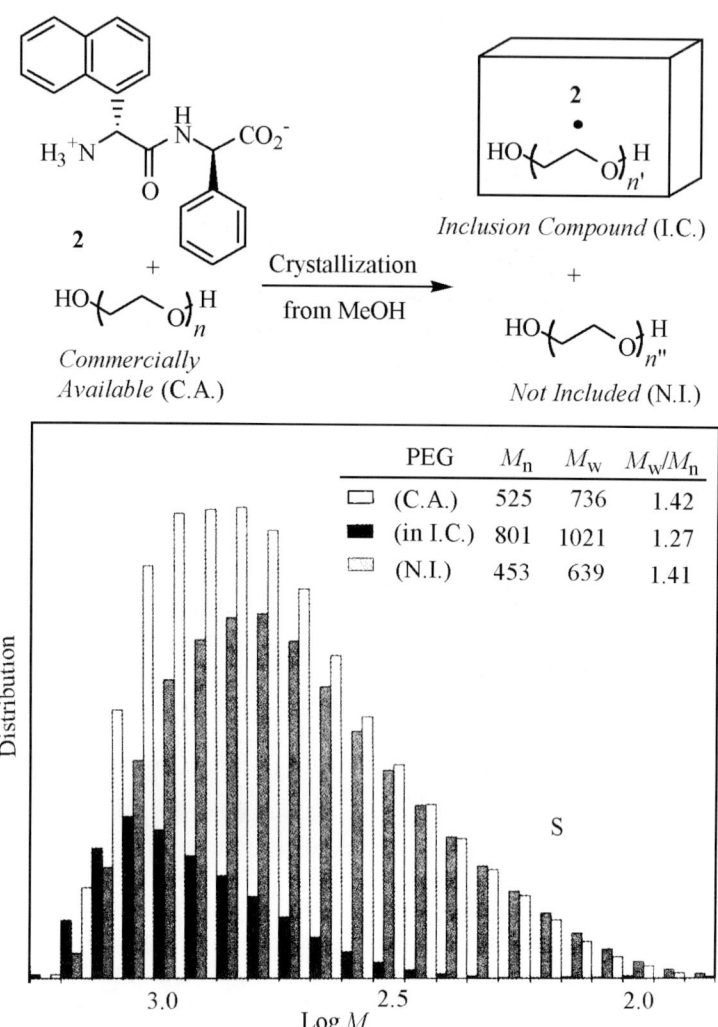

Figure 12. Molecular weight distribution: The employed polymer (white), included polymer (black), and not-included polymer (gray).

We succeeded in the fractionating a mixture of PEGs with various molecular weights by the mean of its inclusion into **2** crystals. Preliminarily, we investigated the fractionation of PEG DM(4) and PEG DM(12). After a mixture (82.0 mg) of PEG DM(4) and PEG DM(12) (49:51, wt/wt) was dissolved in methanol containing **2**, the resulting solution was allowed to stand at room temperature. After about one week, the deposited crystals were collected by filtration. The ratio of PEG DM(4): PEG DM(12) in the crystals was 14:86 (the total amount of PEG DM(4) and PEG DM(12) = 20.2 mg). In a similar manner, PEGs with a broad distribution of molecular weight were fractionated. The PEG mixtures was prepared by mixing commercially available PEGs. Sample A is a 1:1 mixture of commercial PEG #1,000 and PEG #20,000. Sample B is a 1:1:1 mixture of PEG #600, PEG #1,000, and PEG #1,500. In the case of Sample A, the ratio of the included PEGs #1,000 and #20,000 was 35:65, while the not-included one was 56:44. The favorable inclusion of the PEG with a higher molecular weight was also observed in the inclusion of Sample B, as summarized in Figure 12.

Since analogous phenomena were reported in the formation of urea-poly(tetrahydrofuran) inclusion compounds,[20,21] it seems to be generally conclusive that, in the crystallization process, the inclusion of the polymer with higher molecular weight is more favorable. This is reasonably explained in terms of entropic advantage in the inclusion: inversely speaking, the inclusion of the polymer with higher molecular weight is smaller in entropy loss.

5. Chiral Recognition of Sulfoxides

As mentioned before, isopropyl phenyl sulfoxide was included in the crystals of **1** with a high enantioselectivity of its (S)-form. Generally, the sulfoxide-included compounds were prepared by two methods: (a) insoluble **1** was simply stirred in the presence of an alkyl phenyl sulfoxide and water [Method A: "sorption"] and (b) **1** is recrystallized in the presence of the sulfoxide [Method B: "crystallization"].[7,12] The results are summarized in Table 4, which shows enantioselectivity and efficiency in the inclusion of various alkyl phenyl sulfoxides. The efficiency means the mol percentage of the guest molecule based on **1** molecule in the inclusion compound.

By Methods A and B, isopropyl phenyl sulfoxide was included in crystalline **1** with high (S)-enantioselectivity (86 and 87% ee, respectively). Ethyl phenyl sulfoxide formed no inclusion compound by Method A, but the inclusion compound of its (S)-enantiomer was obtained by Method B. The inclusion crystal of (S)-ethyl phenyl sulfoxide is isostructural with that of (S)-isopropyl phenyl sulfoxide (Figure 3). As mentioned above, (S)-ethyl phenyl sulfoxide was not included by Method A. The lack of one methyl group may make enthalpy (interaction with the inclusion cavity) and entropy disadvantageous in crystal packing to result in no inclusion of ethyl phenyl sulfoxide via Method A.

As shown in Figure 4 of section 2, the dipeptide molecules are arranged in a parallel β-sheet-like structure which is constructed by ionic pairing of carboxyl and amino groups *via* a hydrogen bonding network: one terminal COO^- bridged two $^+NH_3$ of adjacent dipeptides, and the $^+NH_3$ also bound two adjacent COO^- groups. The phenyl groups are placed perpendicular to the sheet. The neighboring two phenyl rings stacked each other with the aid of the edge-to-face interaction to form a wall.[16] Interestingly, a typical intermolecular hydrogen bonding among amide groups, which is essential to the formation of oligopeptide crystals,[22] did not contribute to the parallel β-sheet formation. The dipeptide backbones of the parallel mode are aligned with an interchain spacing of 5.51 Å, which corresponds to the c-axis length. The value is larger than the interchain spacings of general parallel β-sheet structures (4.85 Å ideally, 4.90-5.08 Å observed).[12]

TABLE 4. Enantioselective inclusion of sulfoxides by crystalline **1**

Entry	Sulfoxide	A(sorption)		B(cystallization)	
		ee/%	Ef.[a]/%	ee/%	Ef.[a]/%
1	⟨phenyl⟩–S–CH(CH₃)CH₃, O	86(*S*)	100	87(*S*)	100
2	⟨phenyl⟩–S–CH₂CH₃, O	not included		91(*S*)	95
3	⟨phenyl⟩–S–CH₂Cl, O	not included		not included	
4	⟨phenyl⟩–S–CH₃, O	92(*R*)	95	93(*R*)	100
5	⟨o-CH₃-phenyl⟩–S–CH₃, O	92(*R*)	54	94(*R*)	98
6	⟨o-Cl-phenyl⟩–S–CH₃, O	48(*R*)	100	99(*R*)	97
7	⟨phenyl⟩CH₂–S–CH₃, O	Rac.[b]	100	Rac.[b]	100

[a] Ef. is an abbreviation of "Efficiency". Efficiency means mol% of guest based on **1** in the inclusion compound. [b]Rac. means the enantioselectivity is less than 10%.

The sulfoxide molecules are accommodated in the void between the layers via three-point interaction: hydrogen bonding between $^+NH_3$ and the sulfinyl group, the "tilted T"-shaped interaction between two phenyl groups,[17] and the CH/π interaction of phenyl and isopropyl groups.[24] In proteins, π-π interactions on their side chains have already been investigated and the edge-to-face interaction between two benzene rings was revealed to take an important role in their interaction.[16] It is noteworthly that the inclusion compound of the dipeptide (**1**) with ethyl phenyl sulfoxide has the crystal structure similar to that of isopropyl phenyl sulfoxide inclusion crystal. Interestingly, the behavior of methyl phenyl sulfoxide in the inclusion of crystalline **1** is in sharp contrast with those of the above alkyl phenyl sulfoxides;   Methyl phenyl sulfoxide was absorbed into crystalline **1** by Methods A and B to give the inclusion compound with high (*R*)-enantioselectivity (92 and 93% ee, respectively).[12]   A question is why (*R*)-methyl phenyl sulfoxide is included in crystalline **1**.   Since a single crystal with good quality could not be obtained for the (*R*)-methyl phenyl sulfoxide inclusion compound, we designed a new guest molecule, bis[2-(methylsulfinyl)benzyl] ether, that has two methyl phenyl sulfoxide units linked by a -CH$_2$OCH$_2$- bridge(Figure 13a).

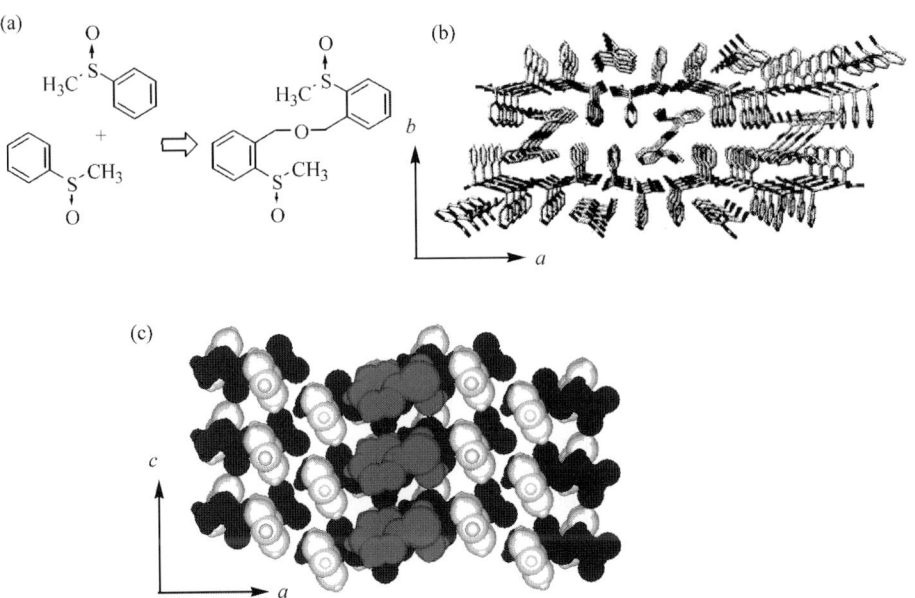

Figure 13.   (a) Design of a new guest linking two methyl phenyl sulfoxide molecules (b) Perspective views from a-b plane of the inclusion compound of **1** (c) Topview of CPK model from a-c plane of the inclusion compound of **1**(the half of guest molecule is shown). For clarity, phenyl groups of **1** and guest are colored white and gray, respectively.

Consideration using CPK molecular model suggested that the designed disulfoxide   is enough large in size to spread over two continuous cavities between the layers of **1**. This may make it easier for the inclusion compound of the disulfoxide to crystallize. Fortunately, its a single crystal with good quality was obtained by Method B.   As expected, the disulfoxide molecules were intercalated between two layers of **1**.   From a perspective side view (Figure 13b) and a top view (Figure 13c) of its crystal structure, the following distinct features became apparent: the layer structure is same to those of the inclusion compound of (*S*)- isopropyl phenyl sulfoxide *except for the phenyl-phenyl stacking mode*. The phenyl groups on the dipeptide layer change their conformation from the "tilted T"-shaped mode to "parallel stacking and displaced" mode.[17] (*R*)-(Methylsulfinyl)phenyl moiety interacted with two phenyl groups of the layer. One interaction is "T"-mode phenyl-phenyl stacking[17] and a CH/π interaction[24] works between methyl and phenyl groups.   For the interaction with phenyl group, the methyl group that binds directly to a highly polarized sulfinyl group seems to have advantage of ethyl group in entropy and enthalpy. In fact, strong CH/π interaction was observed between a benzene ring and the methyl group on a cationic nitrogen.[25] The same inclusion mode was also observed in the inclusion crystals of (R)-2-chlorophenyl methyl sulfoxide as shown in Figure 14.

(a)

(b)

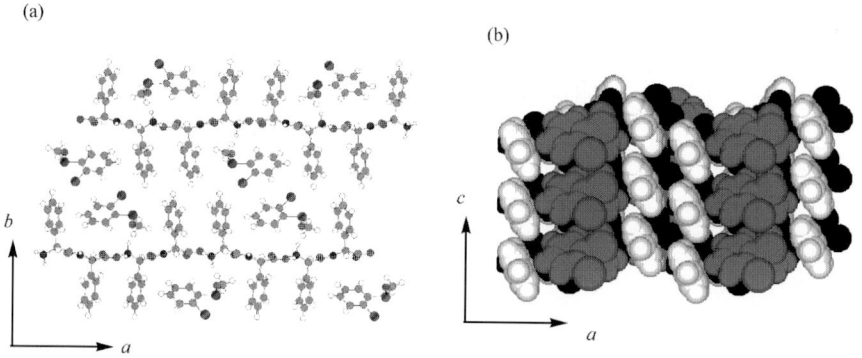

Figure 14. Inclusion compound of (R)-2-chlorophenyl methyl sulfoxide by **1**. (a) Layer sturcture. (b) Packing (CPK model) of the inclusion compound. For clarity, phenyl groups of **1** and guest are colored white and gray, respectively

Thus, the shape of the sulfoxide guest was shown to induce the conformational change of the phenyl groups on the dipeptide layer (Figure 15). These behavior, which is analogous to induced-fit phenomena at the active site of enzymes, seems to be based on an allostreric character of the crystalline dipeptide host.

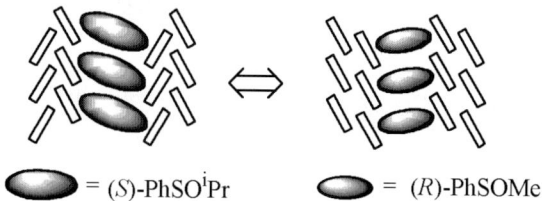

= (S)-PhSO^{i}Pr   = (R)-PhSOMe

Figure 15. Conformational change of phenyl groups of **1** depending on the guest size.

It should be noted that benzyl methyl sulfoxide was included without the recognition of their chirality (entry 7 in Table 4).[26] By single crystal X-ray analyses of this inclusion compound, it was elucidated that the molecules of **1** form the layer structure similar to the one of the above alkyl phenyl sulfoxide-inclusion crystals, and the sulfoxides are included between these layers. There are two different recognition cavities on the upper side for its S enantiomer and on the lower side for its R enantiomer. The upper side cavity can be illustrated by motif A, and the lower side one by motif B in Figure 16.

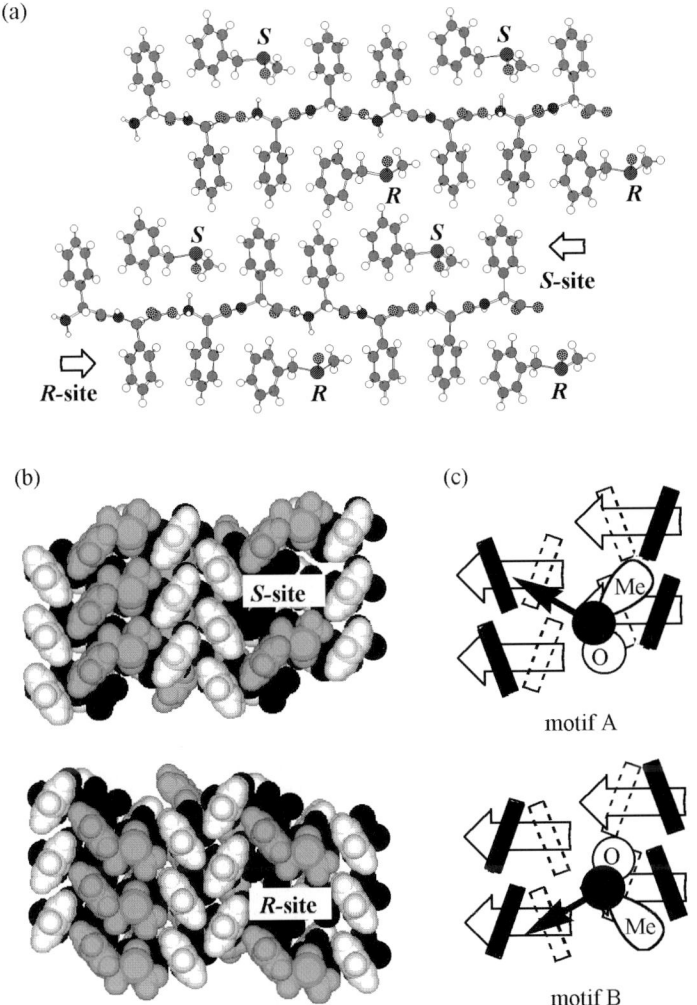

Figure 16. Inclusion compound of benzyl methyl sulfoxide by **1**. (a) Layer sturcture. (b) Packing (CPK model) and schematic representaion of both recognition sites. For clarity, phenyl groups of **1** and guest are colored white and gray, respectively

These cavities are just like a mirror image to each other in shape. In these cases, (*R*)- and (*S*)-benzyl methyl sulfoxides directed their methyl group and benzyl groups to the C-terminal phenyl group and to the N-terminal phenyl group, respectively. This arrangement makes the phenyl groups of **1** and benzyl methyl sulfoxide pack tightly in the usual herringbone motif.[27] The relative position between the ammonio proton and the C-terminal phenyl group in one molecule of the dipeptide determined the stereochemistry of the methyl sulfinyl groups that are recognized.

Finally, we would like to describe the enantiomeric separation of racemic methyl phenyl sulfoxide by the dipeptide (**1**). Optically active sulfoxides are utilized as a useful reagent for synthesizing various chiral compounds.   For preparing these optically active sulfoxides, asymmetric oxidation of the corresponding sulfides,[28] the reaction of *l*-mentyl arenesulfinate with organometallic reagent,[29] and chiral separation of racemic sulfoxide[5,] have been developed.   Among various optically active sulfoxides reported to date, (*R*)-alkyl *p*-tolyl sulfoxide has been often employed.   This is probably because it can be easily prepared by the reaction of alkyl magnesium halide and mentyl *p*-toluenesulfinate (Andersen Method),[29] one diastereomer of which shows enough good crystallinity to be isolated.   In contrast, diastereomerically pure mentyl benzenesulfinate is not obtainable because of its poor crystallinity.   Thus, good synthetic route of optically active methyl phenyl sulfoxide was not available.   Hence, we investigated whether the present enantiomeric recognition of crystalline dipeptide (**1**) is applicable to the preparative resolution of racemic methyl phenyl sulfoxide.

TABLE 5.   Optical Resolution of Mehyl Phenyl Sulfoxide by the Dipeptide (**1**) and Its Reuse.

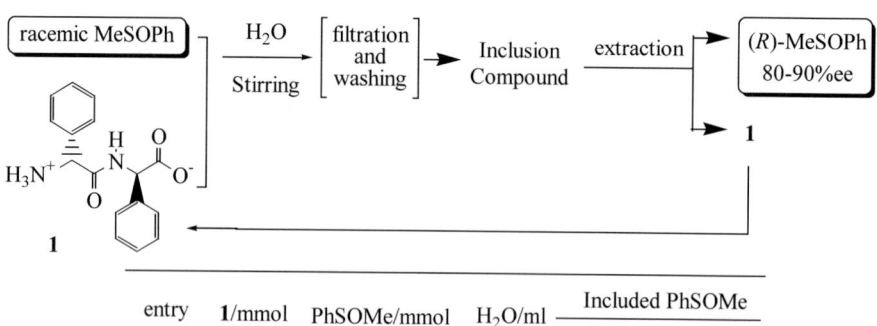

entry	1/mmol	PhSOMe/mmol	H_2O/ml	Included PhSOMe	
				yield/%	e.e./%
1	7.4	18.3	18.0	69	76
2	6.4 [a]	15.8	16.0	83	87

a The dipeptide (**1**)that was recovered in entry 1 was used.

A suspension of the dipeptide (**1**) (1 mmol) in water (2 ml) was stirred together with racemic methyl phenyl sulfoxide (2 mmol) at room temperature for one day.   The formed inclusion compound was collected by filtration and washed with water (20 ml) and dichloromethane (20 ml).   From the inclusion compound, we recoverd the included methyl phenyl sulfoxide by extraction with dichloromethane to give (*R*)-methyl phenyl sulfoxide.   The dipeptide (**1**) remained as a solid.   When the recovered dipeptide (**1**) was again subjected to the formation of inclusion compound with racemic methyl phenyl sulfoxide, (*R*)-methyl phenyl sulfoxide was obtained as summarized in Table 5.   Thus, it was shown that the dipeptide (**1**) can be used repeatedly for enantiomeric separation of methyl phenyl sulfoxide.

      We also examined the efficient separation of two enantiomers from racemic methyl phenyl sulfoxide by the use of the dipeptide (**1**) and its enantiomer (**1***).   According to the above procedure for racemic methyl phenyl sulfoxide and the dipeptide (**1**), we obtained (*S*)-rich methyl phenyl sulfoxide from the washings with dichloromethane.

This (S)-rich methyl phenyl sulfoxide was subjected to the inclusion with the enantiomer (1*) as summarized in Scheme 1. In this procedure, (S)-methyl phenyl sulfoxide was obtained with high enantioselectivity. It is noteworthy that racemic methyl phenyl sulfoxide can be recovered from the washings. Thus, two enantiomers of methyl phenyl sulfoxide can be obtained by the use of the dipeptide (1) and its enantiomer (1*). It should be noted that the present procedure can be applicable to other alkyl phenyl sulfoxides.

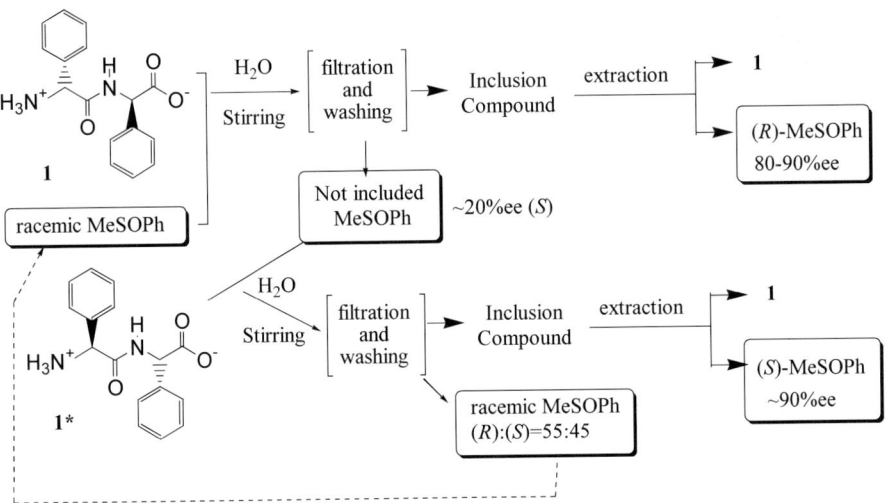

Figure 17. Separation of methyl phenyl sulfoxide using (R)-phenylglycyl-(R)-phenylglycine (**1**) and (S)-phenylglycyl-(S)-phenylglycine (**1***).

6. Summary

We have described the molecular recognition of various guests with (R)-phenylglycyl-(R)-phenylglycine (**1**) and (R)-(1-naphthyl)lglycyl-(R)- phenylglycine (**2**) to form inclusion crystals. Their crystal structures are changeable according to the functionalities and size of the guest. Since these dipeptides have a straight glycylglycine backbone, a two-dimensional layer is generally constructed by intermolecular salt formation between COOH and NH_2 groups. Guest molecules are included between the layers via host-guest interactions such as hydrogen bonding, aromatic-aromatic interaction, and CH/π interaction. α-Hydroxy esters and sulfoxides were recognized with high enantioselectivity to be included in the chiral cavities. Motifs of the dipeptide aggregation depend on the kinds of guests: sulfoxide and ether guests assemble the dipeptide molecules in parallel, whereas α-hydroxy esters arrange them in antiparallel. It was demonstrated that these simple dipeptide hosts change their conformation according to the shape of the guest. This seems like a flexible model for the recognition of substrates by enzymes. Thus, the present chiral recognition by the dipeptides (**1** and **2**) would contribute to a novel method for the optional resolution of α-hydroxy esters and sulfoxides.

References

1) MacNicol, D. D., Toda, F., and Bishop R. (1996) *Comprehensive Supramolecular Chemistry, Vol. 6,* Elsevier Science Ltd., Oxford.

2) Toda, F. (1989) Studies of Host-Guest Chemistry. −Fundamentals and Applications of Molecular Recognition and their Development to New Organic Solid State Chemistry-, *J. Syn. Org. Chem., Jpn.* **47**, 1118-1131.

3) Sada, K. and Miyata, K. (1996) Inclusion Compounds of Bile Acids and their Derivatives, *J. Syn. Org. Chem., Jpn.* **54**, 113-121.

4) Seebach, D., Beck, A. K. Heckel, A. (2001) TADDOLs, Their Derivatives, and TADDOL Analogues: Versatile Chiral Auxiliaries, *Angew. Chem., Int. Ed.* **40**, 92-138.

5) a) Toda, F., Tanaka, K., and Nagamatsu, S. (1984) Mutual optical resolution of 2,2'-dihydroxy-1,1'-binaphthyl and alkyl aryl or dialkyl sulfoxides by complex formation, *Tetrahedron Lett.* **25**, 4929-4934 ; b) Toda, F., Tanaka, K., and Mak, T. C. W. (1984) Mutual optical resolution of bis-β-naphthol and sulfoxides. Absolute configuration and crystal structure of a 1:1 molecular complex, *Chem. Lett.* **1984**, 2085-2088.

6) Aoyama, Y., Endo, K., Kobayashi, K., and Masuda, H. (1995) Hydrogen-bonded network and enforced supramolecular cavities in molecular crystals: an orthogonal aromatic-triad strategy. Guest binding, molecular recognition, and molecular alignment properties of a bisresorcinol derivative of anthracene in the crystalline state, *Supramolecular Chem.* **4**, 229-241.

7) Ogura, K., Uchida, T., Noguchi, M., Minoguchi, M., Murata, A., Fujita, M. and Ogata, K. (1990) A new system for molecular recognition: Highly specific inclusion of (*S*)-isopropyl phenyl sulfoxide by solid (*R*)-phenylglycyl-(*R*)-phenylglycine, *Tetrahedron Lett.* **31**, 3331-3334.

8) a) Görbitz, C. H. (1997) Cyclic water pentamers in L-leucyl-L-alanine tetrahydrate, *Acta Cryst.* **C53**, 736-739 ; b) Görbitz, C. H. (1999) L-Leucyl-L-leucine 2-methyl-1-propanol solvate, *Acta Cryst.* **C55**, 670-672 ; c) Görbitz, C. H. (1999) L-alanyl-L-phenylalanine-2-propanol (1/2) (α-form), L-valyl-L-phenylalanine-2-propanol (1/1) and L-leucyl-L-phenylalanine-2-propanol (1/1) (β-form), *Acta Cryst.* **C55**, 2171-2177 ; d) Görbitz, C. H. and Torgersen, E. (1999) Symmetry, pseudosymmetry and packing disorder in the alcohol solvates of L-leucyl-L-valine *Acta Cryst.* **B55**, 104-113 ; e) Mitra, S. N., Govindasamy, L. and Subramanian, E. (1996) L-Leucyl-L-alanine dimethyl sulfoxide solvate, *Acta Cryst.* **C52**, 2871-2873.

9) Yaghi, O. M., Davis, C. E., Li, G., and Li, H. (1995) Selective binding and removal of guests in a microporous metal–organic framework, *Nature* **378**, 703.

10) Russell, V. A., Evans, C. C., Li, W., and Ward, M. D. (1997) Nanoporous molecular sandwiches: Pillared two-dimensional hydrogen-bonded networks with adjustable p (cover story), *Science* **276**, 575-579.

11) Akazome, M., Sumikawa, A., Sonobe, R., and Ogura, K. (1996) Optional formation of "parallel or antiparallel" β-sheet-like structures in (*R*)-(1-naphthyl)glycyl-(*R*)-phenylglycine crystals, *Chem. Lett.* **1996**, 995-996.

12) Akazome, M., Noguchi, M., Tanaka, O., Sumikawa, A., Uchida, T., and Ogura, K. (1997) Enantiomeric recognition of alkyl phenyl sulfoxides by crystalline (*R*)-phenylglycyl-(*R*)-phenylglycine, *Tetrahedron* **53**, 8315-8322.

13) There are a few reports for enantiomeric inclusion of hydroxy esters by other crystalline hosts, see: a) Cramer, F., and Dietsche, W. 378 (1959) Occlusion compounds. XV. Resolution of racemates with cyclodextrins, *Chem. Ber.* **92** ; b) Toda, F., and Tohi, Y. (1993) Novel optical resolution methods by inclusion crystallization in suspension media and by fractional distillation, *J. Chem. Soc., Chem. Commun.,* 1238-1240 ; c) Toda, F., and Tanaka, K. (1988) A new chiral host compound 10,10'-dihyrroxy-9,9'-biphenanthryl. Optical resolution of propionic acid derivatives, butyric acid derivatives, and 4-hydroxycyclopent-2-en-1-one derivatives by complexation, *Tetrahedron Lett.* **29**, 1807-1810 ; d) Toda, F., Sato, A., Tanaka, K., and Mak, T. C. W. (1988) The crystal structure of 1:1 complexes of (*S*)-(-)-10,10'-dihydroxy-9,9'-biphenanthryl with (*S*)-(-)-methyl 2-chloropropionate and (*S*)-(-)-methyl 3-chloro-2-hydroxybutyrate, *Chem. Lett.* 781-784 ; e) Mravik, A., Böcskei, Z., Katona, Z., Markovits, I., and Fogassy, E. (1997) Coordination-mediated optical resolution of carboxylic acids with *O,O*'-dibenzoyltartaric acid, *Angew. Chem., Int. Ed. Engl.* **36**, 1534-1536.

14) Akazome, M., Takahashi, T., and Ogura, K. (1999) Enantiomeric Inclusion of α -Hydroxy Esters by (*R*)-(1-Naphthyl)glycyl-(*R*)-phenylglycine and the Crystal Structures of the Inclusion Cavities, *J. Org. Chem.* **64**, 2293-2300.

15) Akazome, M., Yanagita, Y., Sonobe, R., and Ogura, K. (1997) Specific Inclusion of

1,2-Dimethoxybenzene Derivatives by Crystalline (*R*)-Arylglycyl-(*R*)-phenylglycines and Its Structure, *Bull. Chem. Soc. Jpn.* **70**, 2823-2827.

16) a) Burley, S. K., and Petsko, G. A. (1985) Aromatic-aromatic interaction: a mechanism of protein structure stabilization, *Science* **229**, 23-28 ; b) Gould, R. O., Gray, A. M., Taylor, P., and Walkinshaw, M. D. (1985) Crystal environments and geometries of leucine, isoleucine, valine and phenylalanine provide estimates of minimum nonbonded contact and preferred van der Waals interaction distances, *J. Am. Chem. Soc.* **107**, 5921-5927 ; c) Burley S. K., and Petsko, G. A. (1986) Dimerization energetics of benzene and aromatic amino acid side chains, *J. Am. Chem. Soc.* **108**, 7995-8001 ; d) Hunter, C. A., and Sanders, J. K. M. (1990) The nature of π–π interactions, *J. Am. Chem. Soc.* **112**, 5525-5534 ; e) Bisson, A. P., Carver, F. J., Hunter, C. A., and Waltho, J. P. (1994) Molecular Zippers, *J. Am. Chem. Soc.* **116**, 10292-10293.

17) a) Jorgensen, W. L., and Severance, D. L. (1990) Aromatic-aromatic interactions: free energy profiles for the benzene dimer in water, chloroform, and liquid benzene *J. Am. Chem. Soc.* **112**, 4768-4774 ; b) Tsuzuki, S., Honda, K., Uchimura, T., Mikami, M., and Tanabe, K. (2002) Origin of Attraction and Directionality of the Interaction: Model Chemistry Calculations of Benzene Dimer Interaction, *J. Am. Chem. Soc.* **124**, 104-112.

18) Akazome, M., Takahashi, T., Sonobe, R., and Ogura, K. (2001) Inclusion of poly(ethylene glycol)s by crystalline (*R*)-(1-naphthyl)glycyl-(*R*)-phenylglycine, *Supramolecular Chem.* **13**, 109-136.

19) Yokoyama, M., Ishihara, H., Iwamoto, R., and Tadokoro, H. (1969) Structure of Poly(ethylene oxide) Complexes. III. Poly(ethylene oxide)-Mercuric Chloride Complex. Type II, *Macromolecules* **2**, 184-192.

20) Chenite, A., and Brisse, F. (1992) Poly(tetrahydrofuran)-urea adduct: a structural investigation, *Macromolecules*, **25**, 776-782.

21) Schmidt, G., Enkelmann, V., Westphal, U., Dröscher, M., and Wegner, G. (1985) Preparation of urea-poly(tetrahydrofuran) complexes and their application for fractionation of oligomers, *Colloid Polym. Sci.* **263**, 120-127.

22) Ashida, T., Tanaka, I., and Yamane, Y. (1981) β-Pleated sheets in oligopeptide crystals, *Int. J. Protein Res.* **17**, 322-329.

23) a) Tsang, K. Y., Diaz, H., Graciani, N., and Kelly, J. W. (1994) Hydrophobic Cluster Formation Is Necessary for Dibenzofuran-Based Amino Acids to Function as β-Sheet Nucleators, *J. Am. Chem. Soc.* **116**, 3988-4005 ; b) LaBrenz, S. R., and Kelly, J. W. *J.* (1995) Peptidomimetic Host That Binds a Peptide Guest Affording a β-Sheet Structure That Subsequently Self-Assembles. A Simple Receptor Mimic, *J. Am. Chem. Soc.* **117**, 1655-1656 ; c) Schneider, J. P., and Kelly, J. W. (1995) Synthesis and Efficacy of Square Planar Copper Complexes Designed to Nucleate β-Sheet Structure, *J. Am. Chem. Soc.*, **117**, 2533-2546 ; d) Schneider, J. P., and Kelly, J. W. (1995) Templates That Induce α-Helical, β-Sheet, and Loop Conformations, *Chem. Rev.* **95**, 2169-2187 ; e) Nowick, J. S., Smith, E. M., and Noronha, G. (1995) Molecular Scaffolds. 3. An Artificial Parallel β-Sheet, *J. Org. Chem.* **60**, 7386-7387 ; f) Nowick, J. S., Mahrus, S., Smith, E. M., and Ziller, J. W. (1996) Triurea Derivatives of Diethylenetriamine as Potential Templates for the Formation of Artificial β-Sheets, *J. Am. Chem. Soc.* **118**, 1066-1072 ; g) Nowick, J. S., Holmes, D. L., Mackin, G., Noronha, G., Shaka, A. J., and Smith, E. M. (1996) An Artificial β-Sheet Comprising a Molecular Scaffold, a β-Strand Mimic, and a Peptide Strand, *J. Am. Chem. Soc.* **118**, 2764-2765.

24) Nishio, M., Umezawa, Y., Hirota, M., and Takeuchi, Y. (1995) The CH/π interaction: Significance in molecular recognition, *Tetrahedron* **51**, 8665-8701. For a review of CH/π interaction, and references therein.

25) Dougherty, D. A. (1996) Cation-π interactions in chemistry and biology: A new view of benzene, Phe, Tyr, and Trp, *Science* **271**, 163-168.

26) Akazome, M., Ueno, Y., Ooiso, H., and Ogura, K. (2000) Enantioselective Inclusion of Methyl Phenyl Sulfoxides and Benzyl Methyl Sulfoxides by (*R*)-Phenylglycyl-(*R*)-phenylglycine and the Crystal Structures of the Inclusion Cavities, *J. Org. Chem.* **65**, 68-76.

27) a) Swift, J. A., Pal, R., McBride, and J. M., (1998) Using Hydrogen-Bonds and Herringbone Packing to Design Interfaces of 4,4'-Disubstituted *meso*-Hydrobenzoin Crystals. The Importance of Recognizing Unfavorable Packing Motifs, *J. Am Chem. Soc.* **120**, 96-104 ; b) Desiraju, G. R., and Gavezzotti, A. (1989) From molecular to crystal structure; polynuclear aromatic hydrocarbons, *J. Chem. Soc., Chem. Commun.* **1989**, 621-623 ; c) Gavezzotti, A. (1989) On the preferred mutual orientation of aromatic groups in organic condensed media, *Chem. Phys. Lett.* **161**, 67-72.

28) a) Pitchen, P., Duñnach, E., Deshmukh, M. N., and Kagan, H. B. (1984) An efficient asymmetric oxidation of sulfides to sulfoxides, *J. Chem. Soc.* **106**, 8188-8193 ; b) Furia, F. D. Modena, G., and Seraglia, R. (1984) Synthesis of chiral sulfoxides by metal-catalyzed oxidation with t-butyl hydroperoxide, *Synthesis*,

325-326 ; c) Sugimoto, T., Kokubo, T., Miyazaki, J., Tanimoto, S., and Okano, M. (1979) Preparation of optically active aromatic sulfoxides of high optical purity by the direct oxidation of the sulfides in the presence of bovine serum albumin, *J. Chem. Soc., Chem. Commun.* **1979**, 402-404 ; Ogura, K., Fujita, M., and Iida, H.(1980) A two-phase reaction catalyzed by a protein — asymmetric oxidation of formaldehyde dithioacetals with aqueous sodium metaperiodate, *Tetrahedron Lett.* **21**, 2233-2236 ; d) Palucki, M., Hanson, P., and Jacobsen, E. N. (1992) Asymmetric oxidation of sulfides with H_2O_2 catalyzed by (salen)Mn(III) complexes, *Tetrahedron Lett,* **33**, 7111-7114.

29) Andersen, K. K., Bujnicki, B., Drabowicz, J., Mikolajczyk, M., and O'Brien, J. B. (1984) Synthesis of enantiomerically pure alkyl and aryl methyl sulfoxides from cholesteryl methanesulfinates, *J. Org. Chem.* **49**, 4070-4072.

30) Toda, F., Tanaka, K., and Okuda, T. (1995) Optical resolution of methyl phenyl and benzyl methyl sulfoxides and alkyl phenylsulfinates by complexation with chiral host compounds derived from tartaric acid, *J. Chem. Soc., Chem. Commun.* **1995**, 639-640.

OPTICAL RESOLUTION VIA COMPLEX FORMATION
WITH O,O'-DIBENZOYL-TARTARIC ACID

Ferenc Faigl[1,2] and Dávid Kozma[1]
[1]Department of Organic Chemical Technology, Budapest University of Technology and Economics, H-1521 Budapest, Hungary; [2]Research Group for Organic Chemical Technology, Hungarian Academy of Sciences, H-1521 Budapest, Hungary

Abstract: Recent methods for preparation of diastereoisomeric coordination complexes of O,O'-dibenzoyltartaric acid metal salts with hydroxycarboxylic acid esters, hydroxycarboxylic acids and alcohols as well as host-guest complexes of the same chiral dicarboxylic acid with racemic phosphine oxides and racemic alcohols have been reviewed in this chapter. Practical and theoretical aspects of these optical resolution processes have also been discussed.

Key words: optical resolution, diastereoisomeric complexes, O,O'-dibenzoyltartaric acid, hydroxycarboxylic acid esters, hydroxycarboxylic acids, chiral phosphine oxides, racemic alcohols.

1.   INTRODUCTION

Tartaric acid and its derivatives are among the most frequently used chiral reagents in organic synthesis. Hundreds of racemic bases have been resolved with (R,R)-tartaric acid (TA) and its O,O'-dibenzoyl (DBTA) or O,O'-di-*p*-toluyl (DPTTA) derivatives and, of course, with the enantiomers of these compounds. [1, 2, 3] Another series of chiral ligands has been synthetised by transformation of the carboxylic groups of tartaric acid into carbinols, amides or other functions whith parallel slight modification of the two alcoholic hydroxy groups. One of the first and most famous representative of these compounds was *(R,R)*-4,5-bis(hydroxydiphenyl-methyl)-2,2-dimethyl-1,3-dioxacyclopentane (TADDOL, Scheme 1). [4] In the last decades numerous similar compounds have been described in the literature as efficient ligands for enantioselective reactions [5, 6] or host compounds for enantiomer separation via complex formation. [7, 8] Effectiveness of tartaric acid and the above menitoned derivatives in a wide range of chiral recognition processes can be rationalised if the C_2 symmetry and the relatively large number of functional groups

73

F. Toda (ed.), Enantiomer Separation, 73-101.

(coupled directly to chiral centers) of these molecules have been taken into consideration. Carboxyl or carboxylate groups, as well as the hydroxy groups are outstanding donors and acceptors in second order interactions, easily form H-bridges and the aromatic ring containing derivatives (such as TADDOL or DBTA) suffer π electrons for further second order bond formations with suitable guest compounds. [9]

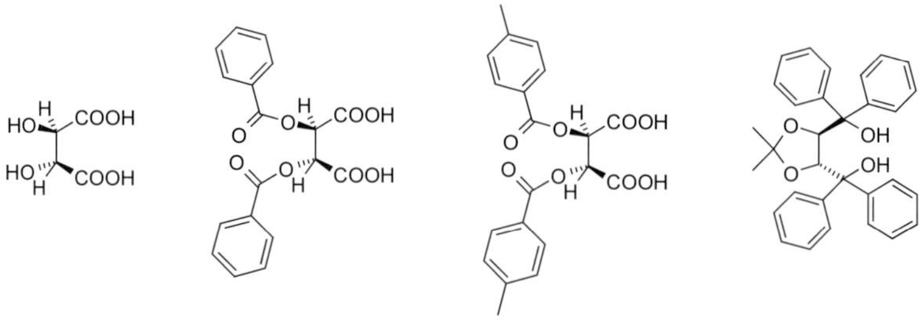

| TA | DBTA | DPTTA | TADDOL |

Scheme 1

Crucial role of second order interactions during chiral recognation of enantiomers was demonstrated by successful application of Ogston's three point interaction model [10] in a quantitative approach to optical resolution processes. [11] Experimental data and molecular modelling calculations confirmed that similar second order interactions determine the efficiency of optical resolutions [12] and enantiomeric enrichment processes. [13] In other words, energy difference between diastereoisomers has been determined by the difference of second order interactions formed between the chiral selector and the *R* or *S* enantiomer, separately, and it is practically independent of the type of the leading bond (the strongest bond: covalent, ion pair, H-bridge, etc.) formed between the resolving agent and the enantiomer. In real solutions and solid phases supramolecular structures are formed in which energy differences among the second order bonds of the diastereoisomers are summarised resulting in macroscopic difference in solubility, partition coefficient or other behaviours of the diastereoisomers. [13]

Single crystal X-ray diffraction measurements have shown the importance of these attractive second order interactions in stabilisation of a relatively constant structure of TA in diastereoisomeric salts [14], DBTA in complexes [15, 16] and of DBTA derivatives in sole or solvated crystals, too. [17] In the most cases, strong H-bridges stick together the TA or DBTA molecules into long chains and chanels having chiral surface for discrimination between enantiomers.

One can conclude from these data that the criteria of a good chiral ligand or resolving agent are basically the same. The compounds should contain suitable groups for different types of second order interactions as much as possible. In this respect, TA, DBTA, DPTTA and the TADDOL analogs fulfil these requirements. However, significant differences can be observed among their solubility (lipophylic character) and this behaviour influences their application. In the last fifteen years intensive development of TADDOL like compounds generated a family of neutral, lipophylic ligands for resolution of racemates like chiral alcohols or carboximides and carboxylic acid anhydrides by formation diastereoisomeric inclusion complexes [8, 18, 19, 20] and for enantioselective reactions. [21] It has to be mentioned, that these ligands are usually prepared via multistep synthesis from tataric acid.

Among the more polar acides (TA, DBTA and DPTTA) O,O'-dibenzoyltartaric acid (DBTA) has been recognized as a good complex forming agent since ten years. [15, 16, 22] This compound can easily be prepared in two steps from natural tartaric acid, it is commercially available as monohydrate or anhydrous crystalline material in large quantities and can be recovered from the reaction mixtures in good yield. Thus, application of DBTA has practical advantages in preparative scale resolutions of non basic racemates via diastereoisomeric complex formation.

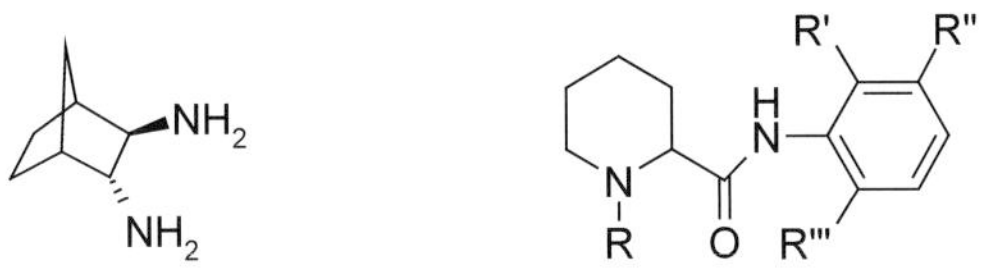

1

2

R: H, Me, Et, Pr, Bu; R': H, Me;
R": H, Me; R"': H, CF$_3$

Scheme 2

Biased classification of DBTA into the group of agents for diastereoisomeric salt formation was breaken through when Hatano and his co-workers recognized that *trans*-bicyclo[2.2.1]heptane-2,3-diamine (**1**, Scheme 2) form diastereoisomeric complexes with DBTA instead of regular salts [15]. Our laboratory has also reported such complex formations of eight N-alkylpipecolic acid-anilide derivatives (**2**, Scheme 2) with DBTA. [22]

Actually, coordination complexes of different metal salts of DBTA with hydroxycarboxylic acid esters, hydroxycarboxylic acids and alcohols as well as host-guest complexes of DBTA with chiral phosphine oxides and racemic alcohols can be prepared and used for separation of optical isomers. In the next subchapters theoretical and practical aspects of these recent resolution processes are summarised.

2.   **OPTICAL RESOLUTIONS VIA DIASTEREOISOMERIC**
        COORDINATION COMPLEXES OF DBTA METAL SALTS

Combination of the coordinating ability of metal ions for hydroxy group containing compounds with the before mentioned exceptional behaviours of DBTA in chiral recognation processes gave us a new possibility for direct resolution of α-alkoxycarboxylic acids, α-hydroxycarboxylic acid esters and series of racemic alcohols via diastereoisomeric coordination complex formation. Theoretically numerous metal ions could be used as central atom of the coordination compounds but successful applications of neutral and acidic calcium, zinc and cooper salts of DBTA have been published until now. [23, 24, 25]

2.1   Resolution with DBTA calcium salts

In practical point of view, use of acidic ((DBTA)$_2$Ca) or neutral (DBTACa) calcium salts of DBTA seems to be the most advantageous choice because of the easy preparation of the resolving agents, low price and toxicity, good compatibility of the calcium salts with different ligands and usually good crystallisation abilities of the coordinative complexes formed.

α-Alkoxycarboxylic acids

Optical resolutions of two α-alkoxycarboxylic acids (**3a** and **3b**, Scheme 3) were accomplished with the neutral calcium salt of DBTA (DBTACa) in aqueous ethanol solution.[24] Preparation of the diastereoisomers is quite simple: a stoichiometric mixture of DBTA monohydrate and calcium oxide had to solve in hot aqueous ethanol to form DBTACa salt and then racemic acid (**3a** or **3b**) had to add into it. On cooling [Ca(H$_2$O)(*R*-**3a**)]DBTA.H$_2$O or [Ca(H$_2$O)(*S*-**3b**)]DBTA.H$_2$O cystallised. Free (*S*)-**3a** or (*R*)-**3b** was obtained from the corresponding filtrate. In the schemes and the formulas (*S*) or (*R*) stands for showing the configuration of the of the major enantiomer in the product. For example, (*S*)-**3a** means that the isolated material contained the *S*-isomer in excess, ee values are given in the Tables.

According to the original recepies, the crystalline diastereoisomers were recrystallized two or three times before liberation of (*R*)-**3a** or (*S*)-**3b** in 51 or 89 % yield, respectively. [24]

Presence of water in the reaction mixture has significant importance because dihydrates crystallized in both cases. Single crystal X-ray analysis of (*S*)-**3b**-DBTACa.2H$_2$O confirmed that one of the water molecules coordinated to the central calcium ion while the other one sticked to the remote carboxylic group by H-bond as it is showed in Scheme 3. [24]

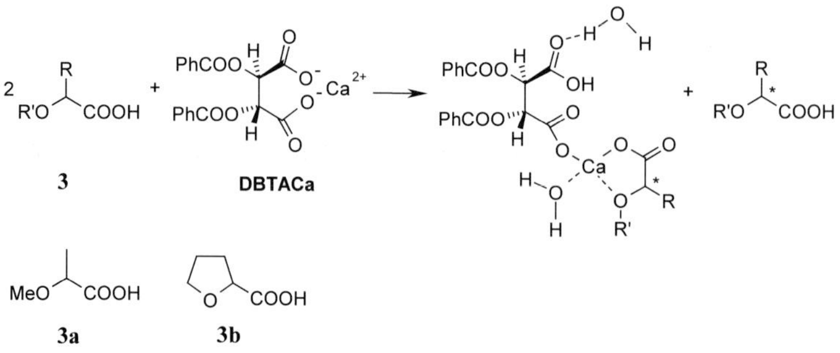

Scheme 3

Crowded structure of these diastereoisomers limitate the scope of this resolution method to relatively small racemic α-alkoxycarboxylic acids. Furthermore, the presence of an alkoxy group in the racemate seems to be essential because its oxygen atom also coordinates to the calcium ion. Luck of such coordination hindered formation of crystalline complex when α-chlorocarboxylic acids were tested. On the other hand, α-hydroxycarboxylic acids precipitated from the reaction mixture as simple calcium salts because of their lowest solubility among the possible combinations of the components of the reaction mixtures.

α- or β-Hydroxycarboxylic acid esters

Transformation of the α- or β-hydroxycarboxylic acids into esters eliminates the acidic character of the racemate, thus one can avoid formation of the above mentioned unsoluble calcium salts. In the same time formation of a mixed calcium salt of DBTA and the racemic acid is impossible therefore coordination bonds should entirely replace it.

Series of racemic α- and β-hydroxycarboxylic acid esters (4 and 5, Scheme 4) could be separated into their enantiomers via coordination complex formation but in these cases the acidic salt, $(DBTA)_2Ca$ had to be used as resolving agent. [23, 24]

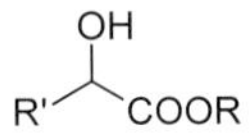

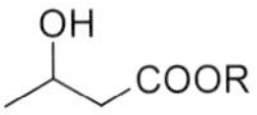

4a: R = Me, R' = Ph; **4b**: R = Et, R' = Ph; **4c**: R = CH$_2$Ph, R' = Ph;
4d: R = CH$_2$CH$_2$Ph, R' = Ph; **4e**: R = Et, R' = Me, **4f**: R = Bu, R' = Me;
4g: R = CH$_2$Ph, R' = Me; **4h**: R = Me, R' = 4-MeOPh; **4i**: R = Me,
R' = CH$_2$Ph; **4j**: R = Me, R' = MeOOCCH$_2$; **4k**: R = Et, R' = EtOOCCH$_2$.

5a: R = Me
5b: R = Et

$$DL + 0.5\ (DBTA)Ca_2 \xrightarrow{\text{solvent mixture}} 0.5\ [Ca(D>L)_n(A_1)_m(A_2)_o](DBTA)_2 + 0.5\ (L>D)$$
$$\mathbf{6}$$

DL: (±)-**4** or (±)-**5**

Scheme 4

Composition of the crystallised complexes corresponds to the general formula **6** where DL stands for compound **4** or **5** (n= 1 or 2), A$_1$ and A$_2$ stand for auxiliary ligands (A$_1$ generally ethyl or propyl acetate (m= 0, 0.5, 1), A$_2$ is water (o= 0, 1, 2), Scheme 4). [23, 24] Amounts of A$_1$ and A$_2$ strongly depend on the solvent mixture used. In some cases solvent molecules, built into the complexes as auxiliary ligands (A$_1$ and/or A$_2$), influence even the configuration and the amount of the chiral ligand (nL) in the coordination complex (see in Table 1). [24] Modification of the alkyl group in the ester function of **4** and **5** influences the efficiency of the resolution and may also cause alteration of the configuration of **4** or **5** in the crystals. Examples are given in Table 1.

TABLE -1. Resolution of hydroxycarboxylic acid esters **4** and **5** with $(DBTA)_2Ca$

Racemate	Co-solvent[a]	n	m(A₁)	o(A₂)	Yield[b] (%)	ee[c] (%)
4a	Acetone	2	-	-	83[d]	99 (R)
4b	Acetone	2	-	1 (H_2O)	41[d]	62 (S)
4b	Ethyl acetate	1	1 (Ethyl acetate)	1 (H_2O)	74[d]	99 (S)
4b	Ethyl propionate	1	1 (Ethyl propionate)	1 (H_2O)	93[e]	78 (S)
4c	Toluene	4	1 (Toluene)	-	136[d]	16 (S)
4c	Ethyl acetate	1	-	1 (H_2O)	42[d]	87 (R)
4d	Acetone	1	-	1 (H_2O)	68[d]	74 (S)
4e	-	2	-	-	98[e]	3 (S)
4f	-	2	-	-	90[e]	27 (S)
4g	-	2	-	-	99[e]	46 (S)
4h	-	2	-	-	90[e]	47 (R)
4i	-	2	-	2 (H_2O)	92[e]	51 (R)
4i	-	1	-	2 (H_2O)	93[e]	65 (R)
4j	Propyl acetate	1	0.5 (Propyl acetate)	1 (H_2O)	86[e]	40 (S)
4k	Ethyl acetate	1	0.5 (Ethyl acetate)	1 (H_2O)	85[e]	28 (S)
5a	Ethyl acetate	1	1 (Ethyl acetate)	2 (H_2O)	68[e]	22 (S)
5b	Propyl acetate	1	1 (Propyl acetate)	2 (H_2O)	63[e]	38 (R)

[a]All resolutions started in 95 % ethyl alcohol then the co-solvent was added or the alcohol was evaporated and replaced with ester type co-solvent. [b]Yield of the recrystallized diastereoisomer is based on the half of the starting racemic **4** or **5**. [c]The ee value and absolute configuration of the liberated **4** or **5** enantiomer. [d]Data from ref. [24]. [e]Data from ref. [23].

Practical advantage of the method is that it does not require dry solvents. The resolving agent can be prepared by simple solution of DBTA monohydrate and half an equivalent amount of calcium oxide in hot 95 % ethyl alcohol. Crystallization of the diastereoisomeric coordination complex can be achieved by cooling and addition of co-solvents (e.g. acetone, toluene, ethyl acetate, etc.) or change ethyl alcohol to an ester type solvent. The enantiomers can be liberated from the crystalline complex by simple acidic workup procedure. [23]

Single crystal X-ray diffraction analysis of $[Ca(R\text{-}4a)_2](DBTA)_2$ showed that the molecules are organized around the central column of calcium ions. [24] Two bidentate ligands ((R)-**4a**) connect with the same metal ion on opposite sides while two carbonyl oxygens of a single DBTA ion are in contact with two different calcium ions. The phenyl rings are situated around the hydrophilic calcium ion chanels. Chiral discrimination betwen **4a** enantiomers can be rationalised if the different space requirements of the hydrogen atom and the phenyl group of **4a** is taken into consideration (the position of thcsc two groups should be changed if (S)-**4a** would be fitted into the same coordination complex instead of (R)-**4a**).

α-Alkoxyalcohols

Mravik et al. published further application of DBTACa salt in optical resolution of different α-alkoxyalcohols. [25] An achiral (**7**) and three racemic (**8, 9, 10**) α-alkoxyalcohols were tested in coordination complex formation reactions (Scheme 5). All the resolutions were carried out in 95 % ethyl alcohol or alcohol/acetone mixtures starting from one mole of DBTACa salt and four (or more) moles of racemic alcohols.

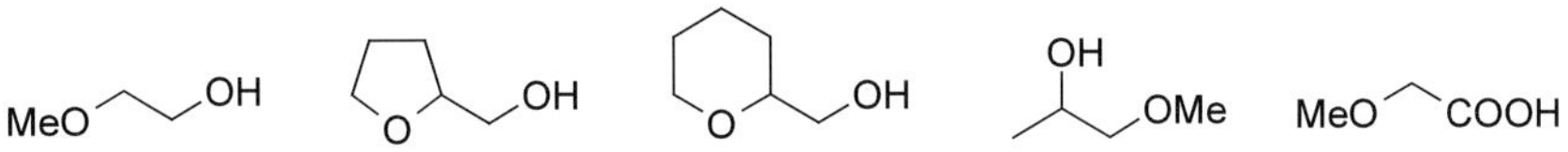

MeO⌒⌒OH   ⟨O⟩⌒OH   ⟨O⟩⌒OH   OH / ⌒OMe   MeO⌒COOH

     7   **8**   **9**   **10**   **12**

Scheme 5

The general structure of the crystallised optically active complexes was [Ca(L)₂]DBTA (**11**: in the formula L stands for **8, 9** or **10**, separately) and they contained two moles of the corresponding (R)-alcohols (L) in all cases (Table 2). [25]

The optically active alcohols can be liberated from the coordination complexes by achiral reagents. One possibility would be the addition of methoxyethyl alcohol **7** which form more stable complex with DBTACa then the resolved chiral alcohols. Practically quantitative recovery of **8, 9** or **10** was achieved by the use of methoxyacetic acid (**12**) that form crystalline mixed salt with DBTACa. The free optically active alcohols were isolated from the filtrate in 90-95 % yields.

TABLE -2. Resolution of alkoxyalcohols **8, 9** and **10** with DBTACa

Racemate	Solvent	Yield[a]	ee[b], (configuration)
8	Ethyl alcohol/Acetone	90 %	67 % (R)
9	Ethyl alcohol/Acetone	89 %	28 % (R)
10	Ethyl alcohol	81 %	69 % (R)

[a]Yield of the complex based on the half of the racemic alcohol. [b]Enantiomeric excess and configuration of the alcohol liberated from the crystalline complex.

It is worth to mention that both coordination compounds, namely [Ca(**7**)₂]DBTA and [Ca(H₂O)](**12**)DBTA form conglomerates with their enantiomers, respectively. Thus, optical resolution of racemic DBTA by preferential crystallization of these coordination complcxes is also possible. [26, 27]

On the basis of the above mentioned workup procedure a new simultaneous resolution method was developed using two moles of racemic tetrahydrofuroic acid (**3b**) instead of **12** for decomposition of complex **11** (Scheme 6, L= **8**). In a methyl alcohol/water system the (S)-**3b** isomer built up a new crystalline coordination complex (**13**) while two moles of (R)-**8** liberated and one mole of (R)-**3b** remained in the solution. [25]

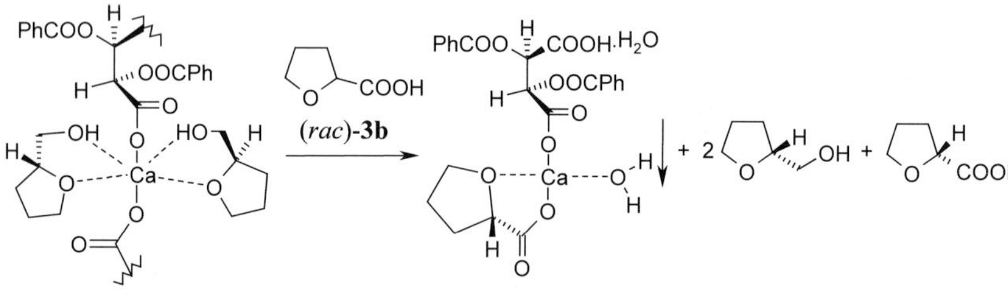

11	**13** (DBTACa(S-**3b**).H$_2$O)	(R)-**8**	(R)-**3b**

Scheme 6

2.2   Resolution with the zinc salt of DBTA

Coordination complex formation of alkoxyalcohols (**8**, **9**, **10**, **14**), 2-butanol (**15**) and 1,3-butanediol (**16**, Scheme 7) with the zinc salt of DBTA was investigated, too. [25] The resolving agent (DBTAZn) was prepared again from DBTA monohydrate and zinc oxide or zinc acetate in aqueous ethyl alcohol solution.

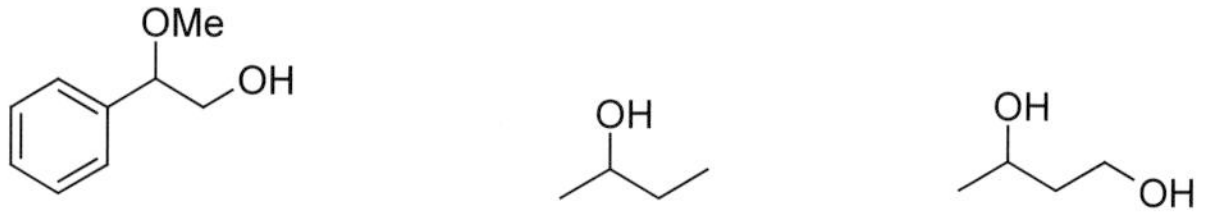

14	**15**	**16**

Scheme 7

Comparison of the zinc containing coordination complex crystalls of **8** and **10** with that of the calcium complexes showed significant differences. First, the opposite enantiomers crystallized with DBTAZn and with DBTACa. Second, compositions of the solids were also different, one mole of the alkoxyalcohol but 2-5 moles of water and one mole of ethyl alcohol or other cosolvent were found in the zinc salt containing crystals. Composition of the coordination complexes (general formula: [Zn(A$_1$)$_n$(A$_2$)$_m$(L)]DBTA, amounts of the achiral ligands are given in Table 3) strongly depended on the quality of the alcohol resolved and, in some cases, it changed during recrystallizations. As an example, Scheme 8 shows the backbone of [Zn(H$_2$O)$_2$(EtOH)(S-**8**)]DBTA.H$_2$O complex according to the single crystal X-ray measurements. [25]

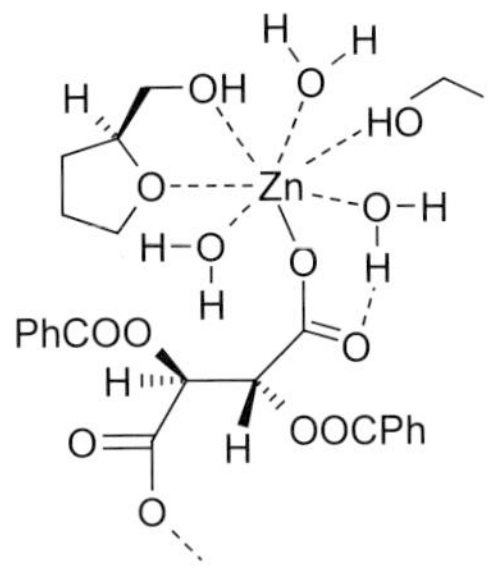

[Zn(H$_2$O)$_2$(EtOH)(*S*-**8**)]DBTA.H$_2$O

Scheme 8

Recovery of the resolved alcohols was achieved by extraction or distillation. In special cases ligand exchange (similar to the before mentioned workup procedures of the calcium compounds) was used to liberate the previously complexed alcohol enantiomer. Results of the optical resolutions made by DBTAZn are summarized in Table 3 where the overall numbers of the coordinated auxiliary ligands (nA$_1$ and mA$_2$) are also given on the basis of X-ray analysis data. [25]

TABLE -3. Resolution of different alcohols with DBTAZn

Racemate	Solvent	Crystallised comlex			
		n A$_1$	m A$_2$	Yield[a] (%)	ee[b] (%)
8	Ethyl alcohol/Water	3 water	1 Ethyl alcohol	78	39 (*S*)
8	Ethyl alcohol/Acetic acid	2 water	1 Ethyl alcohol 1 Acetate anion	84	60 (*S*)
9	Ethyl alcohol/Water	3 water	-	41	33 (*S*)
10	Ethyl alcohol/Water	3 water	1 Ethyl alcohol	71	81 (*S*)
14	Ethyl alcohol/Water	3 water	-	92	57 (*S*)
15	Water	5 water	-	82	30 (*S*)
16	Water/Acetic acid	4 water	1 Acetate anion	93	12 (*S*)

[a]Yield of the complex based on the half of the racemic alcohol. [b]Enantiomeric excess and configuration of the alcohol liberated from the crystalline complex.

2.3   Resolution with the cooper salt of DBTA

The cooper(II) complexes of DBTA are also suitable for partial resolution of alkoxyalcohol **8** and **9** (in general: L) as it was reported in the article of Mravik et al. [25]

General formula of the crystallised complexes is $[Cu(L)_n(H_2O)_m]DBTA(A^-)$. In addition, these cooper containing coordination complexes were used for simultaneous resolution of several α-halogeno-carboxylic acids **17**, **18** and **19** (Scheme 9, in general HA, anion: A⁻) on the analogy of the method developed earlier with complex **11** (L = **8**) for the resolution of **3b**. However the reported yields and enantiomeric excesses are usually low (Table 4). [25]

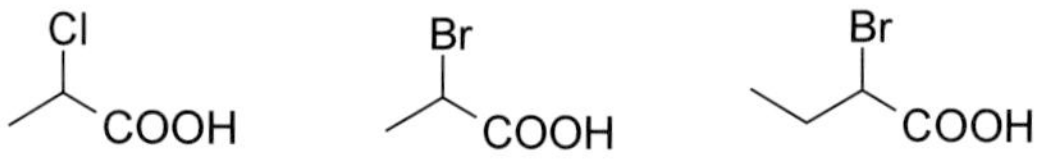

Scheme 9

TABLE -4. Simultaneous resolution of **8** or **9** and **17**, **18**, or **19** with DBTACu

L	HA	Yield[a] (%)	ee_L[b] (%)	ee_HA[c] (%)
8	Acetic acid	74	11 (S)	-
Ethyl alcohol	**17**	86	-	32 (S)
8	**17**	35	8 (S)	33 (S)
8	**18**	60	55 (S)	19 (R)
8	**19**	41	44 (S)	11 (R)
9	Acetic acid	75	31 (R)	-
9	**18**	81	62 (S)	29 (R)
Ethyl alcohol	**18**	66	-	3 (R)
Ethyl alcohol	**19**	66	-	14 (R)

[a]Yield of the complex based on the half of the racemic ligand. [b]Enantiomeric excess and configuration of the chiral alcohol liberated from the crystalline complex. [c]Enantiomeric excess and configuration of the chiral acid liberated from the crystalline complex.

2.4   Optical resolution of diphosphine oxides

Synthesis of chiral atropisomeric diphosphine type ligands is a current challenge in chemical research because their late transition metal complexes usually provide high enantioselectivity in homogenous catalytic reactions. [28] In practical point of view, preparation and optical resolution of racemic diphosphine oxides followed by the reduction of the separated enantiomers are usually more advantageous than an expensive enantioselective synthesis of one diphosphine enantiomer. [29]

In 1986, Minami and his coworkers described the optical resolution of the oxide of *trans-bis*-1,2-(diphenylphosphino)cyclobutane (**20**, Scheme 10.) with stoichiometric amount of DBTA in methanol solution. [30] In the same year, Noyori and his coworkers published the synthesis of optically active BINAP via resolution of BINAP oxide (**21**) with DBTA. [31] In this process, hot chloroformic solution of racemic **21** was mixed with a molar equivalent amount of DBTA previously solved in ethyl acetate. Optically pure (*S*)-**21** was isolated in 79 % yield from the recrystallized diastereoisomeric complex by treating it with aqueous potassium hydroxide solution. Noyori's resolution method has been successfully used since 16 years for separation of the optical isomers of numerous diphosphine oxides such as **22**, [32] **23**, [29] **24** [33] and **25** [34] (Scheme 10).

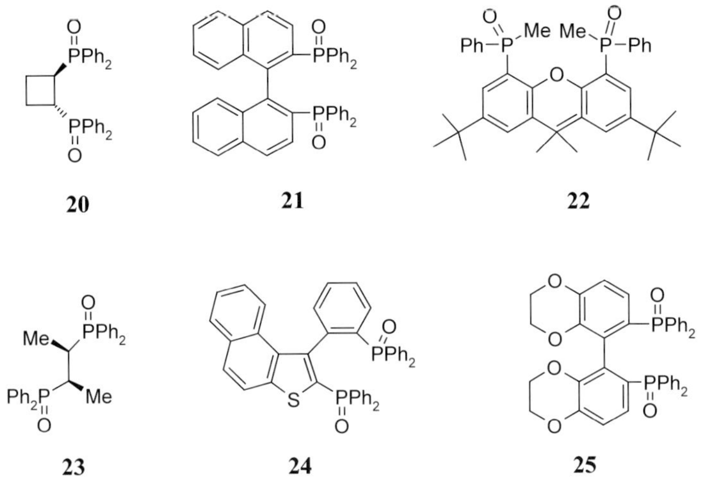

20   **21**   **22**

23   **24**   **25**

Ph: phenyl group

Scheme 10

Good crystallization ability of the (*S*)-**25**-DBTA complex gave a possibility for preparation (*S*)-**25** in a „through process". In that case, diastereoisomeric complex formation was accomplished starting from the crude reaction mixture of (±)-**25** and the residue of the monomeric phosphine oxide was recycled into the previous oxidative coupling reaction from the filtrate of the crystalline diastereoisomeric complex. [34]

All in the above mentioned cases the crystalline complexes contained the corresponding diphosphine oxide enantiomer and DBTA in 1:1 molar ratio. Single crystal X-ray analysis of (*S*)-**25**-DBTA showed the presence of supramolecular chains of alterating (*S*)-**25** and DBTA molecules in the crystal lattice. In one pair of the ligand and the resolving agent strong H-bond was observed between the oxygen atom of a phosphine oxide group and the carboxylic group of DBTA. [34]

2.5   Optical resolution of alcohols

Recently, complex formation reactions of (*R,R*)-tartaric acid (TA) and its O,O'-dibenzoyl derivative (DBTA) with a series of chiral alcohols (**26**, **27**, **8** and **28**, Scheme 11) were investigated in our laboratory using Toda's suspension method. [35] As an example, resolution of **28** is outlined in Scheme 12.

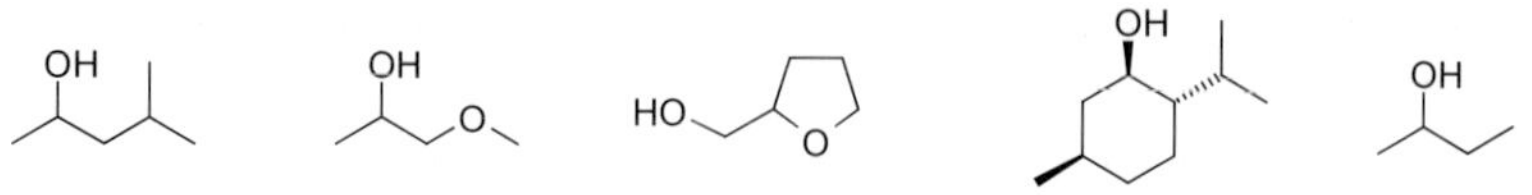

26 **27** **8** **28** **29**

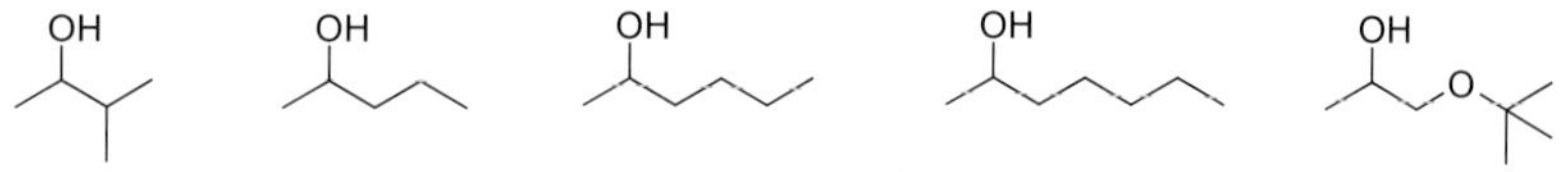

30 **31** **32** **33** **34**

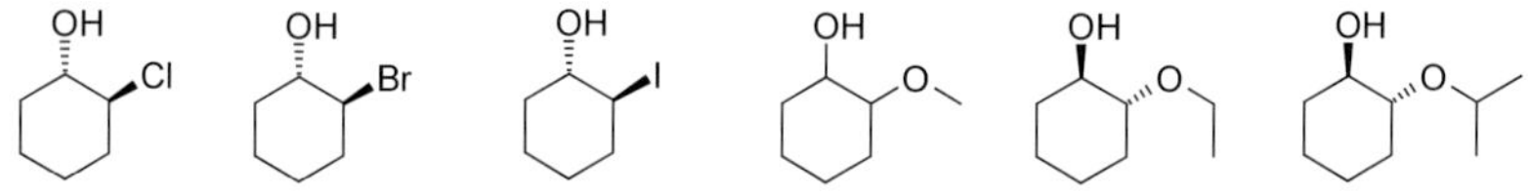

35 **36** **37** **38** **39** **40**

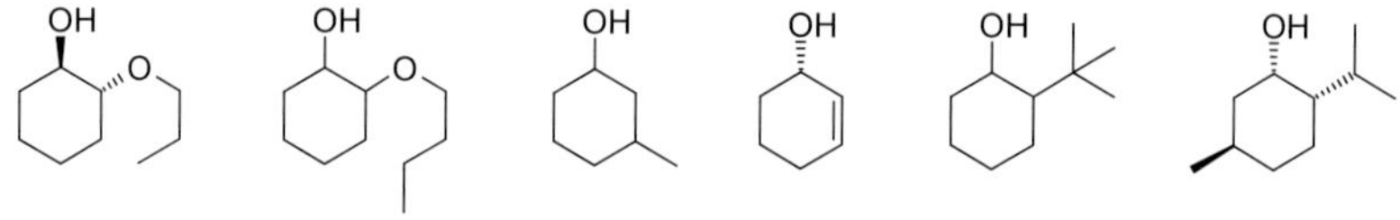

41 **42** **43** **44** **45** **46**

Scheme 11

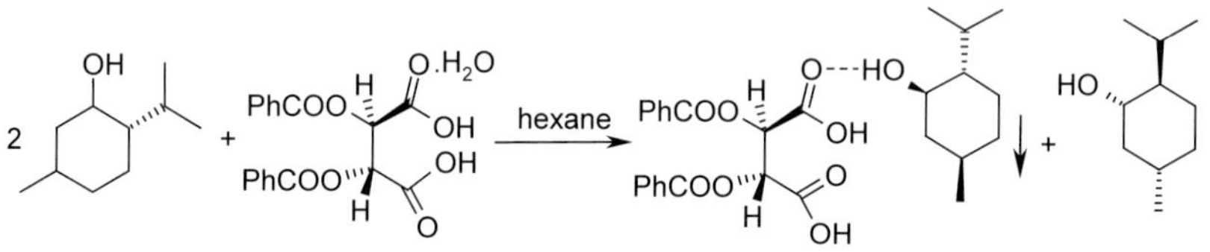

(±)-**28** DBTA.H_2O (*1R,2S,5R*)-**28**.DBTA (*1S,2R,5S*)-**28**

Scheme 12

The experimental results let us to conclude that TA does not form complexes with any of the model compounds but DBTA gives crystalline host-guest complexes with all the four model compounds. [16] The results are summarised in Table 5 where the efficiency of the resolutions is characterised by the S value calculated as the production of the yield and the enantiomeric excess. [11] Application of DBTA monohydrate in hexane suspension provided more efficient separation of **26** and **28** enantiomers than the use of anhydrous DBTA under the same conditions. Data in Table 5 refer to the resolutions made with DBTA monohydrate in all cases.

Thermoanalytical measurements confirmed that the crystallised complexes do not contain any water molecules, consequently the alcohol replaced water during complex formation giving approximatively 1:1 molar ratio of the chiral alcohol and DBTA in these crystals. [36, 37] Diastereoisomeric complex formations of further 18 chiral alcohols (compounds **29-46**) with DBTA were also studied to determine the scope and limitations of this resolution method. The results are collected in Table 5, [38] compositions of the formed complexes are discussed in point 3.5.

The experimental protocol was always the same: the racemic alcohol was solved in hexane and a half molar equivalent of DBTA was suspended in that solution. [16, 38] Without stirring, complex formation reactions were quite slow under these heterogenous conditions, one had to wait for 7-20 days until the process had finished (Figure 1).

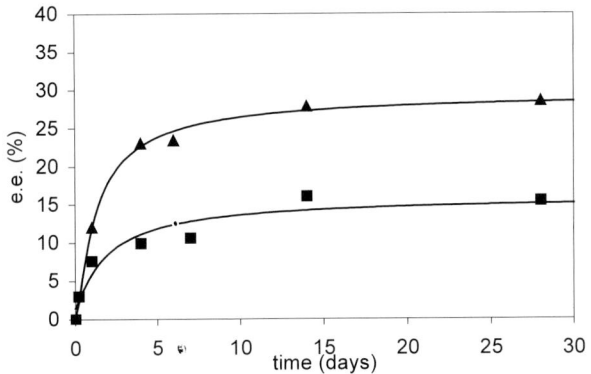

Figure 13. Enantiomeric excess of **8** (■)and **26** (▲) in their DBTA complexes as a function of the complexation time

After filtration, the alcohol enriched in one enantiomer was obtained from the filtrate, the other one was isolated from the solid complex. Thermal methods such as distillation or sublimation were used for liberation of the optically active alcohols from the complexes. [38] Time-scaled experiments showed continous increasing of the enantiomeric excess (ee) in the crystalline diastereoisomeric complexes of **26** and **8** for 14 days, then these ee values practically did not change within the next 16 days (Figure 1).

In several cases (compounds **38**, **43**, **45** and **46**) DBTA remainded unchanged during the long incubation time, no complex formation occured with these alcohols. In one case partial resolution of a chiral primary alcohol could be achieved (compound **8**). All the other examples show complex formation of chiral secondary alcohols with DBTA.

Practical advantage of this resolution method is that most of the investigated racemates could be solved in hexane but DBTA and the complex was insoluble in it. Consequently, complete separation of the solid diastereoisomeric complex and the remained free enantiomeric mixture of the alcohol is simple and chemists can avoid the exhausting work of solvent selection and concentration optimization.

Comparison of the results (Table 5) obtained under the same conditions for structurally similar compounds shed light several tendencies as follows [38]:

a) Among the chiral aliphatic secondary alcohols 2-pentanol derivatives (**26** and **31**) can only be resolved with DBTA, enantiomer separation of shorter or longer chain alcohols could not be achieved.

b) Supramoleular compound formation with DBTA is *trans* selective in the cases of 2-alkylcyclohexanols (good resolution of menthol **28**, no complex formation with *iso*-menthol **46**). Large *tert*-butyl group hinders complex formation in the case of **45**.

c) In the series of 2-alkoxycyclohexanols the enantiopreference of DBTA depends on the size of the alkoxy group. Thus, the methoxy derivative (**38**) does not form crystalline supramolecular compound with DBTA, enrichment of the (*1R,2R*) enantiomers can be observed in the cases of **39** and **40**, (*1S,2S*) isomer is in excess of the solid complex of **41** and racemic mixture containing complex crystallizes from the mixture of **42** and DBTA.

d) (*1S,2S*)-2-Halogenocyclohexanols (**35**, **36**, **37**) form stable supramolecular compounds with DBTA and the selectivity increases with the size of the halogen atom.

TABLE -5. Resolution of alcohols via diastereoisomeric complex formation with DBTA

Racemic alchol	Product from the crystalline phase			
	Configuration	ee (%)	Yield (%)	S^a
26	R	28	48	0.255
27	S	7	66	0.046
8	S	10	34	0.034
28	1R,2S,5R	83	45	0.374
29	R,S	0	48	0
30	R,S	0	59	0
31	R	20	63	0.126
32	R	5	19	0.009
33	R,S	0	11	0
34	R,S	0	93	0
35	1S,2S	35	74	0.259
36	1S,2S	56	63	0.353
37	1S,2R	61	71	0.433
38	-	No solid pase		
39	1R,2R	50	74	0.370
40	1R,2R	15	60	0.090
41	1S,2S	44	66	0.290
42	R,S	0	46	0
43	-	No solid phase		
44	S	21	55	0.115
45	-	No complex formation		
46	-	No complex formation		

[a] Efficiency of the resolution: S = yield x ee. [11]

Efficiency of the above discussed resolutions changed in a wide range: $0 \leq S \leq 0.433$ (Table 5). Menthol (**28**) and 2-halogenocyclohexanol (**35**, **36**, **37**) enantiomers were the best ligands of DBTA for chiral recognition during host-guest complex formation, therefore these model compounds were used for further elaboration of the resolution processes (point 3.4.).

2.6   Combined separation of *cis/trans* and optical isomers of cyclohexanol derivatives

As it was mentioned above, DBTA prefers *trans*-alkylcyclohexanols as guests in the crystalline complexes. This phenomenon was used for achieving enrichment one isomer in the crystalline and the other in the liquid phase, starting from *cis-trans* mixtures of several achiral and chiral alkylcyclohexanols (**47-51** in Scheme 13). [39]

Experimentally the method is simple: *cis-trans* mixtures of 4- or 2-alkylcyclohexanols were solved in hexane and a half or a quarter molar equivalent amount of solid DBTA monohydrate was added to the solutions. After a while the complexes were filtered and the isomer compositions of the filtrates (and after decomposition, that of the solids) were determined by gas chromatography.

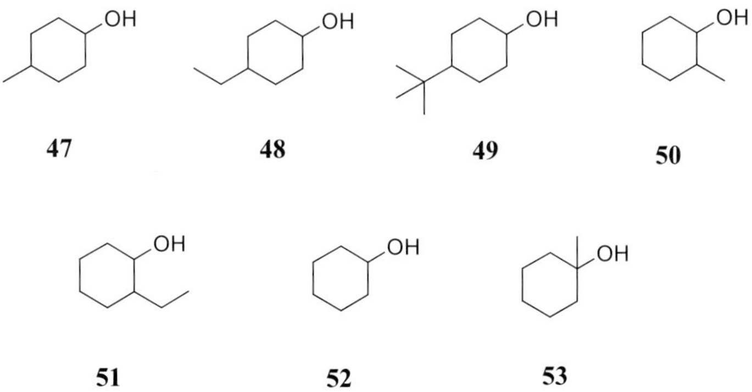

Scheme 13

TABLE -6.

No.	Starting alcohol composition	Alcohol / DBTA molar ratio	Alcohol composition[a] from the complex	from the filtrate
47	$c/t = 33/67$	2/1	$c/t = 25/75$	$c/t = 55/45$
48	$c/t = 34/66$	2/1	$c/t = 7/93$	$c/t = 73/27$
49	$c/t = 9/91$	2/1	$c/t = 8/92$	$c/t = 15/85$
(±)-50	$c/t = 40/60$	4/1	$c/t = 24/76$ $ee_t = 42\%$, $ee_c = 0\%$	$c/t = 45/55$ $ee_t = 16\%$, $ee_c = 0\%$
(±)-51	$c/t = 24/76$	4/1	$c/t = 38/62$ $ee_t = 80\%$, $ee_c = 0\%$	$c/t = 16/84$ $ee_t = 37\%$, $ee_c = 0\%$

[a]ee_t, ee_c : Enantiomeric excess of the *trans* and the *cis* isomers. Enrichment of (*1S,2S*)-**50** and (*1S,2S*)-**51** were observed in the complexes, respectively.

Data in Table 6 show accumulation of the *trans* isomers of 4-alkylcyclohexanols (**47**, **48** and **49**) in the crystalline complexes.The same tendency in combination with partial resolution of *trans*-**50** and *trans*-**51** can be observed at 2-alkylcyclohexanols. [39] Furthermore, *cis* isomer content of the solid complexes remained racemic, consequently DBTA is selective resolving agent of *trans*-**50** and *trans*-**51**. Efficiency of this isomer enrichment method is moderate, but it gives a very simple route to break up *cis-trans* and racemic compositions of the products obtained by nonselective synthesis.

2.7   Alternative methods of resolution by DBTA complex formation

Rate of complex formation between chiral alcohols and DBTA monohydrate in hexane suspension is quite slow (see Figure 1) and numerous separation steps are necessarry for isolation of the alcohol isomers (filtration of the diastereoisomeric complex then concentration of the solution, decomposition of the complex, separation of the resolving agent and the enantiomer, distillation of the product). To avoid these problems, alternative methods have been developed for complex forming resolution of secondary alcohols. In a very first example of solid phase one pot resolution [40] the number of separation steps was decreased radically. Another novel method [41] let us to increase the rate of complex forming reaction in melt. Finally, first examples of the application of supercritical fluids for enantiomer separation from a mixture of diastereoisomeric complexes and free enantiomers [42, 43] are discussed in this subchapter.

One pot resolution of 2-iodocyclohexanol (37)
There are only few examples for one pot resolution of racemates in the literature. [44, 45, 46]. In these examples the solid complex forming resolving agent was mixed with the racemic liquid without any solvent and the enantiomers were separated by distillation.

The first example of combined solid phase reaction and one pot resolution has been developed by our laboratory (Scheme 14). [40]

TABLE -7. Results of solid phase one pot resolution of **37** with DBTA

Reaction time	Temperature of sublimation	Sublimated product (37)				
		No.	Yield[a]	Configuration	*ee*	S[b]
7 days	50 °C	1.	30 %	(*1R,2R*)-(-)	68 %	0,20
	50 °C	2.	58 %	(*1R,2R*)-(-)	86 %	0,50
	100 °C	3.	53 %	(*1S,2S*)-(+)	80 %	0,42
14 days	50 °C	1.	56 %	(*1R,2R*)-(-)	75 %	0,42
	50 °C	2.	37 %	(*1R,2R*)-(-)	55 %	0,20
	100 °C	3.	53 %	(*1S,2S*)-(+)	46 %	0,25
21 days	50 °C	1.	10 %	(*1R,2R*)-(-)	76 %	0,08
	50 °C	2.	77 %	(*1R,2R*)-(-)	78 %	0,61
	100 °C	3.	18 %	(*1S,2S*)-(+)	72 %	0,13

[a]Yield was calculated for the half of the racemic compound **38**. [b]Efficiency of the resolution (S = yield x ee. [11])

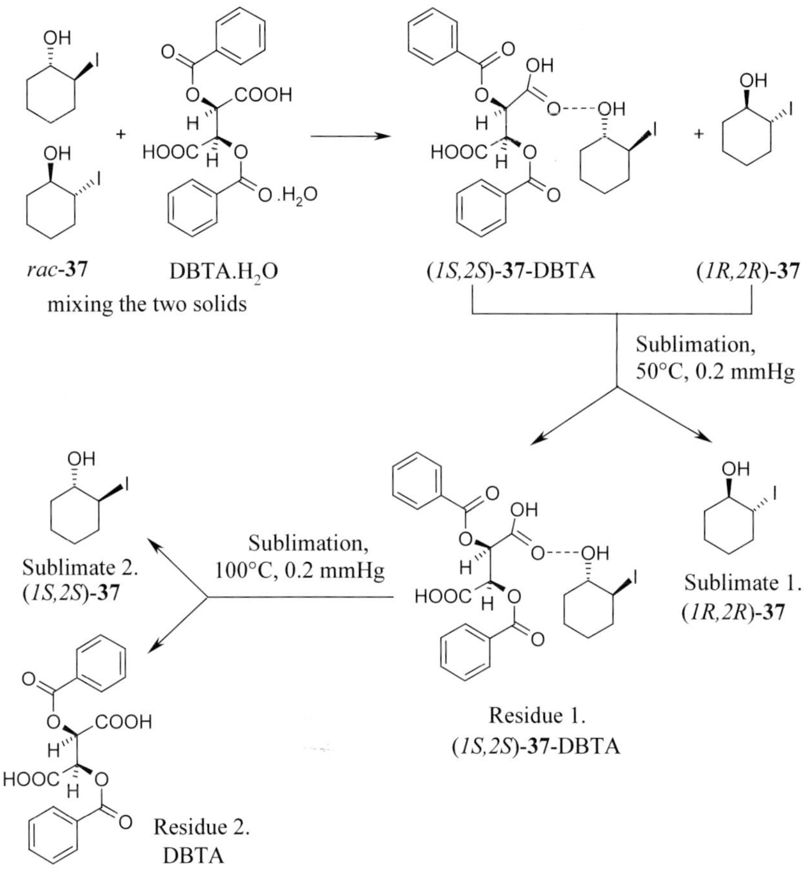

Scheme 14

The melting point of racemic **37** is 43 °C, therfore this racemate and DBTA monohydrate was mixed in 2:1 molar ratio as two powders at ambident temperature, then the mixture was subjected to fractionated sublimation in vacuum (Scheme 14). At 50 °C, the unreacted alkohol sublimated and it contained (*1R,2R*)-**37** in excess. At higher temperature (up to 100 °C) the diastereoisomeric complex decomposed slowly and the second sublimated fraction was enriched in (*1S,2S*)-**37** isomer. Results of one, two and three weeks reactions are summarised in Table 7.

It is important to mention that the rate of complex formation could be radically increased by intensive stirring and fretting of the solid mixture in a porcelain mortar. This resolution process can be considered as the simplest one ever made.

Diastereoisomeric complex formation in melt

Optical resolution of menthol (**28**) with DBTA in hexane suspension is quite efficient (S = 0.374, Table 5 [38]) but complex formation is slow. In order to increase the reaction rate, the highest possible concentrations of the reactants were reached by solving DBTA monohydrate in the melt of racemic menthol. [41]

The melting point of racemic **28** is 33 °C therefore clean liquid could be obtained by gentle heating of the mixture. As far as the stirred reaction mixture cooled down to 25 °C, the unreacted optically active menthol could be withdrawn from the solid by simple extraction [41] (Scheme 15). The yield and enantiomeric enrichment depends on the molar ratio of the racemate and DBTA (Table 8). The most efficient resolution of **28** was accomplished with half an equivalent DBTA (S = 0,456) and this result is significantly better then the selectivity of the original resolution in hexan (S = 0,37).

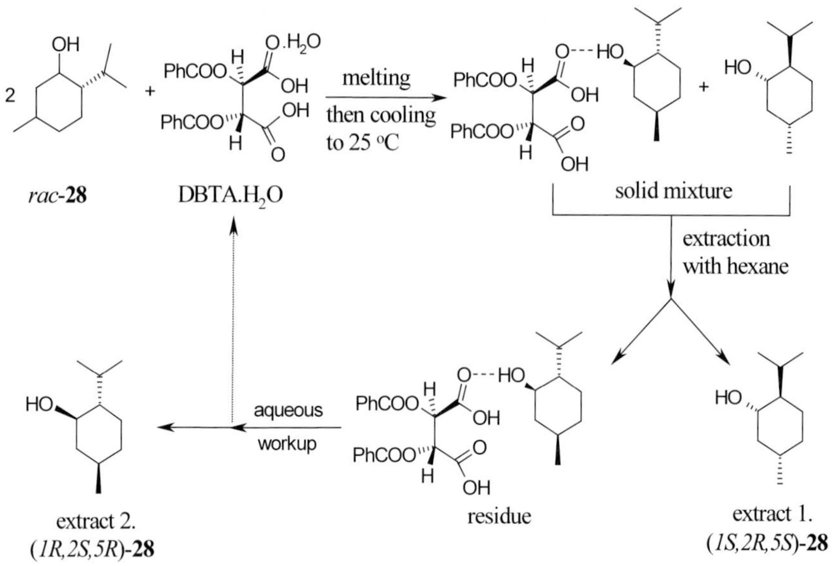

Scheme 15

TABLE -8. Results of optical resolution of **28** in melt

28 / DBTA molar ratio	Unreacted (1S,2R,5S)-28		(1R,2S,5R)-28 from the complex		
	Yield (%)	ee (%)	Yield (%)	ee (%)	S[a]
5 / 1	129	8	12	89	0,107
3.3 / 1	156	16	30	85	0,255
2.5 / 1	138	28	52	76	0,394
2 / 1	113	44	73	63	0,456
	155[b]	24[b]	45[b]	83[b]	0,370[b]
1.6 / 1	124	34	56	67	0,378
1.4 / 1	109	0,3	74	44	0,326

[a]Efficiency of the resolution S = yield x ee. [11] [b]Result of the resolution accomplished in hexane.

Optical resolution did not provide pure enantiomers of **28** in any cases (Table 8) but repeated resolution of the complex forming enantiomeric mixture yielded (*1R,2S,5R*)-**28** in 92-94 % ee. Treatment of the (*1S,2R,5S*)-**28** containing enantiomeric mixture with the same resolving agent (*R,R*-isomer of DBTA) did not affect dramatic change of the enantomeric ratio. Pure (*1S,2R,5S*)-**28** could be prepared by the use of the mirror image isomer of DBTA during a repeated resolution (Table 9).

In order to shed light details of this new complex forming resolution in melt, the process was studied by DSC (Differential Scanning Calorimetry), too. [41] On the basis of these measurements temperature profile of the consecutive physical and physico chemical steps could be determined. During slow heating of a solid mixture of DBTA and (-)-**28** the first endothermic peak at 31.5 °C corresponds to the melting of menthol, the second one at 67 °C belongs to the evaporation of water (coming from DBTA monhydrate). Complex formation starts at that temperature and the formed supramolecular compound cyrstallizes at 90 °C. Further heating results in melting of the complex at 116 °C and menthol evaporation starts at 145 °C. One can conlude from these data that water eliminates from the melt at about 65-70 °C and the diastereoisomeric complex might be destroyed by heating it above 145 °C.

TABLE -9. Repeated resolutions of enantiomeric mixtures of **28** in melt

28 / DBTA molar ratio	Starting mixture of **28**		Product **28** from the complex		
	Configuration	ee (%)	Yield (%)	Configuration	ee (%)
2 / 1	(1R,2S,5R)	64	46	(1R,2S,5R)	94
1.7 / 1	(1R,2S,5R)	64	60	(1R,2S,5R)	92
1.6 / 1	(1S,2R,5S)	29	21	(1R,2S,5R)	16

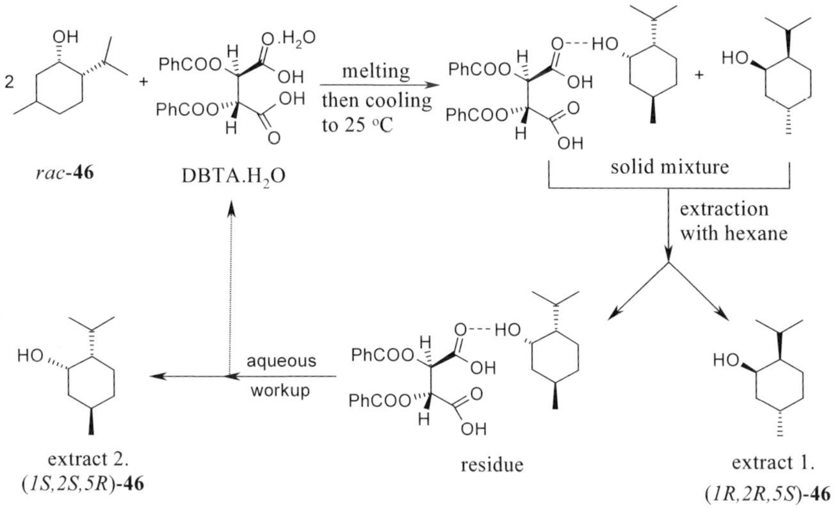

Scheme 16

It is worth to mention that *neo*menthol (**46**) does not form complex with DBTA in hexane but it does in melt. [41] When a mixture of racemic **46** and half an equivalent amount of DBTA monohydrate was melted then cooled down to 25 °C, usual extractive workup provided (-)-**46** in 73 % yield and 49 % ee (Scheme 16).

Optical resolution via consecutive complex formation and supercritical fluid extraction

Supercritical carbon dioxide is widely used in the food industry for separation aroma compounds from drugs of flowers. Big advantages of this separation method are the facts that carbon dioxide is cheap and it can be recycled into the system, easy to evaporate it from the extract by reducing the pressure and the density and solubility power of the fluid can be influenced by simple change of the temperature and the pressure. Disadvantage of the method is that high pressure (100-200 bar) and increased temperature (30-50 °C) are necessary to achieve supercritical state of carbon dioxide. Consequently the cost of such an equipment is high, but it can be used for solving series of separation problems.

Supercritical carbon dioxide is an apolar solvent, thus it is able to replace hexane during separation of the unreacted enantiomer from the diastereoisomeric complex containing reaction mixture. This idea was successfully applied in the complex forming resolution of *trans*-2-halogenocyclohexanols (**35**, **36**, **37**) and menthol (**28**). [42, 43] Diastereo-isomeric complex formation reaction was carried out in the mixture of the hexane solution of the racemic ligand and less then an equivalent amount of pulverised DBTA monohydrate.

In addition, a supporting material (such as Perfil 100[TM]) was also added into the reaction mixture then the solvent was evaporated and the solid residue was extracted with supercritical carbon dioxide at 150 bar pressure. As an example, resolution of **37** by supercritical extraction is outlined in Scheme 17 and the results are sumarised in Table 10.

After a long incubation time the unreacted menthol (**28**) was withdrawn from the solid material in good yields but low ee values. In another experiment two consecutive extractions were accomplished at 32 °C and then at 50 °C after three hours reaction time. The first extract contained (*1S,2R,5S*)-**28** in 24 % ee but the second fraction was practically racemic because the (*1R,2S,5R*)-**28** enantiomer partially escaped from the complex at 50 °C.

TABLE -10. Enantiomer separation via consecutive complex formation and supercritical fluid extraction.

Racemate	Racemate / DBTA molar ratio	Reaction time	Product from the extract		
			Config.	Yield	ee
28	2 : 1	4 days	(*1S,2R,5S*)-**28**	85 %	11 %
28	4 : 3	3 hours	(*1S,2R,5S*)-**28**	1. extract[a] 50 %	24 %
				2. extract[a] 39 %	6 %
35	4 : 2.2	3 hours	(*1R,2R*)-**35**	89 %	63 %
36	3 : 2	3 hours	(*1R,2R*)-**36**	69 %	74 %
37	4 : 2.9	3 hours	(*1R,2R*)-**37**	66 %	95 %

[a]The first extraction was carried out at 32 °C, the second at 50 °C.

(*1S,2S*)-Enantiomers of the *trans*-2-halogenocyclohexanols (**35, 36, 37**) form more stable complexes with DBTA than (*1R,2S,5R*)-**28**, therfore much better separations were achieved in these cases by supercritical fluid extraction. The results are better than that of the hexane extraction method (see in Table 5). [42, 43]

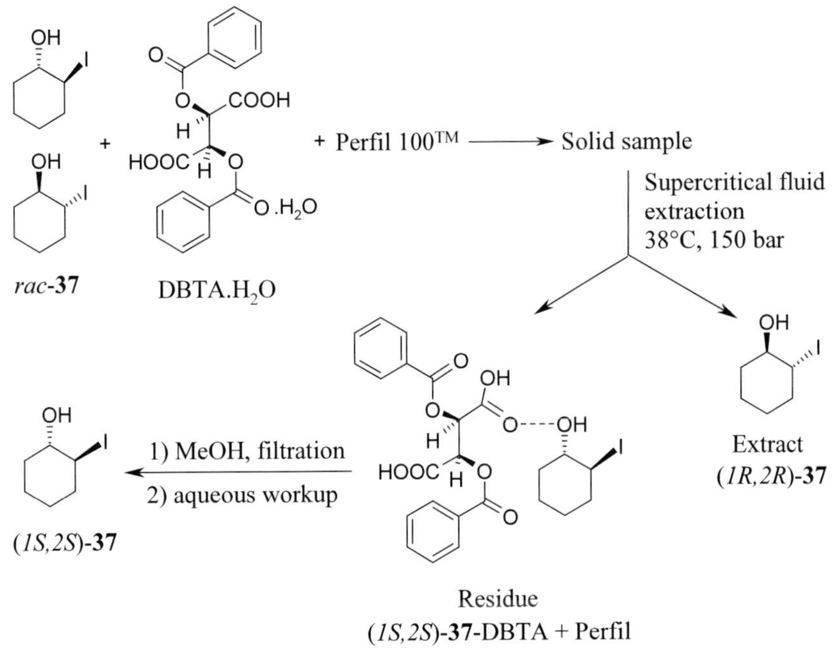

Scheme 17

2.8   Thermoanalytical and single crystal X-ray diffraction analysis of alcohol-DBTA complexes

Combined thermoanalytical (TA,TGA and DSC) investigations of DBTA complexes of chiral alcohols may serve valuable data on the composition and stabilities of these solid supramolecular compounds. DSC monitoring of menthol resolution with DBTA in melt has already discussed in point 3.4. Thermal stability of further ten complexes of DBTA with alcohols 26, 29-31, 35-37 and 39-41 was also determined. [47] The melting points of the DBTA complexes (Table 11) increase in the 2-alkanol, 2-alkoxycyclohexanol, 2-halogenocyclohexanol order.

Within the series of 2-alkanols (**26, 29-31**) the branched alkanols form more stable complexes with higher melting point than the linear 2-alkanols. According to the thermogravimetric measurements, the host-guest molar ratios in the complexes were approximatively 1:1 in these cases. In the more stable complexes of 2-alkoxycyclohexanols (**39-41**) at about two DBTA molecules are complexed with one guest compound and better ee values were obtained than that of the 2-alkanol series. Correlation between the thermal stability and the effectiveness of chiral recognition could also be observed at the complexes of 2-halogenocyclohexanols (**35-37**). According to the thermogravimetric measurements, these solid supramolecular compounds contained less guest molecules than the theoretical amount calculated from the 1:1 host:guest molar ratio.

TABLE -11. Melting points and host-guest ratios of selected DBTA complexes

Alcohol No.	Diastereoisomeric complex			Alcohol
	Alcohol mass ratio	mp. config.	ee	
29	19.5 %	55 °C	0 %	-
30	18.2 %	59 °C	0 %	-
31	27.7 %	75 °C	20 %	(*R*)-**31**
26	23.4 %	105 °C	28 %	(*R*)-**26**
39	17.1 %	87°C	50 %	(*1R,2R*)-**39**
40	20.5 %	112 °C	15 %	(*1R,2R*)-**40**
41	23.1 %	77 °C	44 %	(*1S,2S*)-**41**
35	24.6 %	126 °C	35 %	(*1S,2S*)-**35**
36	26.3 %	137 °C	56 %	(*1S,2S*)-**36**
37	30.1 %	144 °C	61 %	(*1S,2S*)-**37**

Single crystal X-ray diffraction data of nine DBTA complexes of chiral and ahiral alcohols were also collected and analysed with the aim of finding similarities and characteristic differences among the structures of the supramolecular compounds. [48] The more one knows on the constitution of the DBTA complexes the larger the chance to expand this complex forming resolution method for other, new racemates. Molecular structures, packing diagrams and other details on the discussed complexes can be seen and read in the original articles. [16, 48] Here we just mention the general conclusions of the comparative study of those structures.

The nine crystal structures were arranged into three classes: A, B and C (Table 12). In class A 1:1 complexes of chiral alcohols and DBTA can be found. Class B is the group of 2:1 complexes of achiral alcohols. Finally, in class C one can find the 2:1 complex of racemic 2-methylcyclohexanol (50, Scheme 13) and racemic DBTA. Common feature of all crystalline complexes is the lack of water or other solvate molecule even all of them was prepared from DBTA monohydrate. This obsrevation fits well to the thermoanalytical measurements.

The different classes have different space group symmetries (Table 12). On the other hand, similar intermolecular second order interactions and isostructural constitution of the complexes could be observed within the classes.

TABLE -12. Selected data of crystalline DBTA complexes

Guest molecule	Space group[a]	a (Å)	b (Å)	c (Å)	β (°)	D_{calc}[b] (g/cm^3)
			Class A			
(1R,2S,5R)-28	P2$_1$	8.738(3)	12.753(5)	13.036(5)	97.66(3)	1.187
(R)-31	P2$_1$	8.523(2)	12.362(5)	11.699(3)	100.93(2)	1.225
(R)-26	P2$_1$	8.638(1)	12.522(1)	11.694(1)	99.844(11)	1.227
(1S,2S)-47	P2$_1$	8.643(4)	11.821(4)	12.507(6)	102.28(4)	1.257
(1S,2S)-48	P2$_1$	8.7651(17)	11.979(3)	12.643(1)	101.085(12)	1.240
			Class B			
52	C2	18.917(4)	6.455(8)	15.063(4)	122.909(12)	1.201
47	C2	18.793(4)	6.463(4)	14.456(4)	109.047(18)	1.174
53	C2	18.446(2)	7.626(3)	14.2980(17)	123.637(7)	1.164
			Class C			
(±)-trans-50	C2	14.638(12)	9.911(7)	23.027(4)	101.03(3)	1.188

[a]The crystal systems were monoclinic, α and γ=90° in all cases. [b]Calculated density.

The cystals of class A containe head to tail DBTA chains connected by H-bond between the carboxylic groups and the chains are cross-bonded by the hydroxy groups of the chiral alcohol molecules. The oxygen atom of the alcoholic hydroxy group is the acceptor of the carboxy hydrogen of one DBTA while the hydrogen atom is connected to the carbonyl oxygen of another carboxylic group. In this way, hydrophobe holes are formed for the alkyl chains of the guest molecules. Data in Table 12 show small change of the unit cell parametes by changing the size of the guest compounds but all the five complexes have isosturctural crystals in which the conformation of DBTA is practically the same. In class B the DBTA molecules are not connected to each other directly but the hydroxy groups of the achiral alcohols stick them together by H-bonds. Class C is represented by the only structure of racemic trans-2-methylcyclohexanol (50) and racemic DBTA. The unit cell contains four molcules, the mirror image pairs of the host and the guest compounds.

In spite of the different compositions and space groups of symmetry, hydrophobe holes are formed by the parallel benzoyl groups of DBTA in all crystals of classes A and B. These holes play crucial role in complex formation by accepting the alkyl or cycloalkyl part of the guest molecules. Tartaric acid (TA) can not offer such hydrophobe surface for week second order interactions, therefore TA does not form complexes with the investigated guest compounds.

Computer aided superposition of the molecular complexes of class A and B demonstrated that the host molecule (DBTA) situates practically in the same conformation in every cases and its carboxyl, carbonyl and phenyl groups take part in intermolecular interactions recognizing the enantiomers of chiral alcohols. [48] For clarity, superposition of only four complexes of DBTA is presented in Figure 2.

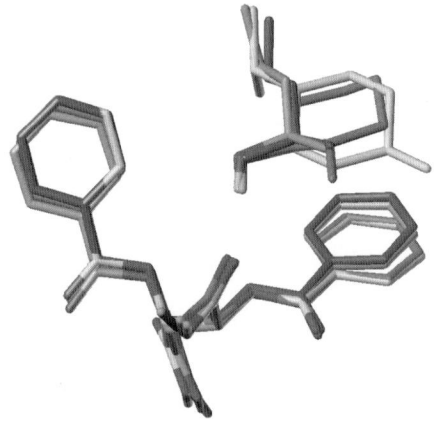

Figure 2. Superpositions of **26-**, **28-**, **31-** and **50**-DBTA complexes in conformations found in their crystals.

2.9   Scope and limitations: resolution trials with DPTTA

Compared with DBTA, there are only two methyl groups more in DPTTA but this small change in the structure causes significant differences in the outcome of complex forming resolutions. Three model compounds (**26**, **28**, **37**) among the previously investigated chiral alcohols were selected for test reactions with DPTTA. Optical resolutions of these compounds served the best results with DBTA. The results of the comparative resolution experiments with DPTTA and DBTA are collected in Table 13. [49]

TABLE -13.

Racemate	Resolving agent[a]	Alcohol from the filtrate		Alcohol from the solid complex		
		Yield[b]	ee	Yield[b]	ee	S[c]
28	DPTTA	128 %	0 %	16 %	0.5	0
	DBTA	155 %	24 %	45 %	% 83 %	0,37
26	DPTTA	195 %	9 %	0 %	0 %	0
	DBTA	109 %	23 %	91 %	28 %	0,25
37	DPTTA	75 %	28 %	114 %	22 %	0,25
	DBTA	100 %	48 %	71 %	61 %	0,43

[a]Half equivalent amounts of resolving agents were used in all cases. [b]Yields were calculated for the half of the starting racemate. [c]The efficiency (0<S<1) of the optical resolution. [11]

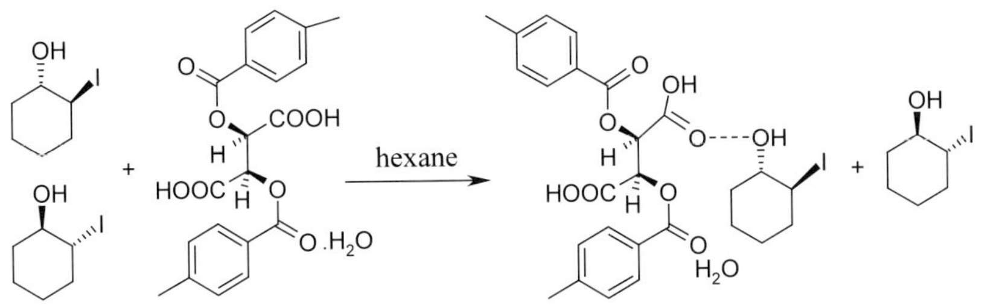

rac-**37** DPTTA.H_2O (*1S,2S*)-**37**-DPTTA.H_2O solid (*1R,2R*)-**37** in solution

Scheme 18

Under the same conditions compound **26** did not form complex with DPTTA and the crystalline **28**-DPTTA contained the racemic alcohol, while both compounds could be resolved with DBTA. Partial resolution of **37** served 22% enrichment of the (*1S,2S*)-**37** enantiomer in the crystalline DPTTA complex (Scheme 18) but much higher ee (61 %) was achieved with DBTA.

These experimental data can be rationalised if the crystal structures of the DBTA complexes and the thermoanalytical data of (*1S,2S*)-**37**-$(DPTTA)_2.H_2O$ complex are taken into consideration. [49] DSC and TG analysis of (*1S,2S*)-**37**-$(DPTTA)_2.H_2O$ complex have shown that it contains water which easily eliminate at about 40 °C. In spite of the fact that both resolving agents were used as monohydrate, the DBTA complexes never contained water. Further heating of the DPTTA containing sample results in decomposition of the complex and sublimation of **37** at lower temperature (~ 100 °C) than in the case of the corresponding DBTA complex. According to the thermogravimetric analysis, two DPTTA molecules are necessarry for complexation of one 2-iodocyclohexanol molecule (**37**) while the molar ratio in the DBTA complex is 1:1.

As it was shown in point 3.5., parallel benzoyl groups of DBTA molecules form hydrophobic holes in the crystals and the side chains of the chiral alcohols fit in this holes. The *para*-methyl groups in DPTTA probably hinder such tight arrangement of the aromatic rings, thus the energy gain of attractive week interactions between the phenyl rings and the side chains of the alcohols diminishes. Furthermore, there are several examples in the literature where the presence or lack of weak second order interaction between the *para* hydrogen of a phenyl ring with oxygen atom of another molecule influences the efficiency of the resolution dramatically. [50, 14]. In the present cases *para* methyl groups of the phenyl rings in DPTTA also eliminate the possibility of formation such type of week second order bonds. Lack of the before mentioned interactions and bulkiness of the methyl groups are supposed to be the main reasons of the radically decreased stability of the studied DPTTA comlpexes.

Comparison of the results of the parallel resolution experiments with DBTA and DPTTA confirmed that small variation in the structure of DBTA may cause radical change in its complex forming ability. On the other hand, DBTA itself is a versatile resolving agent, it can be used for resolution of a wide range of non basic racemates via coordination compounds or host-guest complexes. Enantiomer separation of chiral alkoxycarboxylic acids, hydroxycarboxylic acid esters, different classes of substituted alcohols and phosphine oxides have been accomplished until now. The flexibility and the low price of DBTA together with the simple experimental conditions of the reviewed complex forming resolution processes promise of further development and practical application of these methods in the near future.

ACKNOWLEDGEMENTS

The authors are grateful to Professor Elemér Fogassy for helpful discussions. The research work has partially been supported by the Hungarian Scientific Research Foundation (OTKA T-042805).

REFERENCES

1. Jacques, J., Collet, A., and Wilen, S.H.: *Enantiomers, Racemates and Resolutions.* Wiley, New York, 1981, pp 259-260.
2. Newman, P.: *Optical Resolution Procedures for Chemical Compounds*, Vol. 1-3. Optical Resolutions Information Center, New York, 1978-1983.
3. Ács, M., Fogassy, E., Kassai, Cs., Kozma, D., and Nógrádi, M.: *CRC Handbook of Optical Resolutions via Diastereoisomeric Salt Formation*, Ed.: Kozma, D., CRC Press, Boca Raton, 2002.
4. Toda, F., and Tanaka, K.: Design of a new chiral host compound, trans-4,5-bis(hydroxydiphenylmethyl)-2,2-dimethyl-1,3-dioxacyclopentane. An effective optical resolution of bicyclic enones through host-guest complex formation, *Tetrahedron Lett.* **1988**, *29*, 551-554.
5. Weber, B., and Seebach, D.: Enantiomeric tertiary alcohols by TADDOL-mediated additions to ketones - or, how one makes a Grignard reagent enantioselective, *Angew. Chem.* **1992**, *104*, 96-7.
6. Seebach, D., Pichota, A., Pinkerton, A.K., Litz, T., Karjalainen, J., and Gramlich, V.: Preparation of new TADDOL derivatives for new applications, *Org. Letters*, **1999**, 1, 55-58.
7. Toda, F., and Tanaka, K.: Efficient optical resolution of 2,2'-dihydroxy-binaphtyl and related compounds by complex formation with novel chiral host compounds derived from tartaric acid, *J. Org. Chem.* **1988**, *53*, 3607-3609.
8. Tanaka, K., Honke, S., Urbanczyk-Lipkowska, Z., and Toda, F.: New chiral hosts derived from dimeric tartaric acid: efficient optical resolution of aliphatic alcohols by inclusion complexation, *J. Org. Chem.* **2000**, *65*, 3171-3176.
9. Eliel, E.L., Wilen, S.H., Mander, L.N.: *Stereochemistry of Organic Compounds*, Wiley, New York, 1994, pp. 351-359.
10. Ogston, A.G.: Interpretation of experiments on metabolic processes using isotopic tracer elements, *Nature* **1948**, *162*, 963.
11. Fogassy, E., Lopata, A., Faigl, F., Darvas, F., Ács, M., and Tőke, L.: Quantitative approach to optical resolution, *Tetrahedron Lett.* **1980**, 21, 647-648.
12. Fogassy, E., Faigl, F., and Ács, M.: A new method for designing optical resolutions and for determination of relative configurations, *Tetrahedron*, **1985**, *41*, 2837-2840.
13. Fogassy, E., Faigl, F., and Ács, M.: Diastereoisomeric interactions and selective reactions in solutions of enantiomers, *Tetrahedron*, **1985**, *41*, 2841-2845; Faigl, F., Simon, K., Lopata, A., Kozsda, É., Hargitai, R.,

Czugler, M., Ács, M., and Fogassy, E.: A combined DSC, X-ray diffraction and molecular modelling study of chiral discrimination in the purification of enantiomeric mixtures of cis-permethrinic acid, *J. Chem. Soc. Perkin Trans. 2*. **1990**, 57-63.

14.   Fogassy, E., Ács, M., Faigl, F., Simon, K., Rohonczy, J., and Fcsery, Z.: Pseudosymmetry and chiral discrimination in optical resolution via diastereoisomeric salt formation. The crystal structures of (*R*)- and (*S*)-N-methylamphetamine bitartarates, *J. Chem. Soc. Perkin Trans 2*. **1986**, 1881-1885; *Acta Cryst. S.* **1985**, *40A* C81.

15.   Hatano, K., Takeda, T., and Saito, R.: Optical resolution of trans-bicyclo[2.2.1]heptane-2,3-diamine: chiral recognation in the crystal of its complex with (2*R*,3*R*)-O,O'-dibenzoyltartaric acid, *J. Chem. Soc. Perkin Trans 2*. **1994**, 579-584.

16.   Kozma, D., Böcskei, Zs., Kassai, Cs., Simon, K., and Fogassy, E.: Optical resolution of racemic alcohols by diastereoisomeric complex formation with O,O'-dibenzoyl-(2*R*,3*R*)-tartaric acid, the crystal structure of the (-)-1*R*,2*S*,5*R*-menthol.O,O'-dibenzoyl-(2*R*,3*R*)-tartaric acid complex. *J. Chem. Soc. Chem. Commun.* **1996**, 753-754.

17.   Ryhlewska, U., and Warzajtis, B.: Interplay between dipolar, stacking and hydrogen-bond interactions in the crystal structures of unsymmetrically substituted esters, amides and nitriles of (R,R)-O,O'-dibenzoyltartaric acid, *Acta Cryst., Sec. B*. **2001**, *B57*, 415-427.; Isostructuralism in a series of methyl ester/methylamide derivatives of (R,R)-O,O'-dibenzoyl tartaric acid; inclusion properties and guest-dependent homeotypism of the crystals of (2*R*,3*R*)-O,O'-dibenzoyltartaric acid diamide, *Acta Cryst., Sec. B*. **2002**, *B58*, 265-271.

18.   Toda, F., Miyamoto, H., and Ohta, H.: Efficient optical resolution of cis-4-methylcyclohex-4-ene-1,2-dicarboxylic anhydride, cis-4-methylcyclohex-4-ene-1,2-di-carboximide, and their derivatives by complexation with optically active host compounds derived from tartaric acid, *J. Chem. Soc. Perkin Trans 1*, **1994**, 1601-1604.

19.   Miyamoto, H., Yasaka, S., Takaoka, R., Tanaka, K., and Toda, F.: One-pot preparation of optically active sec-alcohols, epoxides, and sulfoxides by a combination of synthesis and enantiomeric resolution with optically active hosts in a water suspension medium *Enantiomer* **2001**, *6*, 51-55.

20.   Miyamoto, H., Sakamoto, M., Yoshioka, K., Takaoka, R., and Toda, F.: Resolution of hydrocarbons by inclusion complexation with a chiral host compound *Tetrahedron: Asymmetry* **2000**, *11*, 3045-3048; Tanaka, K., Honke, S., Urbanczyk-Lipowska, Z., and Toda, F.: New chiral hosts derived from dimeric tartaric acid: Efficient optical resolution of aliphatic alcohols by inclusion complexation, *Eur. J. Org. Chem.* **2000**, 3171-3176.

21.   Seebach, D., Beck, A.K., and Heckel, A.: TADDOLs, their derivatives, and TADDOL analogs: versatile chiral auxiliaries, *Angewandte Chemie Int. Ed. Engl.* **2001**, *40*, 92-138.

22.   Nemák, K., Ács M., Jászay M.Zs., Kozma, D., and Fogassy, E.: Study of the diastereoisomers formed between pipecolic acid N-alkylanilides and 2R,3R-tartaric acid or O,O'-dibenzoyl-2R,3R-tartaric acid. Do the tartaric acids form molecular complexes instead of salts during optical resolutions?, *Tetrahedron* **1996**, *52*, 1637-1642.

23.   Mravik, A., Böcskei, Zs., Katona, Z., Markovits, I., Pokol, Gy., Menyhárd, D.K., and Fogassy, E.: A new optical resolution method: coordinative resolution of mandelic acid esters. The crystal structure of calcium hydrogen (2*R*,3*R*)-O,O'-dibenzoyltartrate-(2*R*)-methyl mandelate, *J. Chem. Soc. Chem. Commun.* **1996**, 1983.

24.   Mravik, A., Böcskei, Zs., Katona, Z., Markovits, I., and Fogassy, E.: Coordination-mediated optical resolution of carboxylic acids with O,O'-dibenzoyltartaric acid, *Angew. Chem. Int. Ed. Engl.* **1996**, *36*, 1534-1536.

25.   Mravik, A., Böcskei, Zs., Simon, K., Elekes, F., and Izsáki, Z.: Chiral recognition of alcohols in the crystal lasttice of simple metal complexes of O,O'-dibenzoyltartaric acid: enantiocomplementarity and simultaneous resolution, *Chem. Eur. J.* **1998**, *4*, 1621-1627.

26.   Mravik, A., Lepp, Zs., and Fogassy, E.: Simple resolution of O,O'-dibenzoyltartaric acid by preferential crystallization of its calcium salt-methoxymethanol complex, *Tetrahedron: Asymmetry* **1996**, *7*, 2387-2390.

27. Elekes, F., Kovári, Z., Mravik, A., Böcskei, Zs., and Fogassy, E.: New access to enantiopure O,O'-dibenzoyltartaric acid: resolution of the mixed calcium methoxyacetate by preferential crystallization, *Tetrahedron: Asymmetry* **1998**, *9*, 2895-2900

28. Kagan, H.B.: In *Comprehensive Asymmetric Catalysis* (Eds.: Jacobsen, E. N., Pfaltz, A., Yamamoto, H.), Vol. I., Chapter 2., Springer, Berlin, 1999, p. 9.

29. Matteoli, U., Beghetto, V., Schiavon, C., Scrivanti, A., and Menchi, G.: Synthesis of the new chiral biphosphine ligand 2,3-bis(diphenylphosphino)butane (CHIRAPHOS), *Tetrahedron: Asymmetry* **1997**, *8*, 1403-1409.

30. Minami, T., Okada, Y., Nomura, R., Hirota, S., Nagahara, Y., and Fukuyama, K.: Synthesis and resolution of a new type of chiral bisphosphine ligand, trans-bis-1,2-(diphenylphosphino)cyclobutane, and asymmetric hydrogenation using its rhodium complex, *Chem. Lett.* **1986**, 613-616.

31. Takaya, H., Mashima, K., Koyano, K., Yagi, M., Kumobayashi, H., Taketomi, T., Akutagawa, S., and Noyori, R.: Practical synthesis of (*R*) -or (*S*)-2,2''-bis(diaryl-phosphino)-1,1'-binaphtyls (BINAPs), *J. Org. Chem.*, **1986**, 51, 629-635.

32. Hamada, Y., Matsuura, F., Oku, M., Hatano, K., and Shioiri, T.: Synthesis and application of new chiral bidentate phosphine, 2,7-di-tert-butyl-9,9-dimethyl-4,5-bis(methylphenylphosphino)xanthene, *Tetrahedron Lett.* **1997**, *38*, 8961-8964.

33. Benincori, T., Gladiali, S., Rizzo, S., and Sannicolo, F.: New modular class of easily accessible, inexpensive, and efficient chiral diphosphine ligands for homogenous stereoselective catalysis, *J. Org. Chem.* **2001**, *66*, 5940-5942.

34. Dupart de Paule, S., Jeulin, S., Ratovelomanana-Vidal, V., Genet, J.P., Champion, N., Dellis, P.: SYNPHOS, a new chiral ligand: synthesis, molecular modelling and application in asymmetric hydrogenation, *Tetrahedron Lett.* **2003**, *44*, 823; Synthesis and molecular modeling studies of SYNPHOS (R), a new, efficient diphosphane ligand for ruthenium-catalyzed asymmetric hydrogenation, *Eur. J. Org. Chem.* **2003**, *10*, 1931-1941.

35. Toda, F., and Tohi, Y.: Novel optical resolution methods by inclusion crystallization in suspension media and by fractional distillation, *J. Chem. Soc. Chem. Commun.* **1993**, 12381240.

36. Illés, R., Kassai, Cs., Pokol, Gy., Fogassy, E., and Kozma, D.: Thermoanalytical study of O,O'-dibenzoyl-(2*R*,3*R*)-tartaric acid supramolecular compounds, part I., *J. Therm. Anal. and Cal.* **2000**, *61*, 745-755.

37. Kassai, Cs., Illés, R., Pokol, Gy., Sztatisz, J., Fogassy, E., and Kozma, D.: Thermoanalytical study of O,O'-dibenzoyl-(2*R*,3*R*)-tartaric acid supramolecular compounds, part II., *J. Therm. Anal. and Cal.* **2000**, *62*, 647-655.

38. Kassai, Cs., Juvancz, Z., Bálint, J., Fogassy, E., and Kozma, D.: Optical resolution of racemic alcohols via diastereoisomeric supramolecular compound formation with O,O'-dibenzoyl-(2*R*,3*R*)-tartaric acid, *Tetrahedron* **2000**, *56*, 8355-8359.

39. Kassai, Cs., Bálint, J., Juvancz, Z., Fogassy, E., and Kozma, D.: Isomer and enantiomer separation of 2-and 4-alkylcyclohexanols by stereoselective complex formation with O,O'-dibenzoyl-(2*R*,3*R*)-tartaric acid, *Synth. Commun.* **2001**, *31*, 1715-1719.

40. Kassai, Cs., Fogassy, E., and Kozma, D.: *unpublished results.*

41. Simon, H., Vincze, K., Marthi, K., Lévai, G., Pokol, Gy., Fogassy, E., and Kozma, D.: Thermoanalytical study of O,O'-dibenzoyl-(2*R*,3*R*)-tartaric acid supramolecular compounds, part IV., *J. Therm. Anal. and Cal.* **2003**, accepted for publication.

42. Székely, E., Simándi, B., Fogassy, E., Kemény, S., and Kmecz I.: Enantioseparation of chiral alcohols by complex formation and supercritical fluid extraction *Chirality* **2003**, 15, 783-786.

43. Székely, E., Simándi, B., Illés, R., Molnár, P., Gebefügi, I., Kmecz I., and Fogassy, E.: Application of supercritical fluid extraction for fractionation of enantiomers, *J. Supercritical Fluids 2003*, accepted for publication.

44. Toda, F.: Solid State Organic Chemistry: Efficient Reactions, Remarkable Yields, and Stereoselectivity, *Acc. Chem. Res.* **1995**, *28*, 480-486.

45. Ács, M., Szili, T., and Fogassy, E.: New method of optical activation for racemic bases, *Tetrahedron Lett.*, **1991**, *49*, 7325-7328.

46. Kozma, D., and Fogassy, E.: Solvent-free optical resolution of N-methylamphetamine by distillation after partial diastereoisomeric salt formation, *Chirality*, **2001**, *13*, 428-430.

47. Illés, R., Kassai, Cs., Pokol, Gy., Fogassy, E., and Kozma, D.: Thermoanalytical study of O,O'-dibenzoyl-(2R,3R)-tartaric acid - Part III. SMC-s formation with chiral secondary alcohols, *J. Therm. Anal. and Cal.* **2002**, *68*, 679-685.

48. Kovári, Z., Böcskei, Zs., Kassai, Cs., Fogassy, E., and Kozma, D.: Investigation of the structural background of stereo- and enantioselectivity of O,O'-dibenzoyl-(2R,3R)-tartaric acid-alcohol supramolecular compound formation, *Chirality* **2003**, submitted for publication. Crystal data are deposited at the Cambridge Crystal Structure Data Base under the following numbers: CCDC 181497, 181498, 181499,181500, 181501, 181502, 181503, 181504, 181505.

49. Simon, H., Marthi, K., Pokol, Gy., Fogassy, E., and Kozma, D.: O,O'-di-*para*-toluoyl-(2R,3R)-tartaric acid as supramolecular resolving agent, *J. Therm. Anal. and Cal.* **2003**, *74*, 155-162.

50. Taylor, R., and Kennard, O.: Crystallographic evidence for the existence of CH...O, CH...N and CH...Cl hydrogen bonds, *J. Am. Chem. Soc.* **1982**, *104*, 5063.

SPONTANEOUS CHIRAL CRYSTALLIZATION OF ACHIRAL MATERIALS AND ABSOLUTE ASYMMETRIC TRANSFORMATION IN THE CHIRAL CRYSTALLINE ENVIRONMENT

MASAMI SAKAMOTO
Department of Materials Technology, Faculty of Engineering, Chiba University
Yayoi-cho, Inage-ku, Chiba 263-8522, Japan

1. Introduction

Recently, the combination of chiral crystallization and the solid-state photoreaction has provided many successful examples of absolute asymmetric synthesis.[1]   In these reactions achiral materials adopted chiral arrangement only by spontaneous crystallization, and optically active products are obtained from the topochemically controlled reaction with high ees (Figure 1).[2,3]   This method incurs a problem in crystallization of achiral molecules in chiral space groups, while rare and unpredictable. However, crystal engineering and the solid-state reaction in recent years to a variety of new systems has progressed to such extent that it can now be regarded as an important branch of organic chemistry. The achievement of an asymmetric synthesis starting from an achiral reagent and in the absence of any external chiral agent has long been an intriguing challenge to chemists and is also central to the problem of the origin of optical activity on Earth.[4]

In this chapter, a general method of chiral crystallization and asymmetric synthesis using the chiral crystals will be described.

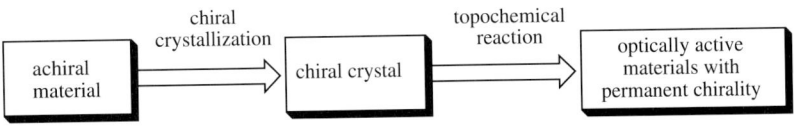

FIGURE 1. Absolute asymmetric synthesis in the chiral crystalline environment

F. Toda (ed.), Enantiomer Separation, 103-133.
© 2004 *Kluwer Academic Publishers, Printed in the Netherlands.*

2.   Chiral Crystallization of Achiral Materials

Optically active molecules must crystallize into chiral space groups, but a racemic mixture in solution may either aggregate to form a nonchiral racemic crystal or undergo a spontaneous resolution where the two enantiomers segregate into a conglomerate of enantiopure crystals. Nonchiral molecules may crystallize into either a nonchiral or a chiral space group. If they crystallize into a chiral space group, the nonchiral molecules reside in a chiral environment imposed by the lattice. Most achiral molecules are known to adopt interconverting chiral conformations in fluid media, which could lead to a unique conformation upon crystallization. Crystals that have chiral space groups are characterized by being enantiomorphrous. They exist in right-handed and left-handed forms that may or may not be visually distinguishable. In spite of impressive work on crystal engineering, predictions on a correlation between crystal symmetry and molecular structures are still hard to make.[5]

Chiral crystals, like any other asymmetric object, exist in two enantiomorphous equienergetic forms, but careful crystallization of the material can induce the entire ensemble of molecules to aggregate into one crystal, of one-handedness, presumably starting from a single nucleus (Figure 2). However, it is not uncommon to find both enantiomorphs present in a given batch of crystals from the same recrystallization.

For achieving asymmetric synthesis we should begin with a compound crystallizing in any one of the chiral space groups (Table 1). Of the 230 distinct space groups the most commonly occurring space groups are $P2_1/c$, P-1, $P2_12_12_1$, $P2_1$, $C2/c$, and Pbca; the chiral ones being $P2_12_12_1$, $P2_1$, P_1 and $C2$ (Table 2).[6]

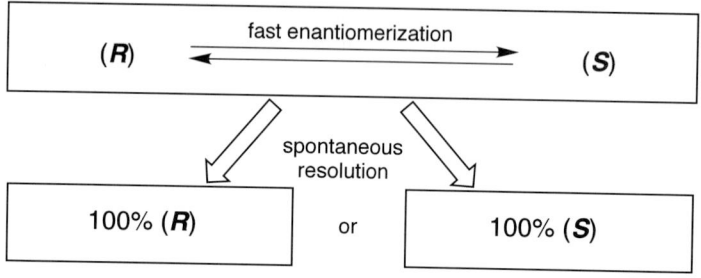

Figure 2. Chiral crystallization with fast enantiomerization

The asymmetric crystallization of achiral compounds is stimulated by autoseeding with the first crystal formed. Although the chiral sense of the spontaneously formed chiral crystals cannot be predicted, seed crystals of the preferred chirality can be added in a more practical procedure to obtain one enantiomorph of a crystal.

Kondepudi and McBride indicate that stirred crystallization is effective to accelerate the enantiomeric excess (ee) of crystals in the recrystallization step.[7] Kondepudi reported that the ee of the crystal greater than 95% can easily be obtained in stirred crystallization of achiral materials that crystallize in chiral form, such as sodium chlorate.   Combination of both seeding and stirred crystallization may lead to assured enantiomeric selectivity with high crystal enantiomeric purity.

TABLE 1 : crystal system and all chiral space groups

crystal system	chiral space group
triclinic	$P1$
monoclinic	$P2, P2_1, C2$
rhombic	$P222, P222_1, P2_12_12, P2_12_12_1, C222_1, C222, F222,$ $I222, I2_12_12_1$
tetragonal	$P422, P42_12, P4_122, P4_12_12, P4_222, P4_32_12, P4_322,$ $P4_32_12, I422, I4_122, P4, P4_1, P4_2, P4_3, I4, I4_1$
trigonal	$P312, P321, P3_112, P3_121, P3_212, P3_221, R3_2, P3,$ $P3_1, P3_2, R3$
hexagonal	$P622, P6_122, P6_522, P6_222, P6_422, P6_322, P6, P6_1,$ $P6_5, P6_2, P6_4, P6_3$
cubic	$P432, P4_232, P432, F4_132, I432, P4_332, P4_132,$ $I4_132, P23, F23, I23, P2_13, I2_13$

TABLE 2 : The most common space groups of organic crystalline compounds based
upon a survey of 29059 crystal structure determinations

order	space group	number	percentage
1	$P2_1/c$	10450	36.0
2	$P\text{-}1$	3986	13.7
3	$P2_12_12_1$*	3359	11.6
4	$P2_1$*	1957	6.7
5	$C2/c$	1930	6.6
6	$Pbca$	1261	4.3
7	$Pnma$	548	1.9
8	$Pna2_1$	513	1.8
9	$Pbcn$	341	1.2
10	$P1$*	305	1.1
11	Cc	277	1.0
12	$C2$*	273	0.9

*Chiral space group.

Actually, it is very hard to determine whether the crystals (or solid) are doped in chiral or achiral. The space group can be determined by only X-ray crystallographic analysis. Now, we can conveniently survey the crystal from the measurement of CD spectra (KBr or nujor method)[8] or activity of SHG (Second Harmonic Generation).[9] However, sense of high accuracy is required for the use of both methods. When we could observe a Cotton effect from the CD spectra, the crystal system should be chiral. On the other hand, there are many examples of chiral crystals that exhibit quite a little Cotton effect. As a matter of course, it is also difficult to recognize chiral crystals when the enantiomeric purity is poor.

In regard to the observation of SHG activity, all crystals without a centrosymmetric space group have the possibility of SHG, whereas the ability is dependent on the packing and the electronic condition of the molecules. Then, the crystals with $Pna2_1$ and Cc, which are achiral, in Table 2 also show SHG. We can not perfectly select out the chiral crystals from these methods; however, we can easily increase the probability to find the chiral crystals without X-ray structural analysis.

3. Absolute Asymmetric Transformation from Nonchiral Molecules in Chiral Crystals

Penzien and Schmidt reported the first absolute asymmetric transformation in a chiral crystal.[10] They showed that enone 4,4'-dimethylchalcone 1, although being achiral itself, crystallizes spontaneously in the chiral space group $P2_12_12_1$ (Scheme 1). When single crystals of this material are treated with bromine vapor in a gas-solid reaction, the chiral dibromide 2 is produced in 6-25% ee. In this elegant experiment, it is the reaction medium, the chiral crystal lattice, that provides the asymmetric influence favoring the formation of one product enantiomer over the other, and the chemist has merely provided a non-chiral solvent (ethyl acetate) for the crystallization and a non-chiral reagent (bromine) for the reaction.

Some other interesting examples of solid-gas reaction using chiral crystals were reported. Reaction of chiral crystals of compound 3 with bromine in connection with rearrangement gave optically active dibromide 4 in 8% ee.[11]

Gerdil et al. reported two examples involving the solid-gas reactions of inclusion complexes of tri-o-thymonide with alkene or epoxycyclopentanone. The complex 5 crystallized in $P3_12_1$, and the reaction with singlet oxygen gave endoperoxide 6; however, the $[\alpha]_D$ value was low.[12] On the other hand, the reaction of chiral crystals of the complex 7 with hydrogen chloride gave two products 8 and 9, ees of which were 9% and 22%, respectively.[13]

These examples are interesting; however, heterogeneous intermolecular solid-state reactions are generally disadvantageous for achieving highly enantioselective reaction, because the reactions generally occur on the crystalline surface while breaking down the crystal lattice in accordance with the loss of chirality.

On the other hand, since the concept of topochemically-controlled reactions was established, various approaches to asymmetric synthesis using solid-state photoreaction have been attempted. Their studies have been concerned with the bimolecular reactions of chiral crystals in the solid state. Research on reactivity in the crystalline state has been extended in recent years to a variety of new systems, and many absolute asymmetric syntheses have been provided.

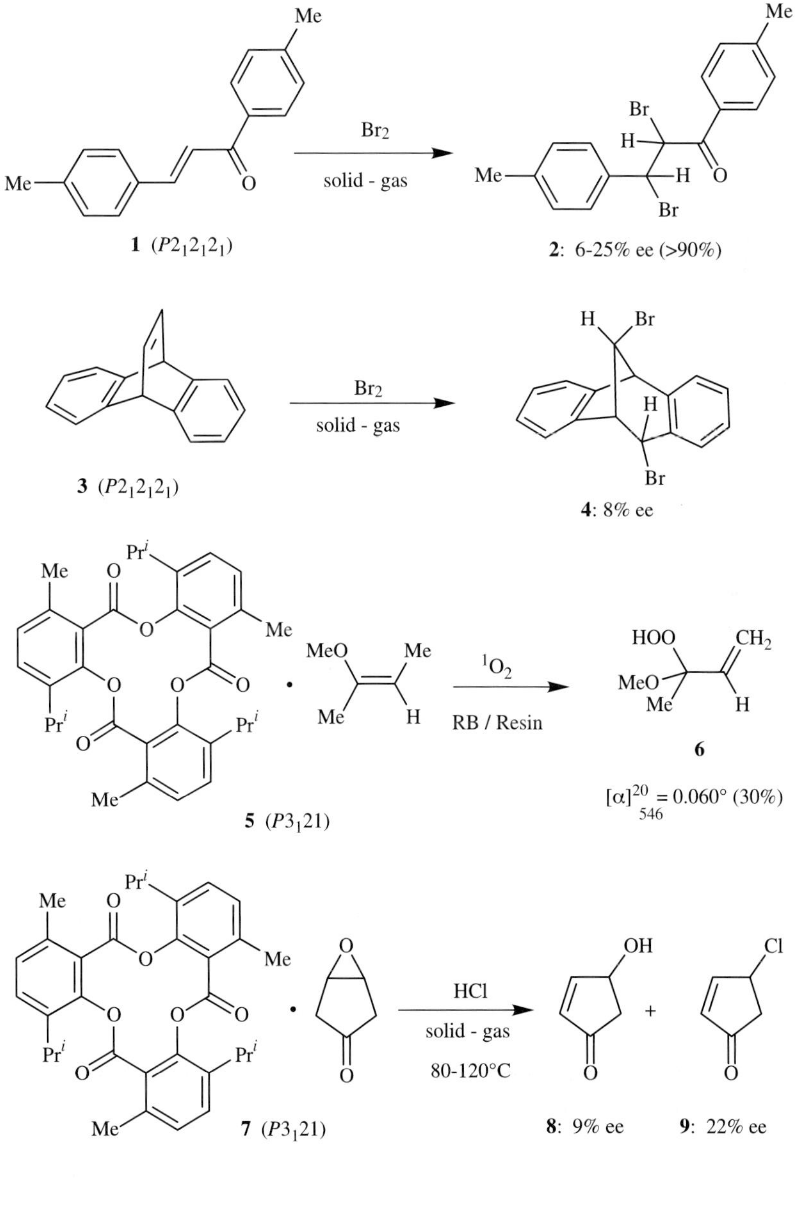

Scheme 1

The first example was promoted for 2+2 intermolecular cyclobutane formation. The strategy and the main results were reported by these investigators for the mixed crystal of 2,6-dichlorophenyl-4-phenyl-trans,trans-1,3-butadiene **10** with the 4-thienyl analog **11** (Scheme 2).[14] These two materials crystallize in two isostructural arrangements in the chiral space group $P2_12_12_1$. Large mixed crystals of the phenyl material containing ~15 % of the thienyl derivative as guest were prepared. The thienyl derivative absorbs light at a longer wavelength. As a result of this selective excitation, the thienyl reacts with the nearer phenyl neighbor to form a mixed cyclobutane dimer **12**. This dimer has been isolated to be optically active with an ee of ~70 %. As expected, some crystals gave left-handed and others right-handed cyclobutanes.

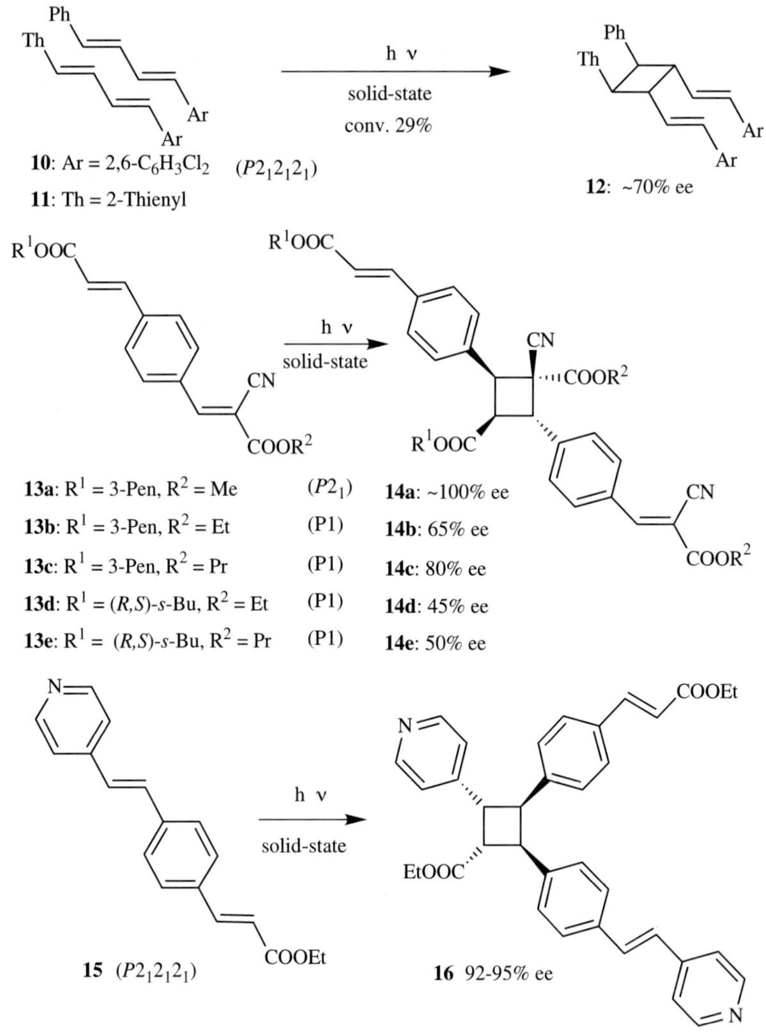

10: Ar = 2,6-C$_6$H$_3$Cl$_2$ ($P2_12_12_1$)

11: Th = 2-Thienyl

12: ~70% ee

13a: R^1 = 3-Pen, R^2 = Me   ($P2_1$)   **14a**: ~100% ee

13b: R^1 = 3-Pen, R^2 = Et   (P1)   **14b**: 65% ee

13c: R^1 = 3-Pen, R^2 = Pr   (P1)   **14c**: 80% ee

13d: R^1 = (R,S)-s-Bu, R^2 = Et   (P1)   **14d**: 45% ee

13e: R^1 = (R,S)-s-Bu, R^2 = Pr   (P1)   **14e**: 50% ee

15 ($P2_12_12_1$)   **16** 92-95% ee

Scheme 2

Several 1,4-disubstituted phenylenediacrylates **13** such as Me, Et, and Pr esters crystallized into chiral structures, and they photodimerized to either (*SSSS*)-cyclobutanes **14** or (*RRRR*)-cyclobutanes ent-**14** with medium to quantitative enantiomeric yields.[15] In addition to these dimers, the corresponding trimers and oligomers were also produced with high enantiomeric yields.

Hasegawa et al. reported another example of a [2 + 2] asymmetric transformation in a chiral crystal.[16] Ethyl 4-[2-(pyridyl)ethenyl] cinnamate **15** crystallizes in a chiral space group $P2_12_12_1$ and upon irradiation yields a chiral dimer **16** with 92-95% ee.

Suzuki et al. reported the photochemical reaction of CT crystals, in which cycloaddition reaction of bis(1,2,5-thiadiazolo)tetracyano-quinodimethane **17** (electron acceptor) and 2-divinylstyrene **18** (electron donor) is efficiently induced (Scheme 3).[17] A structural feature of the CT crystal is the asymmetric nature of the inclusion lattice because of the adoption of a chiral space group, $P2_1$. The [2 + 2] photoadduct **19** was formed via the single crystal-to-single crystal transformation, and the optically active product with 95% ee was obtained.

Most of the bimolecular absolute asymmetric syntheses are limited to 2+2 cyclobutane formation or polymerization of olefins. Koshima et al. reported a unique example of bimolecular reaction whereby acridine **20** and diphenylacetic acid are assembled in a 1:1 molar ratio by hydrogen bonding, and crystallized in a chiral space group, $P2_12_12_1$.[18] Irradiation of the crystals caused stereospecific decarboxylating condensation to give chiral **21** in 33-39% ee.

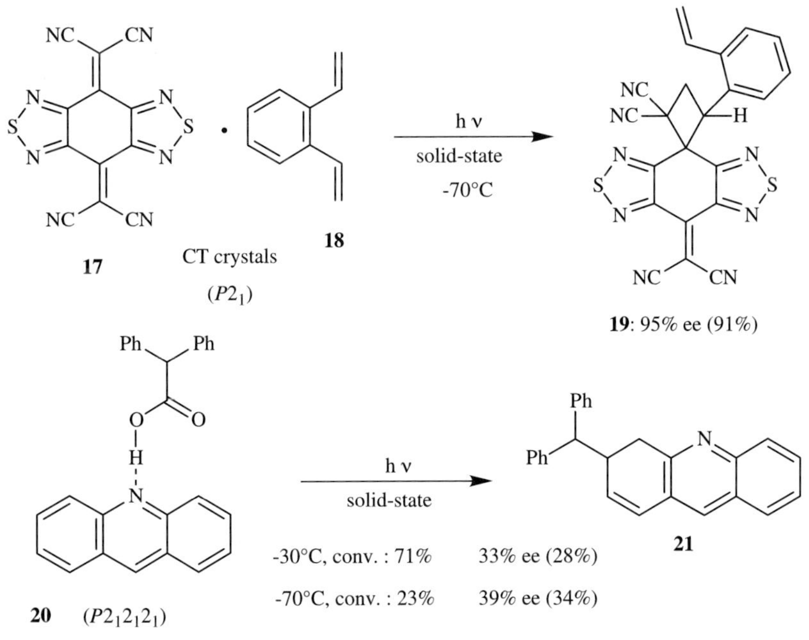

Scheme 3

The concept of absolute asymmetric synthesis using a chiral crystal was applied to unimolecular photochemistry, and now many fine examples are reported. Scheffer et al. reported elegant unimolecular absolute asymmetric transformations (Scheme 4).[19] This group demonstrated that the well-studied solution-phase di-π-methane photorearrangement can also occur in the solid state. Of over 20 symmetrical and unsymmetrical dialkyl 9,10-ethanoanthracene-11,12-decarboxylate 22, only two compounds were found to undergo absolute asymmetric di-π-methane photorearrangement in the solid state. One is the diisopropyl diester 22a and the other is the diethyl analog 22b. The corresponding diisopropylester 22a is dimorphic, and one of the forms grown from the melt is chiral (space group $P2_12_12_1$), and another is nonchiral space group $Pbca$. Irradiation of the chiral crystals gave rearranged product 23a in ee's greater than 95%. Solid-state photolysis of 22b also gave optically active 23b; however, the ee's were variable under the reaction conditions. In general, lower photolysis temperature gave higher ee, and when the crystal melted or disintegrated during photolysis, noticeably lower ee was observed.

Demuth et al. also have reported a solid-state di-π-methane type photorearrangement of 24 to give preparative quantities of two rearranged products.[20] The starting

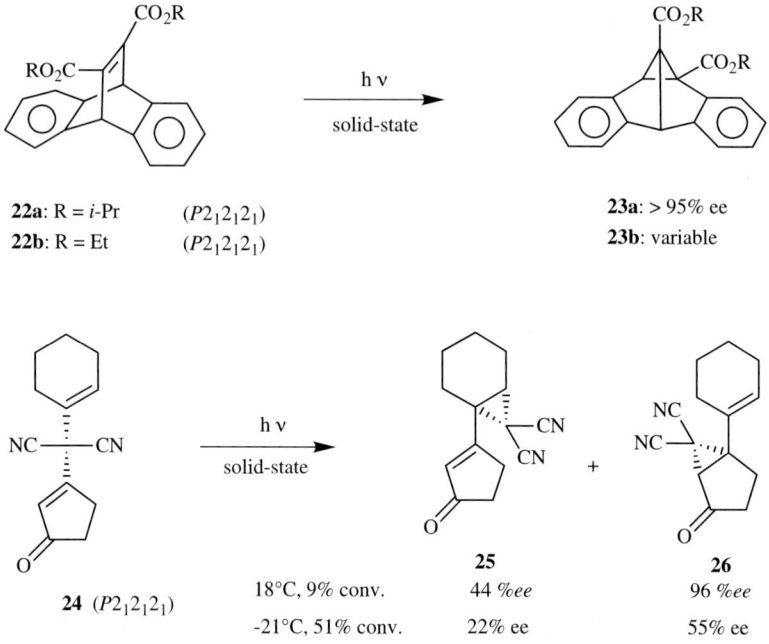

22a: R = i-Pr   ($P2_12_12_1$)
22b: R = Et   ($P2_12_12_1$)

23a: > 95% ee
23b: variable

24 ($P2_12_12_1$)

18°C, 9% conv.
-21°C, 51% conv.

25
44 %ee
22% ee

26
96 %ee
55% ee

Scheme 4

material adopts chiral packing, space group $P2_12_12_1$, and a helical molecular conformation. The rearrangements proceed so that the products, **25** and **26**, are obtained in respective ee of ≤ 44 % and ≤ 96 %, and the ee's decreased proportionately with conversion.

Scheffer et al. provided another unimolecular asymmetric transformation involving the Norrish type II reaction, a well-known excited state process of ketones that is initiated by an intramolecular hydrogen atom transfer from carbon to oxygen through a six-membered transition state (Scheme 5).[19a]   An adamantyl ketone derivative **27** was found to crystallize from ethanol in very large prisms in the chiral space group $P2_12_12_1$. Upon irradiation of these crystals to approximately 10% conversion, the chiral cyclobutanol derivatives **28** were afforded as the major products in 80% ee.

Toda and co-workers reported an example that is similar to the photorearrangement of ketone **27** via initial hydrogen atom transfer from carbon to oxygen followed by coupling of the 1,4-diradical. This involves the photochemical conversion of the achiral α-oxoamide derivative **29** into the chiral β-lactam **30**.[21][22]   When grown from benzene, crystals of **29** occupied the chiral space group $P2_12_12_1$, and irradiation of one of the enantiomorphs was found to give (+)-**30** in 93% ee, while photolysis of the other enantiomorph gave (-)-**30**, also with the same ee. They also prepared seventeen α-oxoamide derivatives and investigated the chiral spontaneous crystallization and the solid-state photoreaction, and found six materials crystallized in a chiral space group.[22c]

Irngartinger et al. reported another example of asymmetric synthesis involving δ-hydrogen transfer followed by cyclization.[23]   Ketone **31** crystallized in the chiral space group $P2_12_12_1$, and the phenyl groups are fixed in a syn conformation. Irradiation of single crystals of **31** gave optically active cis-1,2-diphenylacenaphthen-1-ol **32** as the main product, with up to 97% de. Only a small amount of trans-acenaphthenol is also generated in the solid state, presumably due to an increased thermal motion of the molecules in the crystal during the irradiation (40-45°C). The photoreaction is frozen when the probe is cooled to below 0°C. An excess of one enantiomer of **32** is found in selected single crystals after irradiation.   In optimal cases, 86% ee of the first enantiomer **32** and 49% ee of the other enantiomer **ent-32** were obtained.

Sakamoto et al. provided an example of absolute asymmetric synthesis involving hydrogen abstraction by thiocarbonyl sulfur (Scheme 6).[24] Achiral N-diphenylacetyl-N-isopropylthiobenzamide **33** and N-diphenylacetyl-N-isopropyl(p-chloro)thiobenzamide **33** crystallize in chiral space group $P2_12_12_1$. Photolysis of the chiral crystals in the solid state gave optically active azetidin-2-ones whereas achiral thioketones were obtained as main products. When **33a** was irradiated in the solid state at -45°C followed by acetylation (at -78°C), 2-acetylthio-3,3-dimethyl-1-diphenylacetyl-2-phenylaziridine (**34a**: 39% yield, 84% ee), 4-acetylthio-5,5-dimethyl-2-diphenylmetyl-4-phenyloxazoline (**35a**: 10% yield, 50% ee), 3,3-diphenyl-1-isopropyl-4-phenylazetidin-2-ones (**36a**: 16% yield, 20% ee) were obtained. In the case of the p-chlorothiobenzamide derivative **33b**, the corresponding aziridine (**34b**: 37% yield, 70% ee), oxazoline (**35b**: 13% yield, 40% ee), azetidin-2-ones (**36b**: 19% yield, 13% ee) were obtained.

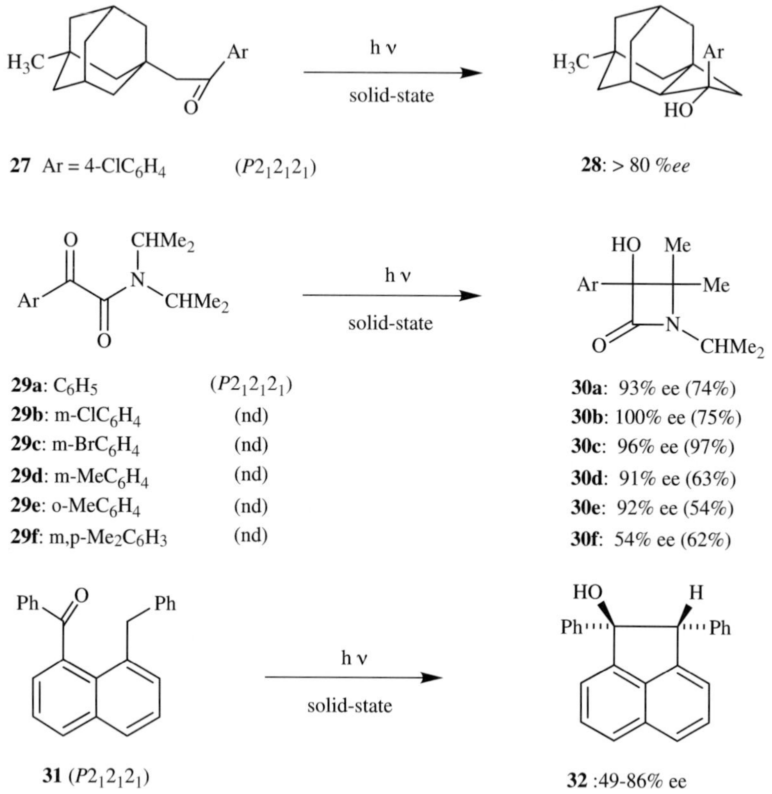

27 Ar = 4-ClC$_6$H$_4$ ($P2_12_12_1$)   28: > 80 %ee

29a: C$_6$H$_5$ ($P2_12_12_1$)   30a: 93% ee (74%)
29b: m-ClC$_6$H$_4$ (nd)   30b: 100% ee (75%)
29c: m-BrC$_6$H$_4$ (nd)   30c: 96% ee (97%)
29d: m-MeC$_6$H$_4$ (nd)   30d: 91% ee (63%)
29e: o-MeC$_6$H$_4$ (nd)   30e: 92% ee (54%)
29f: m,p-Me$_2$C$_6$H$_3$ (nd)   30f: 54% ee (62%)

31 ($P2_12_12_1$)   32 :49-86% ee

Scheme 5

Next, Sakamoto et al. showed an example which involves the hydrogen abstraction by the alkenyl carbon atom.[25] X-Ray crystallographic analysis revealed that the crystals of thioamide **37a** are chiral and the space group is $P2_12_12_1$. The crystals were irradiated at 0°C until the reaction conversion reached 81 % yield. As expected, the asymmetric induction was observed and optically active **38a** (55% ee) was isolated. As a consequence of the suppression of the reaction conversion yield from 81% to 17%, the enantiomeric purity rose up to 74% ee. In the solid-state photoreaction of **37b**, the space group $P2_1$, a more chemoselective reaction occurred, and only β-thiolactam **38b** was obtained almost quantitatively. Furthermore, the X-ray crystallographic analysis and the solid-state CD spectra revealed that the crystals of **37b** are chiral, and the space group is $P2_1$. Of particular importance is the finding that the solid-state photoreaction of **37b** involves a crystal-to-crystal nature where the optically active β-thiolactam **38b** is formed. Irradiation of the chiral crystals of **37b** at 0°C exclusively gave optically active β-thiolactam **38b**, in 81% yield and 81% ee at 100% conversion. This reaction exhibited good enantioselectivity throughout the whole reaction, where a small difference was observed in ee varying from 97 to 81% with increasing conversion from 20 to 100%.

Sakamoto et al. reported an intramolecular [2+2] thietane formation in the solid state (Scheme 7).[26] Achiral *N*-(thiobenzoyl)methacrylamide **39** formed (*E,Z*)-conformation of the imide moiety, crystallized in a chiral space group $P2_12_12_1$, and the photolysis of single homochiral crystals at room temperature resulted in the formation of an optically active thietane-fused β-lactam (40,75%) with 10% ee. The solid-state photoreaction proceeded even at -45°C to give higher ee value, 40% ee (conv. 30%, yield 70%).

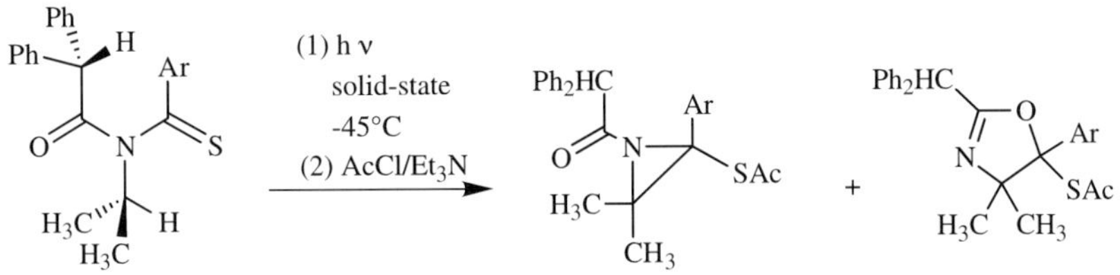

33a: Ar = Ph   ($P2_12_12_1$)
33b: Ar = *p*-ClPh ($P2_12_12_1$)

34a: 84% ee (39%)
34b: 70% ee (37%)

35a: 50% ee (10%)
35b: 40% ee (13%)

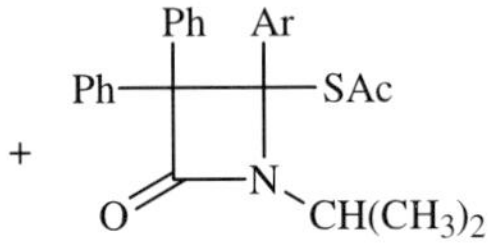

36a: 20% ee (16%)
36b: 13% ee (19%)

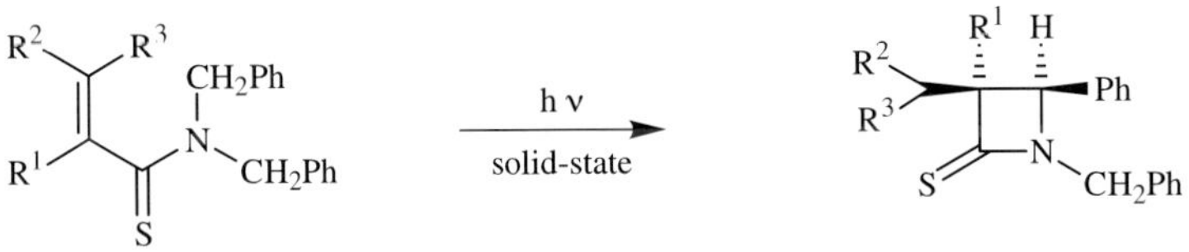

37a : R^1 = H, R^2 = Me, R^3 = Me   ($P2_12_12_1$)

37b: R^1 = R^2 = -(CH$_2$)$_4$-, R^3 = H   ($P2_1$)

38a: conv. : 81%   55% ee (55%)
38a: conv. : 17 %   74% ee (95%)

38b: conv. : 100%   81% ee (81%)
38b: conv. : 20%   97% ee (97%)

Scheme 6

Similar absolute asymmetric synthesis was demonstrated in the solid-state photoreaction of *N*-(β,γ-unsaturated carbonyl)thiocarbamate **41**.[27] Achiral *O*-methyl *N*-(2,2-dmethylbut-3-enoyl)-*N*-phenylthiocarbamate **41** crystallized in chiral space group *P2₁*, and irradiation of these crystals gave optically active thiolactone in 10-31% ee. A plausible mechanism for the formation of **42** is rationalized on the basis that photolysis of **41** undergoes [2 + 2] cyclization to thietane and is subsequently followed by rearrangement to thiolactone **42**.

Sakamoto *et al.* also demonstrated an absolute oxetane synthesis in the solid-state photolysis of *N*-(α,β-unsaturated carbonyl)benzoylformamides **43**.[28] The X-ray analysis of *N*-isopropyl substituted imide **43a** revealed that the crystal system was monoclinic and the space group *P2₁*. Crystals of **43a** were powdered and photolyzed at 0°C. The imide undergoes the [2+2] cycloaddition to afford the bicyclic oxetane **44a**, which is a mixture of diastereomers, namely, *syn-* and *anti-*isomers at the C-7 position. In this reaction optically active *syn*-oxetane **44a** with 37% ee (84% chemical yield) and racemic *anti*-**44a** were obtained. The solid-state photoreaction proceeded even at -78°C, and optically active *syn*-**44a** which showed ee value as high as >95% ee, (conv 100%, chemical yield 89%) was formed in a higher diastereomeric ratio (*syn/anti* = 6.5). Under identical conditions *N*-benzyl substituted **43b** was irradiated in the solid state.

The solid-state photoreaction provided very high *syn/anti* stereoselectivity of **44b** (*syn/anti* ratio 2.1 in solution and 60 in the solid-state, respectively). Furthermore, the solid-state photoreaction was found to give the *syn*-**44b** as a chiral compound in 81% ee; however, the crystal system could not be determined.

Molecules of **43c** adopt chiral packing (space group *P2₁*) and a helical molecular conformation, and crystallize in (*E,Z*) conformation which is unfavorable for the oxetane formation. The solid-state irradiation of **43c** was found to give the oxetane **44c** and a β-lactam derivative **45c**. The β-lactam **45c** was revealed to be enantiomerically enriched to 88% ee, whereas the other photoproduct **44c** was racemic. The occurrence and the mechanism of transformation of **43c** to **45c** involve hydrogen abstraction by the alkenyl carbon atom.

Kohmoto *et al.* reported an example of absolute asymmetric synthesis involving the intramolecular [4 + 4] photocycloaddition of 9-anthryl-*N*-(naphthylcarbonyl)-carboxamide derivatives **46** in the solid state.[29] The crystals of **46a** and **46b** have the chiral space groups *P2₁2₁2₁* and *P2₁*, respectively. Asymmetric induction in the photocycloaddition of **46a** and **46b** in single crystals was observed, and optically active products, **47a** (82% ee) and **48b** (100% ee), were obtained.

Toda *et al.* reported another example of absolute asymmetric photoreaction involving electrocyclization in the solid state; crystallization of 3,4-bis(phenylmethylene)-*N*-methylsuccinimide **48** gave chiral crystals as orange hexagonal plates and two racemic crystals as orange rectangular plates and yellow rectangular plates (Scheme 8).[30] Irradiation of powdered enantiomeric crystals gave optically active photocyclization product **49** of 64 %ee. Photolysis of eight other derivatives led to a racemic cyclization product in quantitative yield.

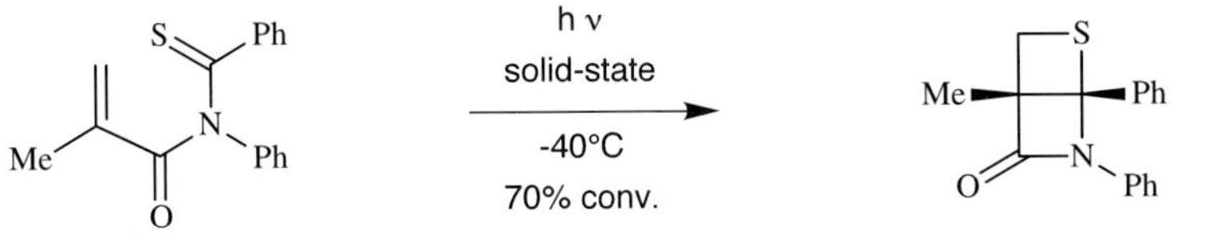

39 ($P2_12_12_1$)

40: 40% ee (70%)

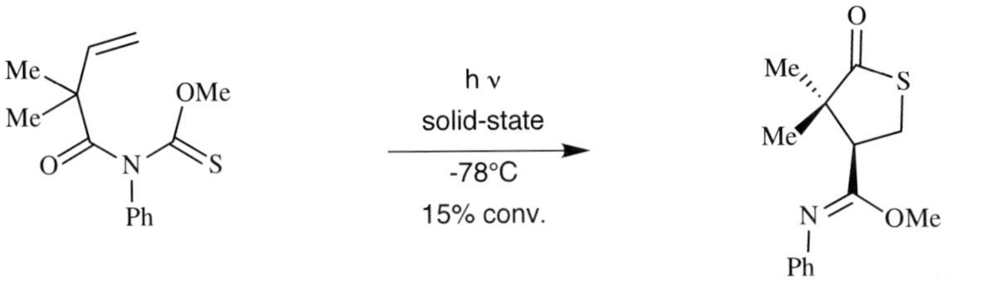

41 ($P2_1$)

42: 31% ee (90%)

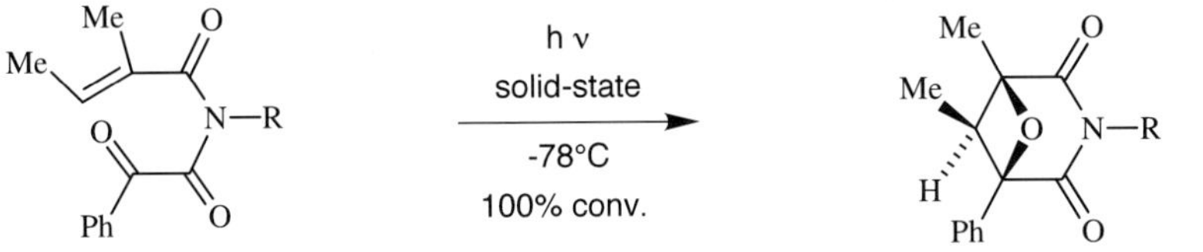

43a: R = Pr^i ($P2_1$)
43b: R = Bn   (nd)

44a: 95% ee (89%)
44b: 81% ee (91%)

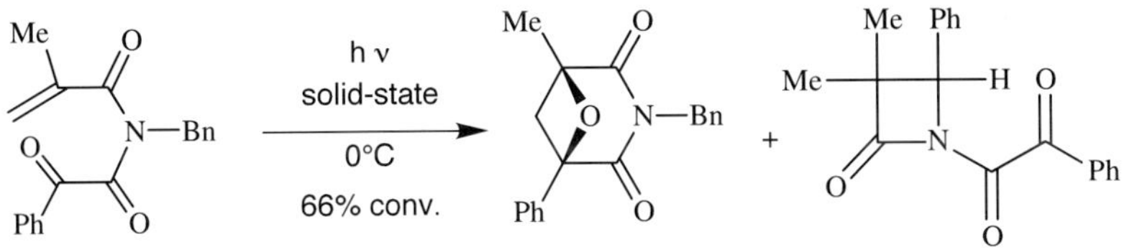

43c ($P2_1$)

44c: racemic (40%)

45c: 88% ee (20%)

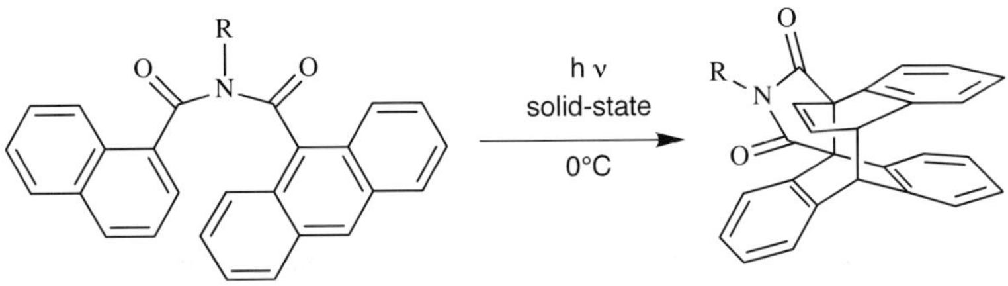

46a: R = CH$_2$Ph
46b: R = o-(CH$_3$)$_3$C$_6$H$_4$CH$_2$

($P2_12_12_1$)
($P2_1$)

47a: 82% ee (100%)
47b: 100% ee (100%)

Scheme 7

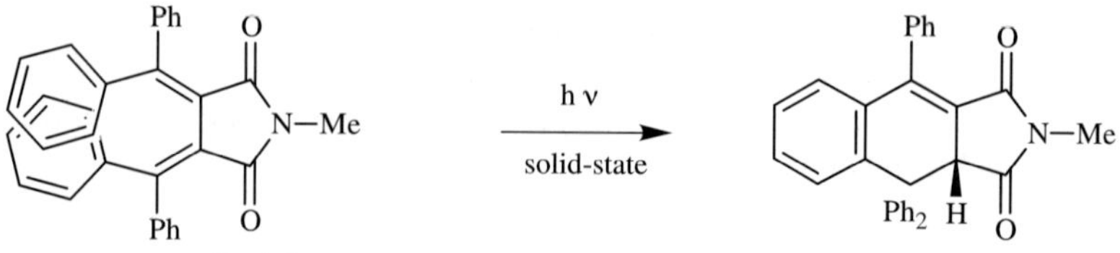

48 (nd)

49: 64% ee

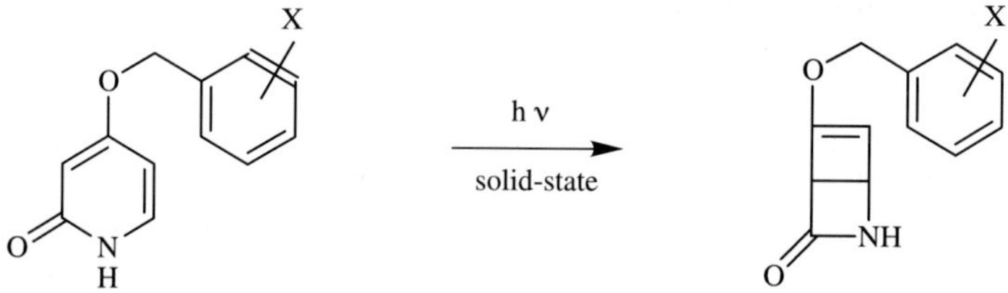

50a: X = m- Cl   ($P2_12_12_1$)

50b: X = m-Me   ($P2_12_12_1$)

50c: X = m-OMe   ($P2_12_12_1$)

50d: X = m-Br   ($P2_12_12_1$)

51a: 69 - 78% ee

51b: 58 - 71% ee

51c: 90 - 100% ee

51d: unstable

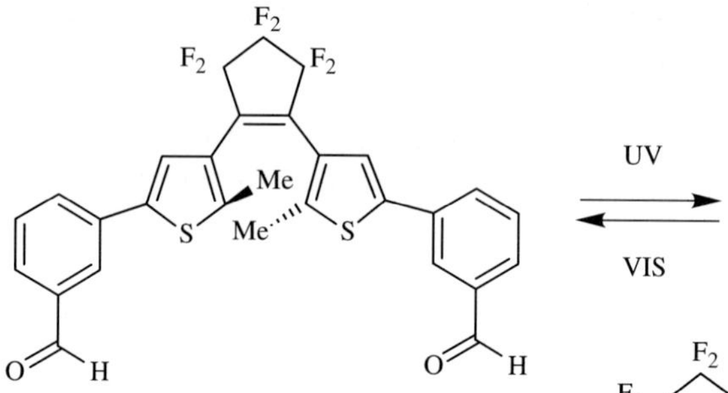

(*P*)-**52**

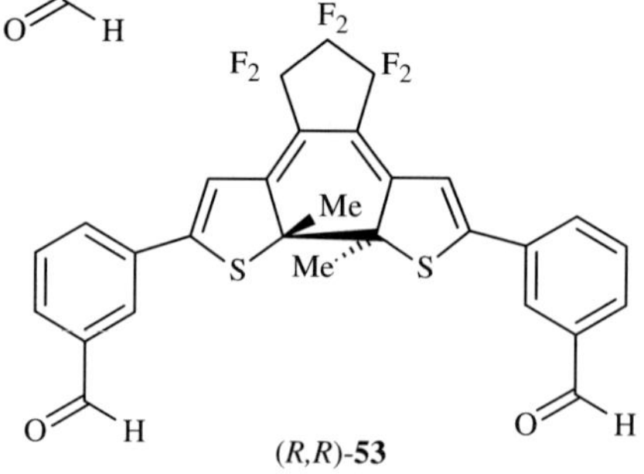

(*R,R*)-**53**

cov. (%)	ee(%)
12.0	94
0.6	99

Scheme 8

In the case of the formation of the β-lactam, it was shown that the ability of a compound to crystallize in a chiral space group was increased by using compounds with meta-substituted aryl groups such as **50a-d** instead of their *ortho-* and *para*-substituted analogues; the absolute asymmetric photocyclization of the pyridones resulted in the formation of **51a-d** with good ees.[31]

Irie *et al.* reported on the absolute reversible photocyclization reactions of achiral diarylethene **52** in chiral crystal leading to **53** in up to 99% ee. The asymmetric photochromic material has potential application in nonlinear optics, for example, in switchable second harmonic generation devices.[32]

Sakamoto et al. found that *S*-phenyl *N*-benzoylformyl-*N*-(*p*-tolyl)thiocarbamate **54** crystallized in chiral space group $P2_1$.[33] Photolysis of the chiral crystals in the solid-state gave optically active 1-benzyl-4-phenyl-4-phenylthiooxazolidine-2,4-dione (**55**, 16% chemical yield, 21% ee) and *cis*-3,4-diphenyl-3-hydroxy-1-(thiophenylcarbonyl)-azetidin-2-one (**56**, 18% chemical yield, 23% ee) in 62 % conversion yield. Better optical purities were observed at low conversion (in 17% cov.), 46% ee for **55**, and 32% ee for **56**, respectively.

Sakamoto *et al.* also reported the solid-state photoreactions of *S*-aryl 2-aroylbenzothioates **57** which led to a remarkable conclusion that the major process involved was an unprecedented reaction sequence, starting with intramolecular cyclization, followed by aryl migration to the photoproducts **58**.[34]

Seven *S*-aryl 2-aroylbenzothioates **57** were subjected to X-ray crystal analysis and it was revealed that three thioesters **57a-c** crystallized in a chiral space group, and the other crystals were racemic. Irradiation of powdered (+)-**57a** at 0°C led to production of 3-phenyl-3-(*o*-tolylthio)-phthalide **58a** as a major product (65% yield) in 100% conversion yield. As expected, the asymmetric generation in **58a** was observed in 30% ee. Pronounced changes in the product profiles were recognized when the reaction was carried out at low (5%) conversion and at low temperature (-78°C). In the former case, a good ee value (77% ee) was obtained with exclusive product formation (>95% conversion). *S*-Phenyl and *S*-(*m*-tolyl) derivatives **57b** and **57c** formed chiral crystals (space groups $P2_12_12_1$) and their solid-state photoreactions also led to optically active products with ee values varying with the reaction conditions.

Absolute asymmetric synthesis was observed in the solid-state photoreaction of benzoylbenzamide **59** to phthalide **60**; however, the reaction mechanism was completely different from that of thioester **57**.[35] Recrystallization of these amides **59a-c** from the chloroform-hexane solution afforded colorless prisms in all cases. X-ray crystallographic analysis revealed that all prochiral amides **59a-c** adopted orthorhombic chiral space group $P2_12_12_1$ and were frozen in chiral and helical conformation in the crystal lattice.

The solid-state photolysis of powdered **59b** gave a quantitative amount of 3-(*N*-methylanilino)-3-phenylphthalide **60b**, whereas **59a** was inert toward topochemical reaction upon irradiation. As expected, the asymmetric induction in **60b** was observed. As a result of the suppression of the reaction conversion yield and by decreasing the reaction temperature to -50°C, the enantiomeric purity rose up to 87% ee (at 25% conversion with quantitative yield). Furthermore, irradiation of single crystals of **59b** gave 97% ee of **60b** when the reaction conversion reached 46%.

When the crystal of **59c** was irradiated at 0°C, a single regioisomer of **60c** was obtained quantitatively. The photoproduct **60c** showed an ee of 42%. Furthermore, a higher ee value of **60c** (87% ee) was obtained in the reaction at -50°C. The radical mechanism has been confirmed in the photo-Fries rearrangement of aromatic amides.

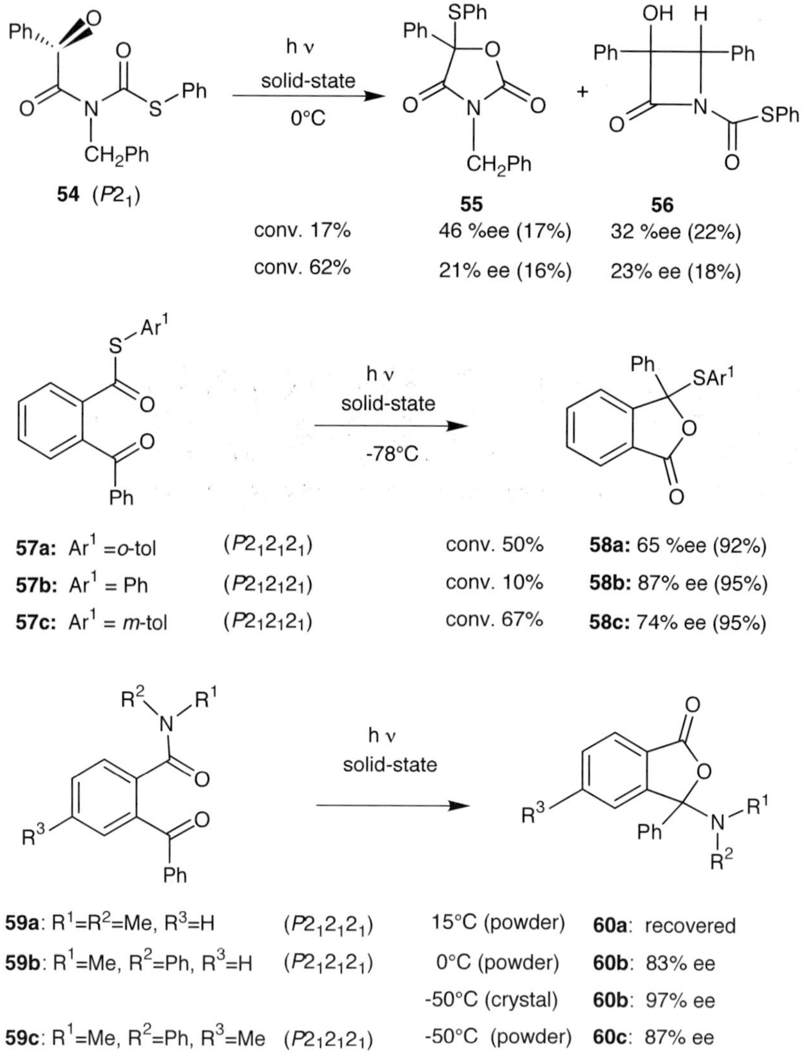

54 (*P2₁*)

conv. 17%

conv. 62%

55

46 %ee (17%)

21% ee (16%)

56

32 %ee (22%)

23% ee (18%)

57a: Ar¹ =o-tol	(*P2₁2₁2₁*)	conv. 50%	**58a:** 65 %ee (92%)
57b: Ar¹ = Ph	(*P2₁2₁2₁*)	conv. 10%	**58b:** 87% ee (95%)
57c: Ar¹ = m-tol	(*P2₁2₁2₁*)	conv. 67%	**58c:** 74% ee (95%)

59a: R¹=R²=Me, R³=H   (*P2₁2₁2₁*)   15°C (powder)   **60a:** recovered

59b: R¹=Me, R²=Ph, R³=H   (*P2₁2₁2₁*)   0°C (powder)   **60b:** 83% ee

                                              -50°C (crystal)   **60b:** 97% ee

59c: R¹=Me, R²=Ph, R³=Me (*P2₁2₁2₁*)   -50°C (powder)   **60c:** 87% ee

Scheme 9

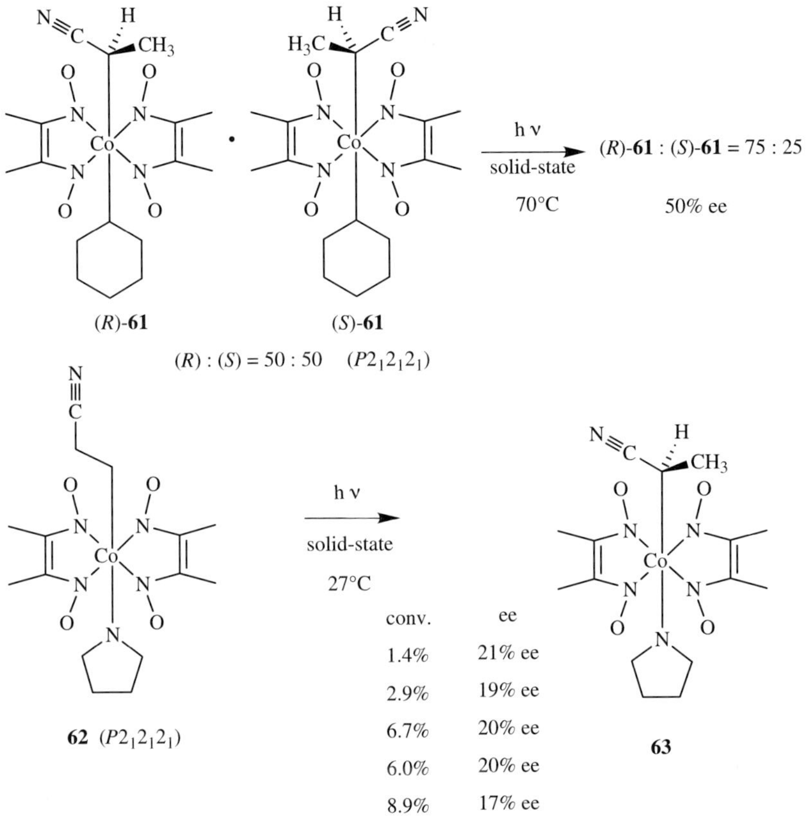

Scheme 10

Ohashi et al. reported a fine example of racemic-to-chiral transformation in cobaloxime complex crystal by irradiation.[36]   A racemic mixture of (1-cyanoethyl)(piperidine)cobaloxime **61**, which contains the chiral 1-cyanoethyl group, crystallized in a chiral space group ($P2_12_12_1$) in which R and S enantiomers occupy crystallographically distinct (diastereoisomeric) positions in the asymmetric unit cell. Single crystals were prepared of the Δ configuration. Irradiating the crystal with X-ray at 343K causes racemization of the S enantiomer while the R enantiomer remains unaltered.   This racemization is thought to be due solely to the different volume constraints on the two enantiomers in their different crystal environments. The result is that the number of R molecules is increased at the expense of S molecules, and so the overall composition of the crystal changes from racemic to enriched in the R enantiomer. (Scheme 10)

Ohashi also reported the asymmetric induction by the irradiation of chiral crystals of achiral (2-cyanoethyl)(pyrrolidine)cobaloxime complex **62**.[37] In this case, the achiral cobaloxime complex crystallized in $P2_12_12_1$ space group, and the solid-state photolysis gave 1-cyanoethyl derivative **63** with optical activity. The ees were variable in accordance with the conversion. The maximum ee (21%) was obtained at low conversion (1.4% yield).

4. Mechanistic studies using the correlation of absolute-absolute configuration in the solid-state reaction

Solid-state photoreaction in the chiral crystal provides not only a useful synthetic method of optically active materials but also mechanistic information on the photochemical process. Two examples of solid-state photoreactions will be described, in which the correlation on the absolute configurations between those before and after the reaction revealed the reaction pathways.

In the first case, achiral *N*-methacryloylthiobenzanilide **39** formed (*E,Z*)-conformation of the imide moiety and crystallized in a chiral fashion. The solid-state photoreaction gave optically active β-lactam **40**. The dynamic molecular rearrangement for cyclization was elucidated on the basis of direct comparison of the absolute configuration of both the starting material and the photoproduct (Scheme 11).[25][38]

The circular dichroism (CD) spectra of the crystals in KBr pellet exhibited specific curves in the region between 300-600 nm, which were mirror images designated as (+) and (-) at the wavelength of 470 nm. The monothioimide **39** has no chiral center; however, the absolute structure of the chiral conformation in the chiral crystal could be established by the X-ray anomalous scattering method for the refinement of Roger's parameter. The absolute structure of the chiral conformation of (*P*)-**39** corresponds to (-)-crystal in the solid-state CD spectra (Figure 3).

X-Ray crystallographic analysis revealed that the crystal of thioamide **39** was chiral and the space group was $P2_12_12_1$. The absolute configuration of (-)-rotatory crystals of **39** was determined by the X-ray anomalous scattering method as (-)-(*M*)-**39** for the helicity. The (-)-rotatory crystals obtained by the seeding method were irradiated at 0°C until the reaction conversion reached 100 % yield. As expected, the asymmetric induction in **40** was observed in 10% ee. By suppression of the reaction conversion to 30% and decrease of the reaction temperature to -45°C, the enantiomeric purity rose up to 40% ee.

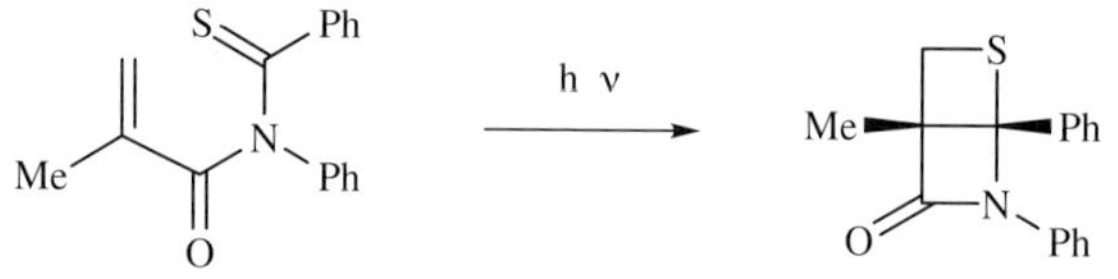

39

space group: $P2_12_12_1$

40

Scheme 11

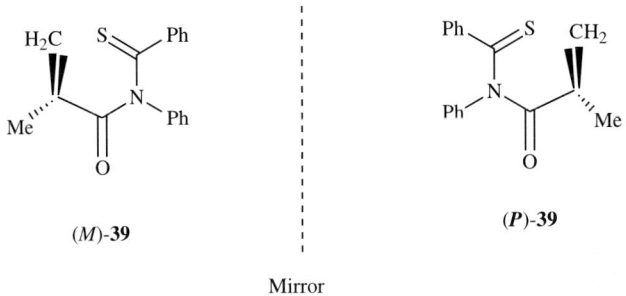

(M)-39   Mirror   (P)-39

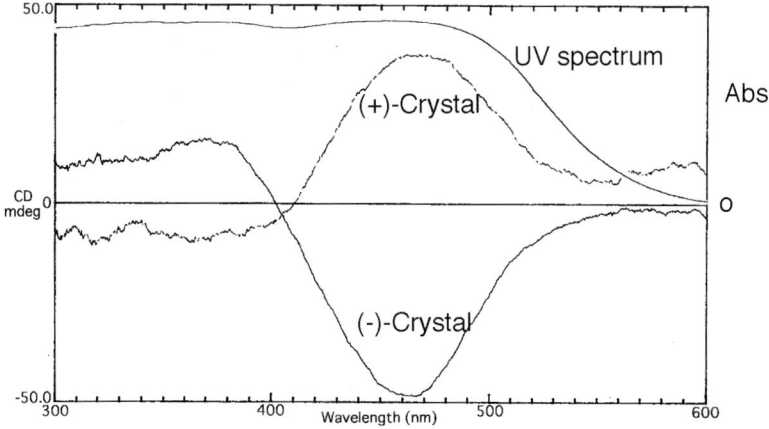

Figure 3. Reflected UV and solid-state CD spectra of enantiomeric crystals
of both antipodes of **1a** in KBr.

Scheme 12 shows the mechanism for the thietane formation, in which six-membered 1,4-biradical **BR** is appropriate. There are two ways of cyclization to thietane **40**, and each pathway gives an enantiomeric structure of thietanes, (1S,4R)- or (1R,4S)-**40**, respectively. What atomic rearrangement is needed for the formation of thietane **40** in the crystal lattice? The answer was provided by a correlation study of the absolute structure before and after the reaction.

The absolute structure of (-)-(M)-**39** and the major isomer (+)-(1S,4R)-**40** was determined by X-ray structural analysis using an anomalous scattering method (Figure 4-a and Figure 4-b). Figure 5 shows the superimposed structure of both absolute structures which was drawn with the overlay program included in CSC Chem3D. The sulfur and the alkenyl carbon atoms are closely placed to make the C-S bond easily, and subsequent cyclization of biradical **BR** needs the rotation of the radical center like path a to yield (1S,4R)-**40**. The molecular transformation from (-)-**39** to (+)-**40** needs considerable molecular rearrangement involving the rotation of the methacryl group.

In this reaction, achiral *N*-methacryloylthiobenzanilide **39** gave chiral crystals by spontaneous crystallization, and absolute asymmetric synthesis was performed by the solid-state photoreaction leading to optically active thietane-fused β-thiolactam. The reaction mechanism for the cyclization was elucidated on the basis of the correlation between the absolute structures of both the prochiral starting monothioimide and the photoproduct.

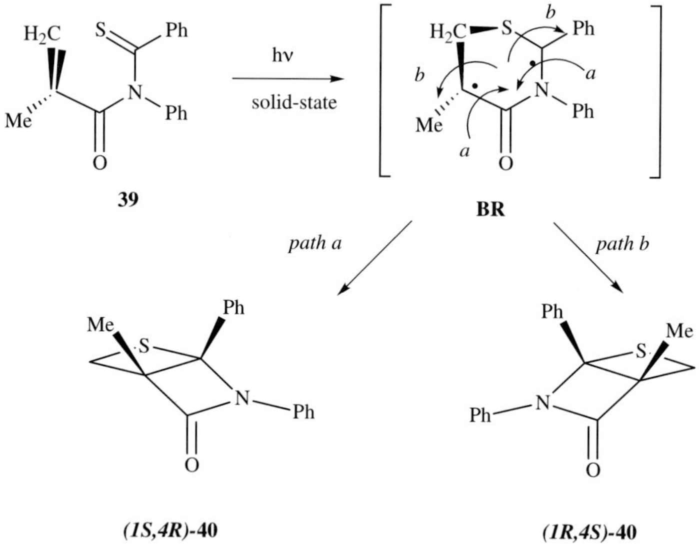

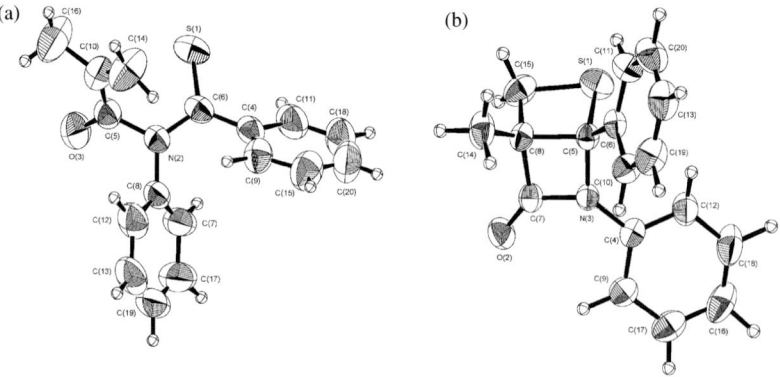

Scheme 12

(a)

(b)

Figure 4. (a) Ortep view of the absolute structure of (-)-(*P*)-**39**. (b) Ortep drawing of the absolute structure of (+)-(*1S,4R*)-**40**

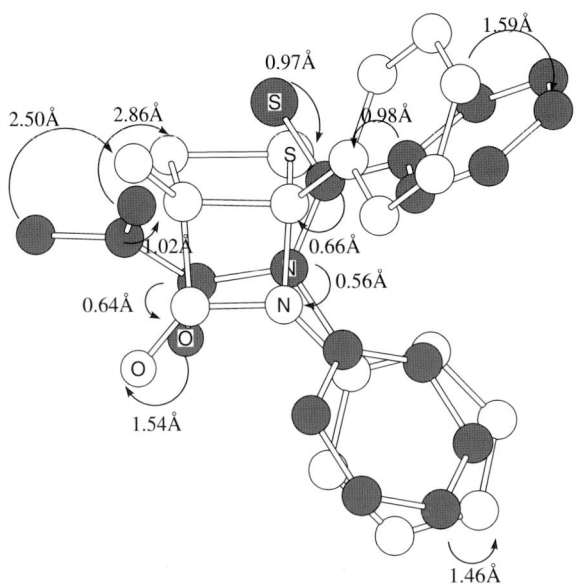

Figure 5. A front view of the superimposed structure of (-)-(*M*)-**39** (black circle) and (+)-(*1S,4R*)-**40** (empty circle) obtained by X-ray structural analysis using Overlay program in Chem3D.

The second example is photochemical reaction of 2-benzoylbenzothioesters **57** in the chiral crystalline environment. Solid-state photoreactions of *S*-(*o*-tolyl), *S*-phenyl, and *S*-(*m*-tolyl) 2-benzoylbenzothioates **57a-c**, which formed chiral crystals by spontaneous resolution, underwent an intramolecular cyclization involving phenyl migration to afford the corresponding optically active 3-phenyl-3-(arylthio)phthalide **58a-c** in good chemo- and enantio-selectivities. (Scheme 11).[34]

Recrystallization of these thioesters from chloroform-hexane solution afforded colorless prisms for all cases. All crystals were subjected to X-ray crystallographic analyses to obtain details on the molecular architecture in the crystals. The constituent molecules adopted orthorhombic chiral space group $P2_12_12_1$ and were frozen in helical conformations. These features were demonstrated through a measurement of the CD spectra of crystalline samples in KBr pellets. Under such conditions conformational interconversion owing to carbon-carbon bond rotations was severely restricted and their optical activities evolving in the crystalline environment were maintained. (Figure 6)

Figure 7 shows CD spectra of two enantiomorphic crystals of **57a-c** in KBr pellets, which were independently obtained by spontaneous resolution. These crystals gave specific curves in the region between 200-400 nm, which were mirror images designated as (+) and (-) at the wavelength of 346 nm within inevitable error.

It is of great significance to correlate the absolute configurations of helical molecules frozen in the chiral crystalline environment with the sign of specific rotations. An attempt to determine the absolute configuration of (+)-**57a** giving rise to a positive

Cotton effect on its CD spectrum was first successfully achieved by X-ray crystallography taking account of anomalous dispersion due to the relatively heavy sulfur atom.   Consequently, the crystals (+)-**57a** were determined to contain *P*-configuration in the molecules with regard to the helix of two carbonyl groups, while its antipode was confidently assigned to possess left-handed *M*-helicity (Figure 8). Furthermore, absolute configurations of both (-)-**57b** and (-)-**57c** were determined to be *M*-configurations by X-ray crystallography.   Compatible with **57a**, both antipodes of **57b** and **57c** gave symmetrical CD absorption curves to their antipodes with almost the same intensities in all regions observed.   Interestingly, a similar tendency in relationships between helicities and absolute configurations for the chiral crystals **57a-57c** was observed, in which minus activity of the crystals was attributed to *M*-helicity while *P*-helicity was responsible for plus activity.

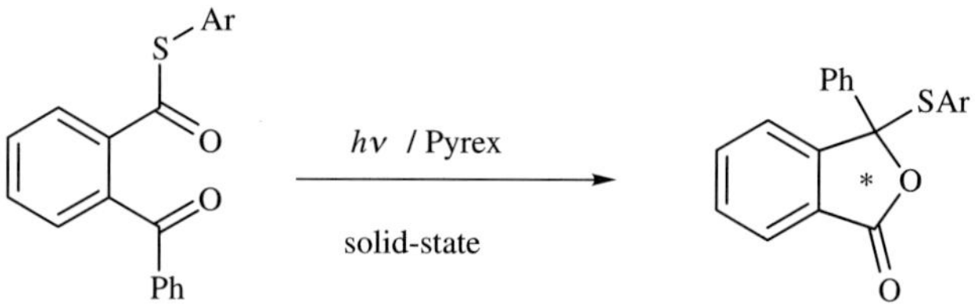

a : Ar = (*o*-tol)

b : Ar = Ph

c : Ar = (*m*-tol)

Scheme 11

Figure 7 includes UV spectra of the crystals of thioesters **57a-c**, indicating the molecules absorbed photons sufficiently in the solid-state beyond 290 nm (Pyrex filtered light).   Irradiation of powdered (+)-**57a** at 0°C led to production of 3-phenyl-3-(*o*-tolylthio)phthalide **58a** as a major product in 65 % yield with complete consumption of the starting material.   As expected, the asymmetric generation in **58a** was realized by an observation of its specific rotation which was +37° which corresponded to 30% ee.

Remarkable changes in the product profiles were recognized when the reaction was carried out at low (5%) conversion or at low temperature (-78°C).   In the former case, a good ee value (77% ee) was obtained with exclusive product formation (>95% as observed by ¹H-NMR of the crude mixture).   Under these conditions, the molecules were supposed to be still under topochemical control leading to the better stereoselective reaction in the crystals.   Additionally, the reaction at -78°C gave similar results (50% yield, 65% ee) for the same reason, where the molecules were strongly frozen in the crystal lattices.   Obviously, the topochemical control was much more effective at lower temperature since 65 %ee was obtained even at 50 % conversion, which value corresponded to about 39 %ee for 0°C on estimation of the curve.   Apart

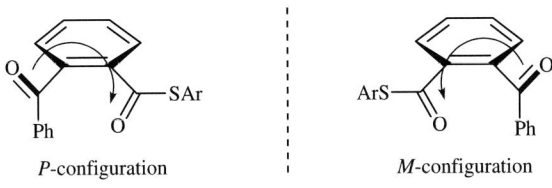

P-configuration   M-configuration

Figure 6

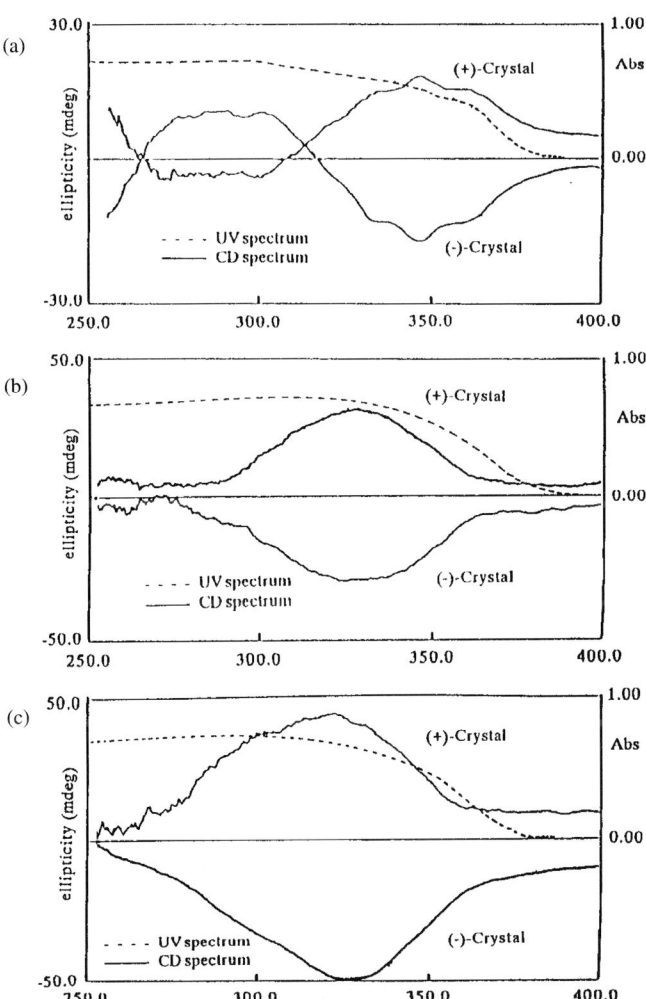

Figure 7. CD and UV spectra of enantiomorphic crystals of both antipodes.
(a) **57a**, (b) **57b**, and (c) **57-c**.

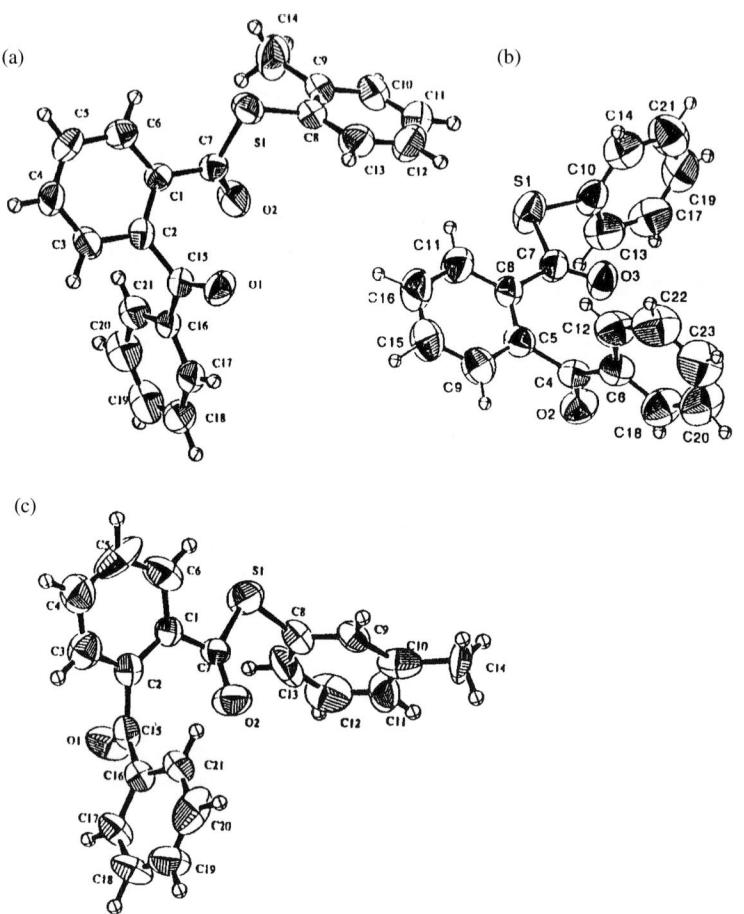

Figure 8. Absolute structure of 2-benzoylthiobenzoate. (a) (+)-(*P*)-**57a**,
(b) (-)-(*M*)-**57b**, and (c) (-)-(*M*)-**57c**.

from **57a**, *S*-phenyl and *S*-(*m*-tolyl) derivatives **57b** and **57c** formed chiral crystals (space groups $P2_12_12_1$) and their solid-state photoreactions also led to optically active products with respective ee values depending on the reaction conditions (Table 3). Under well-optimized conditions, good enantiomeric excesses were eventually obtained in the products as high as 87% for **58b** (10% conversion at -78°C) and 74% for **58c** (67% conversion at -78°C).

In regard to the mechanism for these solid-state photoreactions, there are two possible pathways from the starting thioesters to phthalides (Scheme 12). In the first model (Path A), the reaction is initiated by homolytic dissociation between C-S bonding to form a radical pair intermediate. Such pathway is well recognized as the excitation reaction of thioester compounds. The other model (Path B) consists of direct cyclization and subsequent phenyl migration sequences, where a zwitterionic intermediate would be involved. To solve this mechanistic problem, stereochemical correlation studies before and after the solid-state photoreactions were employed. The optically pure (+)-**58a** was obtained by optical resolution of (+)-enantiomer rich **58a** with HPLC using a chiral cell OJ, whose absolute structure was determined by X-ray analysis using the Bijvoet difference method in the correction process. Finally, the X-ray results indicated that the molecules were crystallized in triclinic chiral space group $P1$ and the unit cell included two discrete rotamers (**58aA** and **58aB**) owing to carbon-carbon bond rotation around the o-tolyl groups. Furthermore, statistic evaluation taking into account the anomalous dispersion determined the single stereocenter in (+)-**58a** to be (R)-configuration (Figure 4). Consequently, this reaction elucidated the stereochemical relationship from P-**57a** to (R)-**58a**.

TABLE 3. Solid-state photoreactions of thioesters **57**

Thioester **57**	Temp.	Conversion (%)	Yield of **58** (%)	ee (%) of **58**
	0°C	100	65	30
a	0°C	5	>95	77
	-78°C	50	92	65
	0°C	20	97	35
b	0°C	10	>95	71
	-78°C	10	>95	87
	0°C	56	96	23
c	0°C	10	95	71
	-78°C	67	>95	74

Assuming Path A was involved in the reaction, this sequence would lead to (S)-**58a** as a major product. This stereochemical outcome is incompatible with the aforementioned results, and this mechanistic interpretation is, therefore, ruled out. On the other hand, Path B may offer a complementary demonstration to satisfy the stereochemical relationship. The course of the initial cyclization step may govern the absolute configuration of the final product, and the phenyl migration step proceeded with steric restrictions resulting in stereoselective formation of a single enantiomer. Thus, the enantioselective transformation from P-(+)-**57a** to (R)-**58a** can be rationalized by Path B involving a convergent sequence of nucleophilic cyclization and phenyl migration.

To clarify the feasibility of stereochemical assignment during the reaction, another correlation study of the absolute configurations of the starting compound and the product was carried out for the reaction of *M*-**57b** to (-)-**58b**. The absolute structure of enantiomerically pure (-)-**58b** was determined as (*S*)-enantiomer by X-ray diffraction solving through refinement of Roger's η parameter (Figure 9). This indicated the aforementioned stereochemical relationship was applicable to this reaction system (*M*)-**57b** to (*S*)-**58b**, supporting the mechanistic proposal for Path B. Unfortunately, additional assignment on the transformation of *M*-**57c** to (-)-**58c** could not be achieved since optically pure (-)-**58c** isolated by HPLC using a chiral cell gave only thin plates unsuitable for X-ray crystallographic studies. On consideration of the aforementioned discussion, the stereochemical relationship between *M*-**57c** and (*S*)-**58c** can be readily predicted.

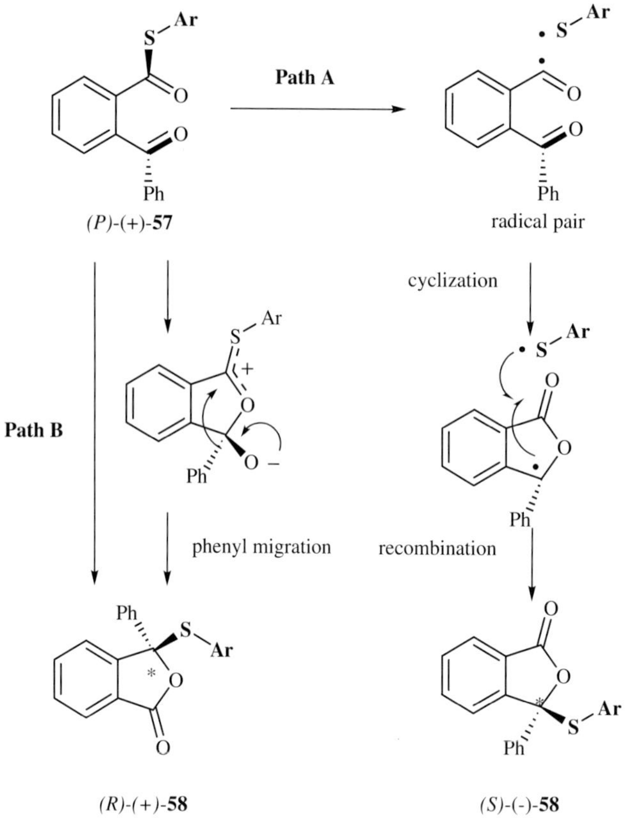

Scheme 12

Investigations of the solid-state photoreactions of *S*-aryl 2-aroylbenzothioates **57** led to the remarkable conclusion that the major process involved was an unprecedented reaction sequence, starting with intramolecular cyclization, followed by aryl migration to the photoproducts **58**. The "absolute" asymmetric generations into the achiral starting molecules were observed with good enantioselectivities, which seriously depended on the conversion and reaction temperature. The mechanistic studies based on stereochemical correlation before and after the reaction gave strong evidence for the putative reaction pathway as well as the geometrical analyses of the starting molecules in the crystal lattices. Thus, the absolute-to-absolute correlation study may offer a powerful tool for manifesting obscure reaction pathways during solid-state reactions.

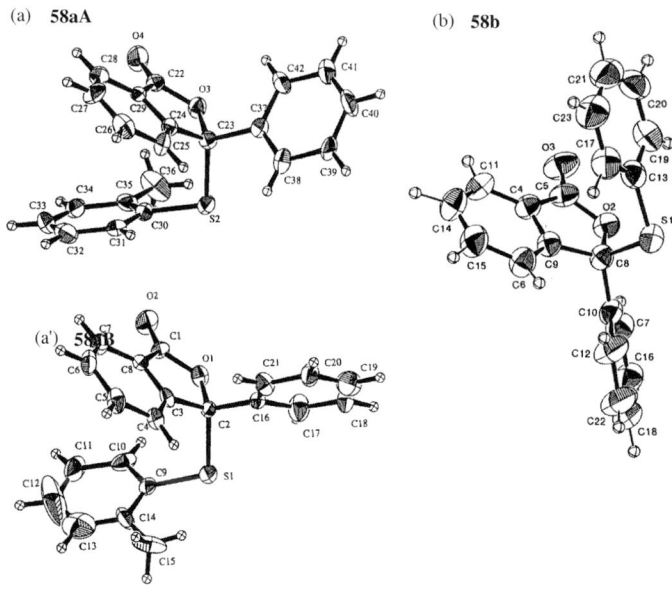

Figure 9. Absolute structure of optically active phthalides **58a** and **58b**. (a) and (a') (+)-(*R*)-**58a**, (b) (-)-(*S*)-**58b**.

5. Conclusion

The chiral crystallization of achiral materials and the asymmetric transformation in the chiral crystal environment are described. Many successful examples are presented; however, it is still rare to find materials which show this behavior. Recently, new asymmetric reactions using chiral crystals in homogeneous conditions have been developed.[39],[40] These reactions used the frozen chirality generated by chiral crystallization, and can be applied to many asymmetric reactions, of course including photochemical reactions. The scientific area of absolute asymmetric synthesis provides not only elucidation of the problem of the prebiotic origin on Earth but also one important branch of asymmetric synthesis.

6.   References

[1] (a) Schmidt, G. M. J. (1971) Photodimerization in the solid state, *Pure Appl. Chem.*, **27**, 647-678.   (b) Green, B. S., Lahav, M., and Rabinovich, D. (1979) Asymmetric synthesis via reactions in chiral crystals, *Acc. Chem. Res.*, **12**, 191-197

[2] For reviews, see (a) Ramamurthy, V. (1986) Organic photochemistry in organized media, Tetrahedron, **42**, 5753-5839.   (b) Scheffer, J. R., Garcia-Garibay, M., and Nalamasu, O. (1987) In: *Organic Photochemistry*, Padwa, A. ed., New York and Basel: Marcel Dekker, Vol. 8, pp 249-338.   (c) Ramamurthy, V. and Venkatesan, K. (1987) Photochemical reactions of organic crystals, Chem. Rev., **87**, 433-481. (d) Vaida, M., Popovitz-Bio, R., Leiserowitz, L., Lahav, M. (1991) In: *Photochemistry in Organized and Constrained Media*, Ramamurthy, V. ed., New York, pp 249-302 (e) Sakamoto, M. (1997) Absolute asymmetric synthesis from achiral molecules in the chiral crystalline environment, *Chem. Eur. J.*, **3**, 384-389.   (f) Koshima, H. and Matsuura, T. (1998) Chiral crystallization of achiral organic compounds. Generation of chirality without chiral environment. 2, *Yuki Gosei Kagaku Kyokaishi*, **56**, 268-477.   (g) Feringa, B. L., and Van Delden, R. (1999) Absolute asymmetric synthesis: the origin, control, and amplification of chirality, *Angew. Chem. Int. Ed.* **38**, 3419-3438.

[3] Asymmetric synthesis using nonchiral crystals was also performed. See [1b] and (a) Chenchaiah, P. C., Holland, H. L., and Richardson, M. F. (1982) A new approach to the synthesis of chiral molecules from nonchiral reactants. Asymmetric induction by reaction at one surface of a single (nonchiral) crystal, *J. Chem. Soc. Chem. Commun.*, 436-437.   (b) Chenchaiah, P. C., Holland, H. L., Munoz, B., and Richardson, M. F. (1986) Synthesis of chiral molecules from non-chiral crystals by controlled reaction at a single surface, *J. Chem. Soc. Perkin Trans. 2*, 1775-1777.

[4] (a) Addadi, L., and Lahav, M. (1979) In: Walker, D. C. ed. *Origin of Optical Activity in Nature*, New York and Basel: Elsevier. (b) Mason, S. F. (1984) Origins of biomolecular handedness, *Nature (London)*, **311**, 19-23.

[5] (a) Jacques, J., Collet, A., and Wilen, S. H. (1991) In: *Enantiomers, Racemates, and Resolutions*, Malabar and Florida, Kreiger publishing Company.   (b) Collet, A., Brienne, M., and Jacques J. (1972) Spontaneous resolution and enantiomeric conglomerates, *Bull. Soc. Chim. Fr.*, 127-142.

[6] Belsky, V. K., Zorkii, P. M. (1977) Distribution of organic homomolecular crystals by chiral types and structural classes, *Acta. Crystallogr. Sect. A*, **33A**, 1004-1006.

[7] (a) Kondepudi, D. K., Kaufma, R., and Singh, N. (1990) Chiral symmetry breaking in sodium chlorate crystallization, *Science*, **250**, 975-976. (b) McBride, J. M., and Carter, R. L. (1991) Spontaneous formation of enantiomorphic crystals with stirring, *Angew. Chem. Int. Ed. Engl.*, **30**, 293-295.   (c) Kondepudi, D. K., Bullock, K. L., Digits, J. A., Hall, J. K., and Miller, J. M. (1993) Kinetics of chiral symmetry breaking in crystallization, *J. Am. Chem. Soc.*, **115**, 10211-10216.

[8] Azumaya, I., Yamaguchi, K., Okamoto, I., Kagechika, H., and Shudo, K. (1995) Total asymmetric transformation of an *N*-methylbenzamide, *J. Am. Chem. Soc.*, **117**, 9083-9084. See also ref. 25(a), 25(b), 34(a), 34(c), and 38.   (b) Toda, F., Miyamoto, H., Koshima, H., and Urbanczyk-Lipkowska, Z. (1997) Chiral Arrangement of *N*-Ethyl-*N*-isopropylphenylglyoxylamide Molecule in Its Own Crystal and in an Inclusion Crystal with a Host Compound and Their Photoreactions in the Solid-State That Give Optically Active β-Lactam Derivatives. X-ray Analytical and CD Spectral Studies, *J. Org. Chem.*, **62**, 9261-9266.

[9] Hermann, J. P.,   Ricard, D., and Ducuing, J. (1973) Optical nonlinearities in conjugated systems. β–Carotene, *Appl. Phys. Lett.*, **23**, 178-180.

[10] (a) Penzein, K. and Schmidt, G. M. J. (1969) Reactions in chiral crystals: an absolute asymmetric synthesis, *Angew. Chem.*, **8**, 608-609. (b) Green, B. S. and Heller. L. (1974) Mechanism for the autocatalytic formation of optically active compounds under abiotic conditions, *Sci.,* **185**, 525-527.

[11] Garcia-Garibay, M., Scheffer, J. R., Trotter, J., and Wireko, F. (1988) Addition of bromine gas to crystalline dibenzobarrelene: an enantioselective carbocation rearrangement in the solid state, *Tetrahedron Lett.*, **29**, 1485-1488.

[12] Gerdil, R., Barchietto, G., and Jefford, C. W. (1984) Heterogeneous chirality transfer on photooxygenation, *J. Am. Chem. Soc.*, **106**, 8004-8005.

[13] Gerdil, R., Huiyou, L., and Gerald, B. (1999) Organic reactions in the solid state. Reactions of enclathrated 3,4-epoxycyclopentanone (= 6-oxabicyclo[3.1.0]hexan-3-one) in tri-o-thymotide and absolute configuration of 4-hydroxy- and 4-chlorocyclopent-2-en-1-one, *Helv. Chim. Acta.*, **82**, 418-434.

[14] (a) Elgavi, A., Green, B. S., and Schmidt., G. M. J. (1973) Reactions in chiral crystals. Optically active heterophotodimer formation from chiral single crystals, *J. Am. Chem. Soc.,* **95**, 2058-2059. (b) Warshel, A. and Shakked, Z. (1975) Theoretical study of excimers in crystals of flexible conjugated molecules. Excimer formation and photodimerization in crystalline 1,4-diphenylbutadiene, *J. Am. Chem. Soc.*, **97**, 5679-5684.

[15] (a) Addadi, L. and Lahav, M. (1978) Photopolymerization of chiral crystals. 1. The planning and execution of a topochemical solid-state asymmetric synthesis with quantitative asymmetric induction, *J. Am. Chem. Soc.,* **100**, 2838-2844. (b) Addadi, L. and Lahav, M. (1979) Photopolymerization in chiral crystals. 3. Toward an "absolute" asymmetric synthesis of optically active dimers and polymers with quantitative enantiomeric yield, *J. Am. Chem. Soc.*, **101**, 2152-2156. (c) Addadi, L. and Lahav, M. (1979) Towards the planning and execution of an "absolute" asymmetric synthesis of chiral dimers and polymers with quantitative enantiomeric yield, *Pure Appl. Chem.,* **51**, 1269-1284. (d) Addadi, L., van Mil, J., and Lahav, M. (1982) Photopolymerization in chiral crystals. 4. Engineering of chiral crystals for asymmetric $(2\pi + 2\pi)$ photopolymerization. Execution of an "absolute" asymmetric synthesis with quantitative enantiomeric yield, *J. Am. Chem. Soc.*, **104**, 3422-3429. (e) van Mil, J., Addadi, L., Gati, E., and Lahav, M. (1982) Useful impurities for optical resolution. 4. Attempted amplification of optical activity by crystallization of chiral crystals of photopolymerizing dienes in the presence of their topochemical products, *J. Am. Chem. Soc.*, **104**, 3429-3434. (f) van Mil, J., Addadi, L., Lahav, M., and Leiserowitz, L. (1982) Asymmetric photopolymerization in chiral crystals. An example of a chiral resolved monomer packing in two quasi-enantiomeric phases, *J. Chem. Soc. Chem. Commun.*, **1982**, 584-587.

[16] (a) Hasegawa, M., Chung, C. M., Murro, N., and Maekawa, Y. (1990) Asymmetric synthesis through the topochemical reaction in a chiral crystal of a prochiral diolefin molecule having a "cisoid" molecular structure and amplification of asymmetry by a seeding procedure, *J. Am. Chem. Soc.,* **112**, 5676-5677. (b) Chung, C. M. and Hasegawa, M. (1991) "Kaleidoscopic" photoreaction behavior of alkyl 4-[2-(4-pyridyl)ethenyl]cinnamate crystals: a crystalline linear high polymer from the methyl ester, an "absolute" asymmetric reaction of the ethyl ester, and two types of dimer formation from the propyl ester, *J. Am. Chem. Soc.,* **113**, 7311-7316. (c) Hasegawa, M. (1983) Photopolymerization of diolefin crystals, *Chem. Rev.*, **83**, 507-518, 1983.

[17] (a) Suzuki, T., Fukushima, T., Yamashita, Y., and Miyashi, T. (1994) An Absolute Asymmetric Synthesis of the [2 + 2] Cycloadduct via Single Crystal-to-Single Crystal Transformation by Charge-Transfer Excitation of Solid-State Molecular Complexes Composed of Arylolefins and Bis[1,2,5]thiadiazolotetracyanoquinodimethane, *J. Am. Chem. Soc.*, **116**, 2793-2803. (b) Suzuki, T. (1996) Photoreactions of crystalline charge-transfer complexes, *Pure Appl. Chem.*, **68**, 281-284.

[18] (a) Koshima, H., Ding, K., Chisaka, Y., and Matsuura., T. (1996) Generation of Chirality in a Two-Component Molecular Crystal of Acridine and Diphenylacetic Acid and Its Absolute Asymmetric Photodecarboxylating Condensation, *J. Am. Chem. Soc.*, **118**, 12059-12065. (b) Koshima,H., Nakata, A., Nagano, M., and Yu H. (2003) Photoreaction of 9-Methylbenz[c]acridine with Diphenylacetic Acid in the Chiral Cocrystal, *Heterocycles*, **60**, 2251-2258

[19] (a) Evans, S. V., Garcia-Garibay, M., Omkaram, N., Scheffer, J. R., Trotter, J., and Wireko, F. (1986) Use of chiral single crystals to convert achiral reactants to chiral products in high optical yield: application to the di-π-methane and Norrish type II photorearrangements, *J. Am. Chem. Soc.*, **108**, 5648-5649. (b) Chen, J., Pokkuluri, P. R., Scheffer, J. R., and Trotter J. (1990) Absolute asymmetric induction differences in dual pathway photoreactions, *Tetrahedron Lett.*, **31**, 6803-6806. (c) Fu, T. Y., Liu, Z., Scheffer, J. R., and Trotter, J. (1993) Supramolecular photochemistry of crystalline host-guest assemblies: absolute asymmetric photorearrangement of the host component, *J. Am. Chem. Soc.,* **115**, 12202-12203. (d) Leibovitch, M., Olovsson, G., Scheffer, J. R., and Trotter, J. (1997) Determination of the Absolute Steric Course of an Enantioselective Single Crystal-to-Single Crystal Photorearrangement, *J. Am. Chem. Soc.*, **119**, 1462-1463,. (e) Leibovitch, M., Olovsson, G., Scheffer, J. R., and Trotter, J. (1997) Absolute configuration correlation studies in solid state organic photochemistry, *Pure Appl. Chem.*, **69**, 815-823.

[20] Roughton, A. L., Muneer, M., and Demuth, M. (1993) Determination of the Absolute Steric Course of an Enantioselective Single Crystal-to-Single Crystal Photorearrangement, *J. Am. Chem. Soc.*, **115**, 2085-2087.

[21] This solid-state reaction was discovered by Aoyama *et al.* and subsequently studied in more detail by Toda *et al.* Aoyama, H., Hasegawa, T., and Omote, Y. (1979) Solid state photochemistry of *N,N*-dialkyl-α-oxoamides. Type II reactions in the crystalline state, *J. Am. Chem. Soc.*, **101**, 5343-5347.

[22] (a) Toda, F. and Soda, S. (1987) Formation of a chiral β-lactam by photocyclization of an achiral oxo amide in its chiral crystalline state, *J. Chem. Soc. Chem. Commun.*, 1413-1414. (b) Sekine, A., Hori, K., Ohashi, Y., Yagi, M., and Toda, F. (1989) X-ray structural studies of chiral β-lactam formation from an achiral oxo amide using the chiral crystal environment, *J. Am. Chem. Soc.*, **111**, 697-699. (c) Toda, F. and Miyamoto, H. (1993) Formation of chiral β-lactams by photocyclization of achiral *N,N*-diisopropylarylglyoxylamides in their chiral crystalline form, *J. Chem. Soc. Perkin Trans. 1*, 1129-1132.

[23] Irngartinger, H., Fettel, P. W., and Siemund, V. (1998) Diastereo- and enantioselective δ-H abstraction in the solid state of 1-benzoyl-8-benzylnaphthalene. Absolute asymmetric synthesis due to a chiral crystal environment, *Eur. J. Org. Chem.*, 2079-2082.

[24] Sakamoto, M., Takahashi, M., Shimizu, M., Fujita, T., Nishio, T., Iida, I., Yamaguchi, K., and Watanabe, S. (1995) "Absolute" Asymmetric Synthesis Using the Chiral Crystal Environment: Photochemical Hydrogen Abstraction from Achiral Acyclic Monothioimides in the Solid State, *J. Org. Chem.*, **60**, 7088-7089.

[25] (a) Sakamoto, M., Takahashi, M., Moriizumi, S., Yamaguchi, K., Fujita, T., and Watanabe, S. (1996) Crystal-to-Crystal Solid-State Photochemistry: Absolute Asymmetric β-Thiolactam Synthesis from an Achiral α,β−Unsaturated Thioamide, *J. Am. Chem. Soc.* **118**, 10664-10665. (b) Sakamoto, M., Takahashi, M., Arai, W., Mino, T., Yamaguchi, K., Watanabe, S., and Fujita, T. (2000) Solid-State Photochemistry: Absolute Asymmetric β-Thiolactam Synthesis from Achiral *N,N*-Dibenzyl-*a,b*-unsaturated Thioamides, *Tetrahedron*, **56**, 6795-6804. (c) Hosoya, T., Ohhara, T., Uekusa, H., and Ohashi, Y. (2002) Crystalline-state photoisomerization of α,β−unsaturated thioamide analyzed by X-rays, *Bull. Chem. Soc. Jpn.*, **75**, 2147-2151.

[26] (a) Sakamoto, M., Hokari, N., Takahashi, M., Fujita, T., Watanabe, S., Iida, I., and Nishio, T. (1993) Chiral thietane-fused β-lactam from an achiral monothioimide using the chiral crystal environment, *J. Am. Chem. Soc.*, **115**, 818. (b) Sakamoto, M., Takahashi, M., Hokari, N., Fujita, T., and Watanabe, S. (1994) Solid-State Photochemistry: Diastereoselective Synthesis of Thietane-Fused β-Lactams from an Acyclic Monothioimide with a Chiral Group, *J. Org. Chem.*, **59**, 3131-3134.

[27] Sakamoto, M., Takahashi, M., Arai, T., Shimizu, M., Yamaguchi, K., Mino, T., Watanabe, S., and Fujita, T. (1998) Solid state photochemical reaction of achiral *N*-(β,γ−unsaturated carbonyl)thiocarbamate to optically active thiolactone in the chiral crystalline environment, *J. Chem. Soc. Chem. Commun.*, 2315-2316.

[28] (a) Sakamoto, M., Takahashi, M., Fujita, T., Watanabe, S., Iida, I., Nishio, T., and Aoyama, H. (1993) Solid-state photochemistry: absolute asymmetric oxetane synthesis from an achiral acyclic imide using the chiral crystal environment, *J. Org. Chem.*, **58**, 3476-3477. (b) Sakamoto, M., Takahashi, M., Fujita, T., Watanabe, S., Nishio, T., and Aoyama, H. (1997) Solid State Photochemical Reaction of *N*-(α,β−Unsaturated carbonyl)benzoylformamides, *J. Org. Chem.*, **62**, 6298-6308.

[29] Kohmoto, S., Ono, Y., Masu, H., Yamaguchi, K., Kishikawa, K., and Yamamoto, M. (2001) Enantioselective Intramolecular Aromatic [4 + 4] Photocycloaddition in Crystalline State: Parameters for Reactivity, *Org. Lett.*, **26**, 4153-4155.

[30] Toda, F. and Tanaka, K., (1994) A new simple chiral helicene 3,4-bis(diphenylmethylene)-*N*-methylsuccinimide in its chiral crystal. A generation of chirality, *Supramol. Chem.*, **3**, 87-88.

[31] Wu, L. C., Cheer, C. J., Olovsson, G., Scheffer, J. R., Trotter, J., Wang, S. L., and Liao, F. L. (1997) Crystal engineering for absolute asymmetric synthesis through the use of mcta-substituted aryl groups, *Tetrahedron Lett.*, **38**, 3135-3138.

[32] Yamamoto, S., Matsuda, K., and Irie, M. (2003) Absolute asymmetric photocyclization of a photochromic diarylethene derivative in single crystals, *Angew. Chem. Int. Ed.*, **42**, 1636-1639.

[33] Sakamoto, M., Takahashi, M., Fujita, T., Nishio, T., Iida, I., and Watanabe, S. (1995) Solid-State Photochemical Reaction of *S*-Phenyl-*N*-(benzoylformyl)thiocarbamates: "Absolute" Asymmetric Synthesis Using the Chiral Crystal Environment, *J. Org. Chem.*, **60**, 4682-4683.

[34] (a) Sakamoto, M., Takahashi, M., Yamaguchi, K., Fujita, T., and Watanabe, S. (1996) Solid State Photochemistry: Absolute Asymmetric Photocyclization of an Achiral *S*-Aryl *o*-Benzoylbenzothioate to an Optically Active Phthalide Involving a Novel Phenyl Migration, *J. Am. Chem. Soc.*, **118**, 8138-8139. (b) Takahashi, M., Fujita, T., Watanabe, S., and Sakamoto M. (1998) Photochemical transformation of *S*-aryl 2-benzoylbenzothioates to 3-(arylthio)-3-phenylisobenzofuranones involving aryl migration, *J. Chem. Soc.*

Perkin Trans. 2, 487-491. (c) Takahashi, M., Sekine, N., Fujita, T., Watanabe, S., Yamaguchi, K., and Sakamoto, M. (1998) Solid-State Photochemistry of *o*-Aroylbenzothioates: Absolute Asymmetric Phthalide Formation Involving 1,4-Aryl Migration, *J. Am. Chem. Soc.*, **49**, 12770-12776.

[35] Sakamoto, M., Sekine, N., Miyoshi, H., and Fujita, T. (2000) Absolute asymmetric phthalide synthesis via the solid-state photoreaction of *N,N*-disubstituted 2-benzoylbenzamides involving a radical pair intermediate, *J. Am. Chem. Soc.*, **122**, 10210-10211.

[36] Osano, Y. T., Uchida, A., and Ohashi, Y. (1991) Optical enrichment of a racemic chiral crystal by X-ray irradiation, *Nature*, **352**, 510-512.

[37] Koura, T. and Ohashi, Y. (2000) Asymmetric Induction in Cobaloxime Complex Crystals Due to Chiral Crystal Environment, *Tetrahedron*, **56**, 6769-6779.

[38] Sakamoto, M., Takahashi, M., Mino, T., and Fujita, T. (2001) Absolute asymmetric β-lactam synthesis via the solid-state photoreaction of acyclic monothioimides and the reaction trajectory in the chiral crystalline environment, *Tetrahedron*, **57**, 6713-6719.

[39] Tissot, O., Gouygou, M., Dallemer, F., Daran, J.-C. B., and Balavoine, G. G. A. (2001) The combination of spontaneous resolution and asymmetric catalysis: a model for the generation of optical activity from a fully racemic system, *Angew. Chem. Int. Ed.*, **40**, 1076-1078.

[40] Sakamoto, M., Iwamoto, T., Nono, N., Ando, M., Arai, W., Mino, T., and Fujita, T. (2003) Memory of Chirality Generated by Spontaneous Crystallization and Asymmetric Synthesis Using the Frozen Chirality, *J. Org. Chem.*, **68**, 942-946.

PREFERENTIAL ENRICHMENT: A DYNAMIC ENANTIOMERIC RESOLUTION PHENOMENON CAUSED BY POLYMORPHIC TRANSITION DURING CRYSTALLIZATION

Rui Tamura* and Takanori Ushio[†]
*Graduate School of Human and Environmental Studies, Kyoto University, Kyoto 606-8501 Japan.
[†]Taiho Pharmaceutical Co. Ltd., Hiraishi Ebisuno, Tokushima 771-0194, Japan.

1. Introduction

In connection with the industrial needs of chiral organic substances, in which both pure R and S enantiomers are required for (i) the development of new chiral functional materials such as nonlinear optics and ferroelectric liquid crystals and (ii) the examination of safety-test, pharmacological activity and ADME (absorption, distribution, metabolism and excretion) at the beginning of the exploitation of new drugs,[1] the HPLC techniques using chiral stationary phase columns have made great progress for the last two decades as a convenient enantiomeric resolution method to separate the two enantiomers directly.[2] As a reliable method to resolve a large quantity of racemic samples, however, a conventional crystallization method is still superior to the chromatographic one from the economical standpoint.[3-5]

In this context, the practical methods for enantiomeric resolution of racemates comprising nonracemizable enantiomers by crystallization are classified into two categories; one is a method using an external chiral element such as a diastereomeric salt formation followed by fractional crystallization[4] and a diastereoselective host-guest inclusion complexation,[5] and the other is a straightforward method to separate enantiomers by crystallization in the absence of an external chiral element. As the typical example of the latter category, well known is the "Preferential Crystallization" method to resolve *a racemic conglomerate* composed of a mixture of homochiral R and S crystals, in which by repeating crystallization from the supersaturated solution with the aid of its enantiopure seed crystals the enantiomerically enriched crystals are efficiently deposited in the alternating chirality sense (Figure 1).[6]

However, the racemates existing as *a racemic conglomerate* occupy only less than 10% among the characterized crystalline racemates, and more than 90% of them are supposed to belong to racemic crystals, which are further classified into either *a racemic compound* consisting of a regular packing of a pair of R and S enantiomers (Figure 2) or *a racemic mixed crystal* (in other words, a pseudoracemate or a solid solution) composed of a random alignment

135

F. Toda (ed.), Enantiomer Separation, 135-163.

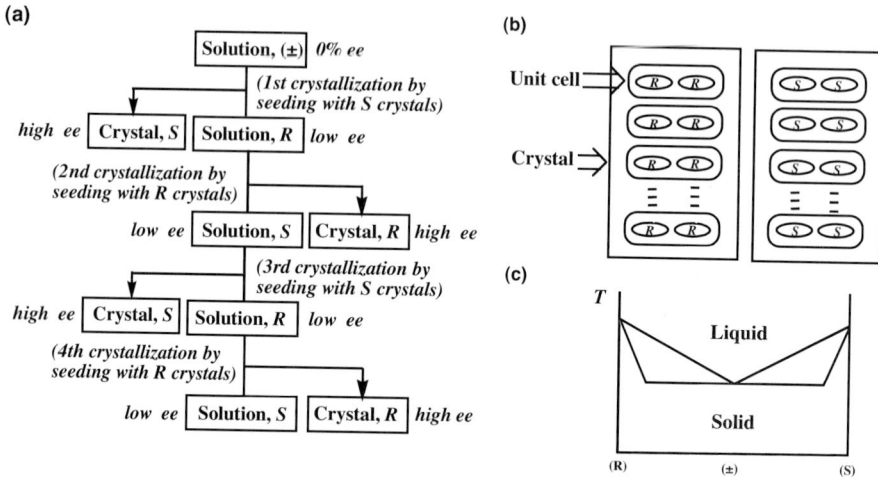

Figure 1. A racemic conglomerate; (a) the principle of Preferential Crystallization, (b) a packing mode of enantiomers in the crystal with $Z = 2$, and (c) a representative binary melting-point phase diagram.

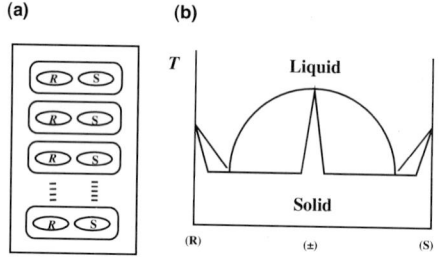

Figure 2. A racemic compound; a packing mode of enantiomers in the crystal with $Z = 2$ and (b) a representative binary melting-point phase diagram.

of the two enantiomers in the defined positions (Figure 3).[7,8]   For this reason, if one could find out an efficient enantiomeric resolution method for these racemic crystals by simple crystallization in the absence of an external chiral element, it would provide a great impact on industrial and academic communities. Preferential Enrichment, found accidentally by us, is the very enantiomeric resoution phenomenon that can be applied to the racemic crystals without any chiral circumstances, and this phenomenon is the completely reverse case of Preferential Crystallization, i.e., it is in the mother liquor that considerable enantiomeric enrichment occurs by recrystallization (Figure 4).[9-19]

        Preferential Enrichment was found in the middle of the development of a certain antiallergic drug, suplatast tosilate (ST).[9]   Whenever the racemic sample of ST was synthesized, the ee value of the solid sample obtained after recrystallization differed, i.e., it was not 0%, but was always distributed in the

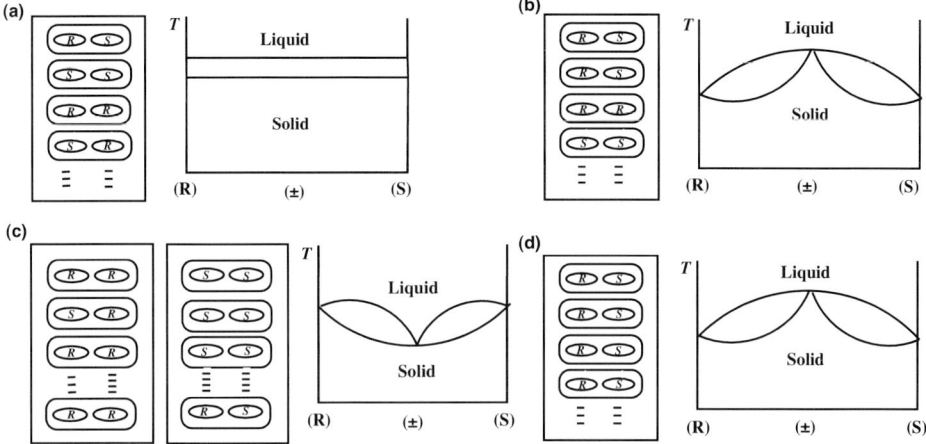

Figure 3. A racemic mixed crystal; a packing mode of enantiomers in the crystal with $Z = 2$ and a binary melting-point phase diagram of (a) disordered type, (b) fairly ordered one (*a racemic compound* type), (c) fairly ordered one (*a racemic conglomerate* type), or (d) highly ordered one.

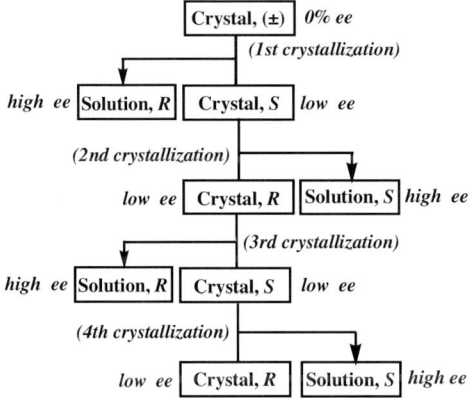

Figure 4. Enantiomeric resolution of *a racemic mixed crystal* using Preferential Enrichment; a case of an enrichment of the *R* enantiomer in solution after the first recrystallization of the racemic sample. Actually, the probability for either the *R* or the *S* enantiomer to be enriched in solution after recrystallization of an exactly racemic sample was 50% (see ref 9a).

range of 0~10%. By the systematic studies on the relationship between the ee values in the supernatant solution and the deposited crystals, a unique dynamic enantiomeric resolution phenomenon turned out to occur. This phenomenon was referred to as Preferential Enrichment.[13]   Later, the appropriate recrystalli-zation conditions to produce exactly racemic ST were devised.

Since Preferential Enrichment is an unprecedented phenomenon, in this chapter we focus on (i) the features of Preferential Enrichment as an enantiomeric resolution phenomenon, (ii) the molecular structural features of the compounds showing this phenomenon, (iii) the crystalline nature of the racemates showing this phenomenon, and (iv) the mechanism of this unusual phenomenon.

2. Features of Preferential Enrichment

Usual standard recrystallization conditions have been applied to the Preferential Enrichment experiment, except that highly supersaturated solutions are employed, because the supersolubility (a solubility obtained by dissolving the sample in a solvent on heating followed by being cooled) of the compounds showing Preferential Enrichment is considerably higher than that of the usual solubility at 25°C.

Preferential Enrichment has the following four features (Figure 4):[9-19]
(i) Repeated crystallization of the racemate and each crop of the resulting deposited crystals from the 4~25-fold supersaturated solution leads to a remarkable alternating enrichment of the two enantiomers up to 100% ee in the mother liquors (*a considerable enantiomeric enrichment in the mother liquor*).
(ii) At the same time, whenever recrystallization is carried out, the resulting deposited crystals always display the opposite chirality (*a slight enrichment of the opposite enantiomer in the deposited crystals*) with a full reproducibility.
(iii) Racemates or nonracemates with low ee values (less than 10% ee) are the more suitable starting materials for Preferential Enrichment than those with higher ee values to achieve a very efficient resolution.   Therefore, only the racemates or nonracemates with low ee values (less than 10% ee) have to be crystalline to carry out the Preferential Enrichment experiment efficiently.   In sharp contrast to Preferential Crystallization of *a racemic conglomerate*, it is no matter whether the enantiomerically enriched materials with high ee values exist as solids or oils.
(iv) No addition of seed crystals is necessary at all.
These are all unusual.   Thus, by collecting the enantiomerically enriched mother liquors with the same handedness, very efficient separation of the two enantiomers (>96% ee) has been easily achieved.   Therefore, to probe if Preferential Enrichment occurs or not for a given compound, one only have to repeat recrystallization of the racemic sample several times at 25, 0, or -20°C and measure the ee value of the supernatant solution after each crystallization.

3. Molecular Structural Features

Typical compounds that have been found to show Preferential Enrichment are summarized in Figure 5, together with the compounds failing to show this phenomenon. These are analogous linear asymmetric secondary alcohols containing a glycerol moiety, an amide group, and a sulfonium sulfonate structure or an ammonium sulfonate one.   Their molecular weights are about 500.

Figure 6 summarizes the structural requirements for the occurrence of Preferential Enrichment:[18]

Compounds Showing Preferential Enrichment:

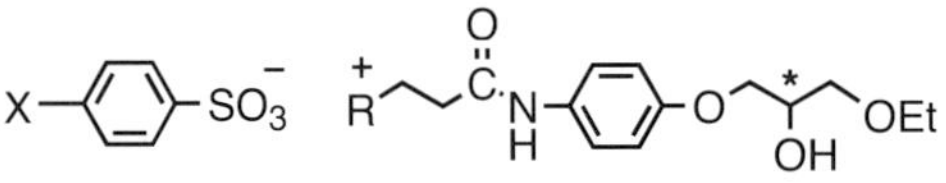

$$X-\langle\bigcirc\rangle-SO_3^- \quad R-\overset{+}{}\sim\overset{\overset{O}{\parallel}}{C}\overset{}{\underset{H}{N}}-\langle\bigcirc\rangle-O\overset{*}{\underset{OH}{\frown}}OEt$$

ST:[9,16] X = CH$_3$, R = Me$_2$S	**NNMe$_3$:**[18] X = NO$_2$, R = Me$_3$N
SB:[19] X = Br, R = Me$_2$S	**NNMe$_2$:**[12] X = NO$_2$, R = Me$_2$NH
SC:[10,16] X = Cl, R = Me$_2$S	**NPMe$_3$:**[18] X = H, R = Me$_3$N
SN:[11] X = NO$_2$, R = Me$_2$S	**NBMe$_3$:**[17] X = Br, R = Me$_3$N
	NCMe$_3$:[15] X = Cl, R = Me$_3$N
	NCMe$_2$:[13] X = Cl, R = Me$_2$NH

Compounds Failing to Show Preferential Enrichment:

$$R^1SO_3^- \quad \overset{+}{R^2}\sim\overset{\overset{O}{\parallel}}{C}\overset{}{\underset{R^3}{N}}-\langle\bigcirc\rangle-O\overset{*}{\underset{OR^4}{\frown}}Y$$

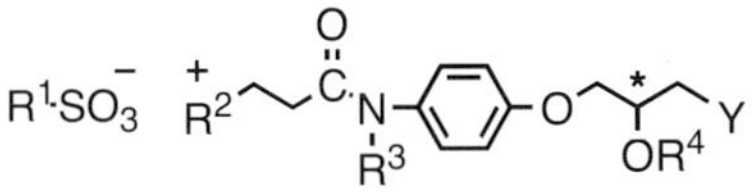

SP:[11]	R^1 = Ph,	R^2 = Me$_2$S, R^3 = R^4 = H,	Y = OEt
SM:[11]	R^1 = Me,	R^2 = Me$_2$S, R^3 = R^4 = H,	Y = OEt
SO:[11]	R^1 = n-C$_8$H$_{17}$,	R^2 = Me$_2$S, R^3 = R^4 = H,	Y = OEt
ST-C:[14]	R^1 = 4-MeC$_6$H$_4$,	R^2 = Me$_2$S, R^3 = R^4 = H,	Y = Pr
SC-C:[14,16]	R^1 = 4-ClC$_6$H$_4$,	R^2 = Me$_2$S, R^3 = R^4 = H,	Y = Pr
ST-NMe:[14]	R^1 = 4-MeC$_6$H$_4$,	R^2 = Me$_2$S, R^3 = Me, R^4 = H,	Y = OEt
ST-OMe:[14]	R^1 = 4-MeC$_6$H$_4$,	R^2 = Me$_2$S, R^3 = H,	R^4 = Me, Y = OEt
NTMe$_3$:[18]	R^1 = 4-MeC$_6$H$_4$,	R^2 = Me$_3$N, R^3 = R^4 = H, Y = OEt	
NTMe$_2$:[12]	R^1 = 4-MeC$_6$H$_4$,	R^2 = Me$_2$NH, R^3 = R^4 = H, Y = OEt	
NNMe$_3$-OPr:[18]	R^1 = 4-O$_2$NC$_6$H$_4$,	R^2 = Me$_3$N, R^3 = R^4 = H, Y = OPr	
NNMe$_3$-C:[18]	R^1 = 4-O$_2$NC$_6$H$_4$,	R^2 = Me$_3$N, R^3 = R^4 = H, Y = Pr	

Figure 5. Sulfonium and ammonium sulfonates showing Preferential Enrichment or failing to show it. The superscripts on the compound names indicate the reference number.

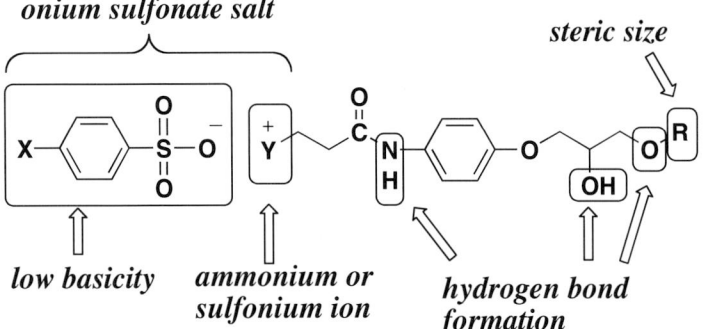

Figure 6. Structural requirements for the occurrence of Preferential Enrichment.

(i) The amide, hydroxy, and terminal alkoxy groups are all indispensable for the formation of hydrogen bonds.
(ii) The terminal alkoxy group must be methoxy or ethoxy; longer or bulky alkoxy groups fail to show Preferential Enrichment.
(iii) The onium sulfonate salt structure is advantageous, inducing a polymorphism owing to the mobility of the anion in the crystal lattice.
(iv) The selection of the sulfonate ion is very important; the sulfonate ions with low basicity which can form weak hydrogen bonds so as to allow a rearrangement of the hydrogen bonds in the crystal lattice, i.e., a polymorphic transition, are appropriate, whereas highly basic sulfonate ions which can form strong hydrogen bonds fail to induce Preferential Enrichment.

4. Unique Binary Melting Point Phase Diagram

To discuss the crystalline nature of the compounds showing Preferential Enrichment, two kinds of unique binary melting-point phase diagrams obtained actually for two typical compounds, $NNMe_3$ and $NBMe_3$, deserve attention.

In the case of $NNMe_3$, the two phase curves were found to intersect at two points around 35% ee (Figure 7).[18] The X-ray crystallographic analysis of the nearly racemic crystal indicated that the phase curve in the range of 0 ~ 15% ee corresponds to *a highly ordered racemic mixed crystal* which can hardly be distinguished from *a racemic compound* (see the section 6.2). On the other hand, the overall flat phase curve in the range of more than 40% ee proved to correspond to *a less ordered racemic mixed crystal* composed of different amounts of the *R* and *S* enantiomers in the crystal lattice.

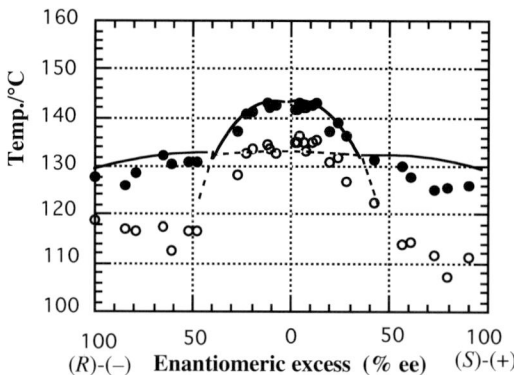

Figure 7. Binary melting-point phase diagram of $NNMe_3$. White and black circles represent the temperatures of the beginning and the end of fusion, respectively. (Reprinted with permission from ref 18. Copyright 2003 American Chemical Society.)

In the case of $NBMe_3$, there are two convex phase curves which do not intersect but are located closely to each other (Figure 8); the upper one proved to belong to *a fairly ordered racemic mixed crystal* by X-ray crystallographic analysis of the racemic and nonracemic single crystals (see the section 6.2), and the other corresponds to a metastable crystalline phase that suddenly disappeared prior to its characterization and was never observed again.[17]

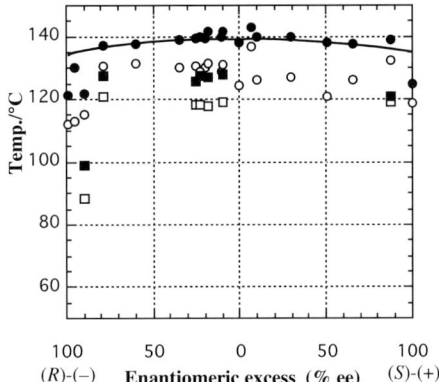

Figure 8. Binary melting-point phase diagram of $NBMe_3$. White and black circles represent the temperatures of the beginning and the end of fusion of the stable polymorph, respectively, while white and black squares are those of the metastable one.

These results indicate that (i) their crystalline nature falls into *a racemic mixed crystal*, and imply that (ii) there must be a polymorphism between the two crystalline phases in both cases and that (iii) the free energy difference between the two polymorphs is small enough to allow a polymorphic transition to proceed at a moderate rate during crystallization. Accordingly, the

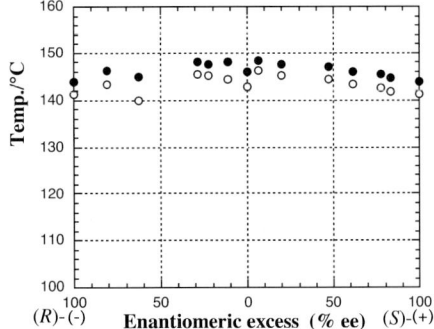

Figure 9. Binary melting-point phase diagram of $NTMe_3$. White and black circles represent the temperatures of the beginning and the end of fusion, respectively. (Reprinted with permission from ref 18. Copyright 2003 American Chemical Society.)

mechanism of Preferential Enrichment must be strongly associated with the relatively slow polymorphic transition of the metastable phase into the stable one that should occur during crystallization from the supersaturated solution of their racemate or nonracemic sample with a low ee value. In fact, the in situ ATR-IR measurements of the crystallization mixture or the DSC traces of the deposited crystals showed appreciable phase changes as the crystallization proceeded (see the section 7), suggesting a high possibility of the occurrence of the polymorphic transition during crystallization.

In contrast, compounds that failed to show Preferential Enrichment such as NTMe$_3$ and NTMe$_2$ have a single convex curve (Figure 9), indicating that these compounds exist only as a single crystalline phase of *a racemic mixed crystal* composed of the two enantiomers over a wide range of ee values.[12,18]

5. Assembly Mode of Enantiomers in Solution

To confirm the occurrence of the polymorphic transition predicted in the section 4 and to elucidate the mechanism, it is primarily necessary to clarify the enantiomeric assembly mode in the first-formed metastable crystal prior to the polymorphic transition and compare it with the stable crystal structure after the polymorphic transition with respect to a compound showing Preferential Enrichment. Since it is very possible that the stable molecular assembly structure in solution would be retained in the crystalline phase first-formed by crystallization from the same solvent,[20] at first, we have investigated the enantiomeric association mode in solutions of the racemates showing Preferential Enrichment. Consequently, in our case, the variable temperature ^{1}H NMR technique proved to be inapplicable to deciding which molecular association mode is more stable in solution, homochiral or heterochiral.[21] Instead, the combined use of the solubility and supersolubility measurements under various conditions and the number-averaged molecular weight measurement by vapor pressure osmometry turned out to become a potent tool for this objective.

5.1 SOLUBILITY AND SUPERSOLUBILITY MEASUREMENTS[18]

For the typical compounds showing Preferential Enrichment such as ST and NNMe$_3$, the solubility (e.g, 11.4 mg/mL for ST of >99.5% ee) of the highly enantio-enriched sample in 2-PrOH at 25°C was higher than that (6.6 mg/mL for ST) of the racemic sample. More interestingly, the supersolubility of the highly enantio-enriched sample in 2-PrOH showed a drastic increase (more than 150 mg/mL for ST of >99.5% ee), whereas that of the racemic sample was only slightly increased (10 mg/mL for ST). These results indicate that their homochiral supramolecular structure in 2-PrOH is very different from that in the highly enantio-enriched crystal and is much more stable than any heterochiral molecular association structure in 2-PrOH.

5.2 NUMBER-AVERAGED MOLECULAR WEIGHT MEASUREMENTS BY VPO[18]

To learn the dimension of the preferential homochiral molecular association in solutions, the number-averaged molecular weights of the racemic and highly enantio- enriched samples of ST and NNMe$_3$ have been measured in CHCl$_3$ (Figure 10). The molecular weight increased in proportion to the solute concentration and there was no appreciable difference between these two samples; these results suggest that the molecular assembly mode is a 1D-chain.

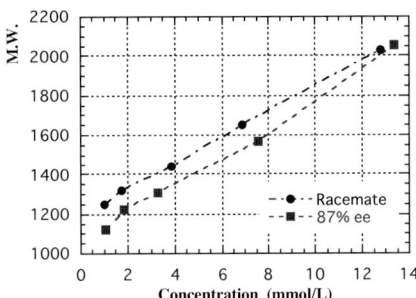

Figure 10. Number-averaged molecular weight measurement of the racemate and nonracemic sample of ST in CHCl$_3$ at 30°C by vapor pressure osmometry. (Reprinted with permission from ref 18. Copyright 2003 American Chemical Society.)

Thus, it was concluded that a homochiral molecular assembly is essentially in preference to a heterochiral one in solution with respect to the racemic samples showing Preferential Enrichment and that the homochiral supramolecular structure must be a 1D-chain. These results were consistent with those obtained by the molecular dynamics simulation of the oligomer models of ST and NNMe$_3$.

6. Characterization of Crystalline Nature

In this section, both the metastable and stable crystal structures that gave a clue to resolve the mechanism of the polymorphic transition are described.

6.1. GENERAL CRYSTALLINE NATURE OF RACEMIC MIXED CRYSTALS COMPOSED OF TWO ENANTIOMERS

The crystalline nature and physicochemical properties of the crystals belonging to *a racemic conglomerate* and *a racemic compound* have been thoroughly investigated.[6-8] In contrast, other types of enantiomers mixture crystals except these two are generally categorized as *a racemic mixed crystal*, namely, *a pseudoracemate* or *a solid solution*.[7,8]   Accordingly, *a racemic mixed crystal* refers to the crystal with a random alignment of the two enantiomers in the defined positions to give both racemic and nonracemic crystals with a variety of ee values flexibly, depending on the molecular structure.   Due to the diversity of its enantiomeric arrangement, the crystalline nature and physicochemical properties of this third class of crystals have not sufficiently been understood.

Whether a given racemic crystal is classified as *a racemic mixed crystal* or not is verified by (i) constructing the binary melting point phase diagram (Figure 3) and (ii) comparing the X-ray diffraction pattern and the solid-state IR spectrum of its racemic sample with those of nearly enantiopure one; almost identical patterns and spectra are obtained for a racemic mixed crystal. Ultimately, it is necessary to observe a molecular structure with the orientational disorder at the position of a substituent on the asymmetric center by the X-ray crystallographic analysis of the racemic and nonracemic single crystals.   Depending on the degree of the order of enantiomeric arrangement, *a racemic mixed crystal* is further categorized into three types; (i) disordered, (ii) fairly ordered (short-range ordered), and (iii) highly ordered (long-range ordered) types, of which the highly ordered type can hardly be distinguished from *a racemic compound* solely by the X-ray crystallographic analysis (Figure 3).[8]   As described in the following section, the crystalline nature of the compounds showing Preferential Enrichment falls into *a highly* or *fairly ordered racemic mixed crystal* (Figure 3b and d).

6.2 STABLE CRYSTAL STRUCTURE

Thus far three types of stable crystal structures, α-, δ- and ε-forms, which are classified as *a highly* or *fairly ordered racemic mixed crystal*, have been obtained for the racemic samples of the compounds showing Preferential Enrichment.   Among them, the δ-form crystal is most commonly found,[10-19] while only one case is observed for each of the α- and ε-forms (Table 1).[9,16,18] It has been confirmed that with respect to ST the formation of the stable α-form crystal is not essential to the occurrence of Preferential Enrichment and is caused by further polymorphic transition of the once-formed δ-form polymorph.[22]   In this section, the crystal structure of the δ-form is described in detail, because this crystal structure provides a crucial information on the mechanism of the polymorphic transition causing Preferential Enrichment. Another important crystal structure of the ε-form is mentioned in detail in the section 7.2. for the same reason.

Highly and fairly ordered crystal structures are obtained for the δ-form crystals of racemic NNMe$_3$ and NBMe$_3$, respectively.[17,18]   Furthermore, the crystal structure of the nonracemic (20% ee) NBMe$_3$ has been confirmed to be virtually isomorphous with that of the racemate and very similar to that of the pure enantiomer.[17,18]

The crystal structure of nearly racemic NNMe$_3$ is characterized by two kinds of centrosymmetric cyclic dimmers, types A and B; the head-to-head

Table 1. Space groups and lattice parameters.[a]

Compound	space group	Z	a (Å)	b (Å)	c (Å)	α (°)	β (°)	γ (°)	V (Å³)
(±)-ST (α)[16,18]	P-1	4	14.885	15.673	12.435	100.17	104.65	108.73	2551
(±)-SB(δ)[19]	P-1	2	9.922	15.184	8.933	98.93	90.68	70.09	1249
(±)-SC (δ)[16]	P-1	2	9.910	14.940	8.901	99.02	91.10	108.11	1234
(±)-NNMe₃(δ)[18]	P-1	2	9.852	15.148	9.040	96.52	92.57	107.19	1276
(±)-NNMe₂ (δ)[12]	P-1	2	10.062	15.365	8.439	97.56	91.48	70.85	1221
(±)-NCMe₂ (δ)[13]	P-1	2	9.896	15.250	8.496	98.20	91.88	71.15	1201
(±)-NCMe₃ (δ)[15]	P-1	2	9.848	14.823	9.147	97.81	92.68	105.92	1267
(±)-NPMe₃ (stable, ε)[25,b]	P-1	2	8.880	17.434	9.311	56.94	77.36	75.40	1163
(±)-NPMe₃ (metastable,γ)	P-1	2	9.497	14.947	9.069	99.301	102.64	89.263	1239
(±)-NBMe₃ (δ)[17]	P-1	2	9.910	15.016	9.13	97.54	92.65	106.84	1284
(±)-SP[11]	P2₁/a	4	11.935	11.759	16.825	-	96.36	-	2346
(±)-SC-C (α)[16]	P-1	2	14.428	15.632	6.283	102.74	97.82	68.742	1286
(±)-NTMe₂[12]	P2₁/c	4	12.699	8.645	22.341	-	91.99	-	2451
(±)-NNMe₃-OPr (γ)[18]	P-1	2	9.337	16.293	9.337	100.57	103.63	80.88	1347
(±)-NNMe₃-C (α)[18]	P-1	2	13.120	16.299	6.5805	96.96	99.59	76.31	1343

[a] Determined by X-ray crystallographic analysis. [b] Solved by the direct-space approach using the Monte Carlo algorithm with the subsequent Rietveld refinement.

cyclic dimer of type A is formed by the hydrogen bonds between the hydroxy groups and the ethoxy oxygen atoms in a pair of R and S molecules, and another head-to-head cyclic dimmer of type B is formed by (i) the hydrogen bond between one oxygen atom of a sulfonate ion and the amide NH and (ii) the electrostatic interaction between another oxygen atom of the same sulfonate ion and the ammonium nitrogen atom in the neighboring long-chain cation (Figure11). By virtue of these intermolecular interactions, a heterochiral 1D-chain is formed. Furthermore, each 1D-chain interacts with two adjacent chains by another electrostatic interaction between the third oxygen atom of the same sulfonate ion and the ammonium nitrogen atom in the adjacent chains, eventually forming a rigid 2D sheet structure.[18]

In the crystals of racemic and nonracemic (20% ee) NBMe₃, the orientational disorder was observed at the position of the hydroxy group on the asymmetric carbon atom (Figure 12); the degree of the disorder per asymmetric unit (one salt) was estimated to be 65:35 and 75:25 (R vs. S, or S vs. R), respectively, by calculating the constrained occupancy factors of the hydroxy group, corresponding to *a fairly ordered racemic mixed crystal*.[17,18] In other respects, the crystal structure is identical to that of the δ-form crystal of NNMe₃. Therefore, it is easily conceivable that such a nonracemic crystal with a low ee value, which is classified as *a fairly ordered racemic mixed crystal* and can accommodate excess enantiomers flexibly in the crystal lattice, is also produced by the Preferential Enrichment experiment in other cases. This result also indicates that the compounds affording *a highly ordered racemic mixed crystal* from the slightly supersaturated solution, such as NNMe$_3$, can give *a fairly ordered racemic mixed crystal* under the Preferential Enrichment experimental conditions using the highly supersaturated solution.

(a)

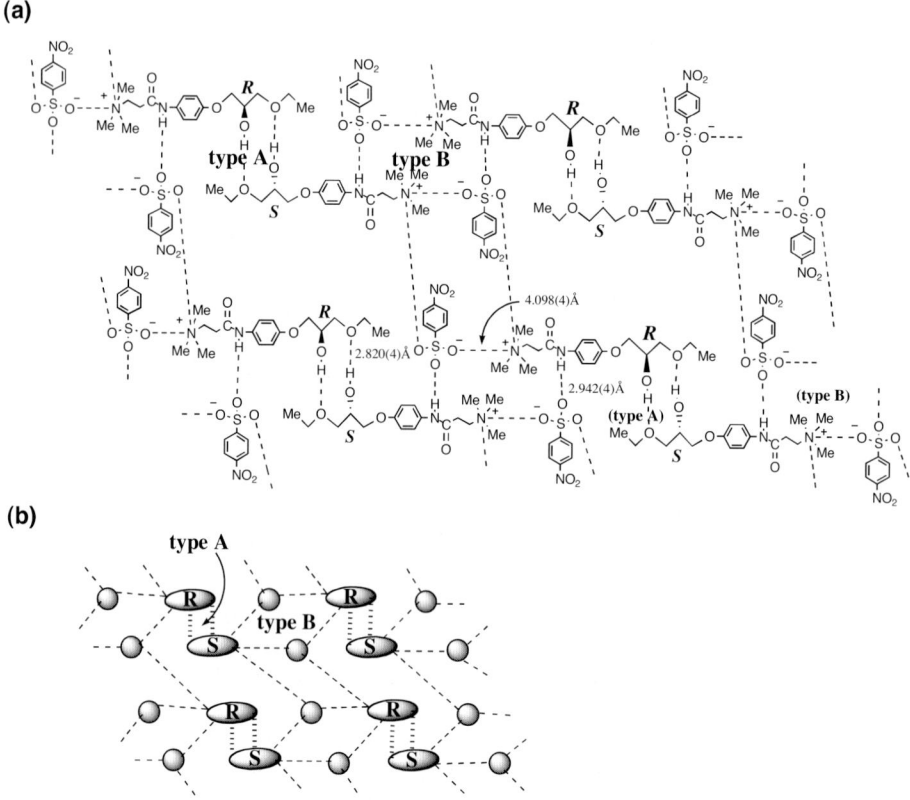

(b)

Figure 11. (a) Intermolecular interactions in the δ-form crystal of racemic NNMe$_3$ and (b) the schematic representation of the heterochiral 2-D sheet structure, in which ellipsoid and circle indicate the ammonium ion and sulfonate ion, respectively. The dashed lines show the intermolecular hydrogen bonds and electrostatic interactions. (Reprinted with permission from ref 18. Copyright 2003 American Chemical Society.)

(a)

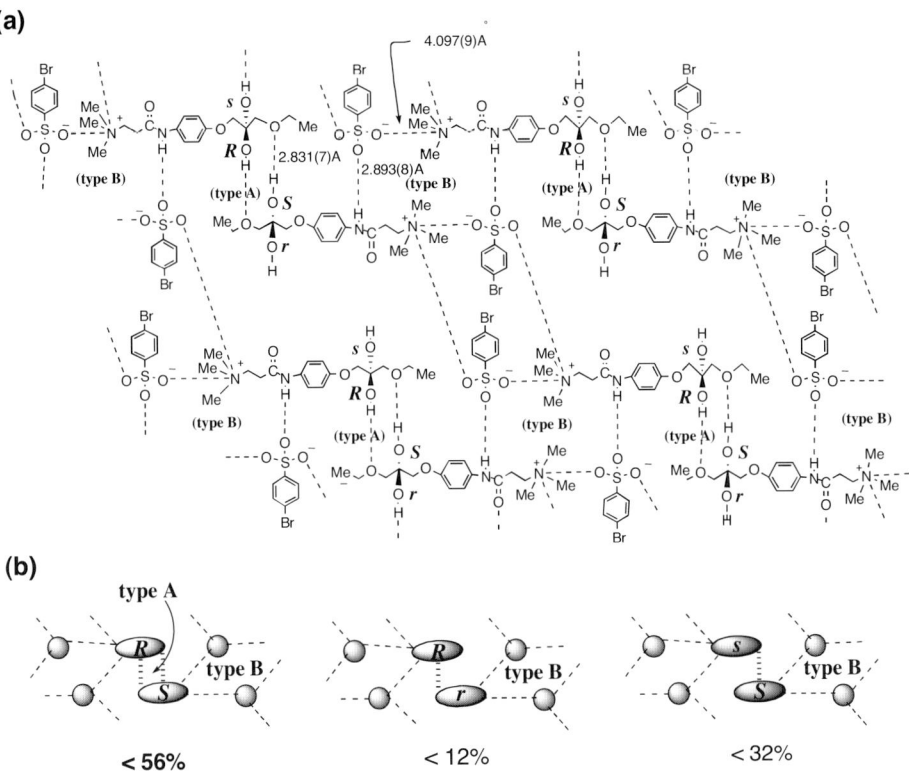

(b)

Figure 12. (a) Intermolecular interactions in the δ-form crystal of *S*-rich NBMe₃ (20% ee) and (b) the schematic representation of the heterochiral cyclic dimer and the homochiral noncyclic ones in the crystal, in which the ellipsoid and circle indicate the ammonium ion and sulfonate ion, respectively.   The hydroxy group on the asymmetric carbon atom is disordered over two positions in panel *a*.   The *R* and *S* enantiomers in the sites with higher occupancy factor (0.75) are designated *R* and *S*, and those in the sites with lower occupancy factors (0.25) are *r* and *s*. The contents of three dimer structures were estimated from the occupancy factors of the orientational disordered hydroxy groups and the ee value (20% ee) of the crystal. (Reprinted with permission from ref 18. Copyright 2003 American Chemical Society.)

6.3 METASTABLE CRYSTAL STRUCTURE

We have searched a compound that shows Preferential Enrichment and possesses a homochiral 1D-chain structure in the metastable polymorphic form, because it is very possible that the homochiral molecular assembly structure in solution would be retained in the crystal first-formed by crystallization from the same solvent.   Finally, racemic NPMe₃ was found to possess the desired γ-form crystal structure that is classified as *a highly ordered racemic mixed crystal* and is composed of alternating alignment of homochiral *R* and *S* 1D-chains in an antiparallel direction with a space group *P*-1 (*Z* = 2) with the second lowest symmetry; each homochiral chain comprises an alternating alignment of the long-chain ammonium ion and the sulfonate ion by two hydrogen bonds (i) between one oxygen atom of the sulfonate ion and the hydroxy group and (ii) between the same oxygen atom and the amide NH (Table 1 and Figure 13).[18] There is no interchain interaction. This γ-form crystal structure is also observed in several other racemic samples such as NNMe$_3$-OPr (Table 1 and Figure 5) which failed to show Preferential Enrichment, indicating that the subsequent polymorphic transition into the δ-form or ε-form is indispensable for Preferential Enrichment to occur. Actually, in the case of NPMe$_3$, the successive polymorphic transition of the γ-form into the ε-form through two other polymorphic forms (ζ- and η-forms) occurred slowly to induce Preferential Enrichment.

7. Observation and Mechanism of Polymorphic Transition Inducing Preferential Enrichment

In general, a polymorphic transition frequently occurs during crystallization from the supersaturated solutions of organic compounds, particularly when the packing mode in the first-formed crystal is unstable as the crystal structure.[23] Although much less is known about the mechanism of the polymorphic transition during crystallization from solution, it has been believed that the phase transition should proceed through either a solvent-mediated dissolution-recrystallization mechanism according to the 'Ostwald's law of stages'[24] or a solid-to-solid transformation one with the free energy change,[25] and that the rate of the polymorphic transition primarily depends on the free energy difference between the two crystalline phases.[23]

Thus far, two types of solvent-assisted solid-to-solid transformations of a kinetically formed metastable crystalline phase into a thermodynamically stable one in a process of crystal growth, relevant to the occurrence of Preferential Enrichment, have been observed: One is a relatively fast polymorphic transition noted in the case of (±)-NNMe$_3$,[18] and the other is a slow one in the case of (±)-NPMe$_3$.[18,26]

7.1 FAST PHASE TRANSITION

The time profiles of the changes in the ee values in the deposited crystals as well as in the mother liquor during crystallization of (±)-NNMe$_3$ from EtOH are shown in Figure 14.[18] The relatively fast phase transition which is complete within 90 min after the beginning of the crystallization could be followed by visual observation and the in situ ATR-FTIR measurement using ReactIR spectroscopy. The photographs shown in Figure 15 indicate that during the first 90 min of crystallization, one of the two enantiomers continues to redissolve predominantly from the just-made crystals into solution, as can be seen as a lot of convective streams evolved from the crystal surface.[18] In fact, the ee values in solution in the proximity of the crystals were always higher by

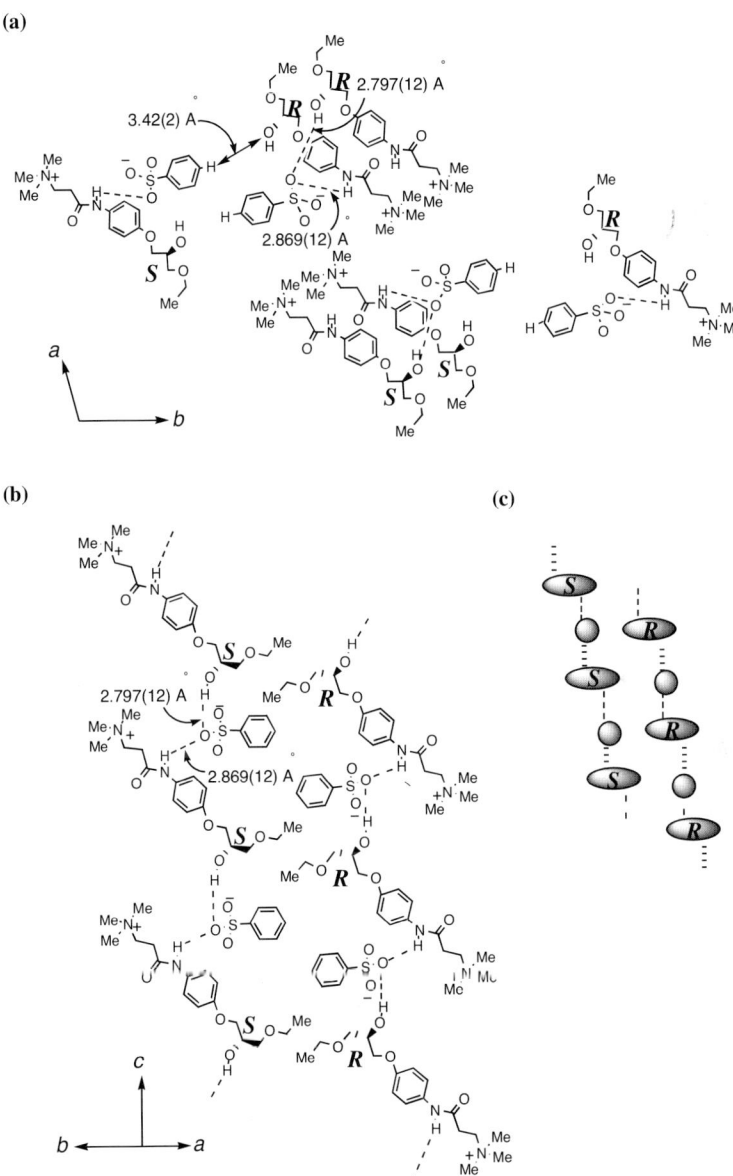

Figure 13. Intermolecular interactions in the γ-form crystal of racemic NPMe₃ viewed (a) down the *c* axis and (b) along the *c* axis, and (c) the schematic representation of the homochiral 1-D chain structure, in which the ellipsoid and circle indicate the ammonium ion and sulfonate ion, respectively. The dashed lines show the intermolecular hydrogen bonds and electrostatic interactions. (Reprinted with permission from ref 26. Copyright 2003 American Chemical Society.)

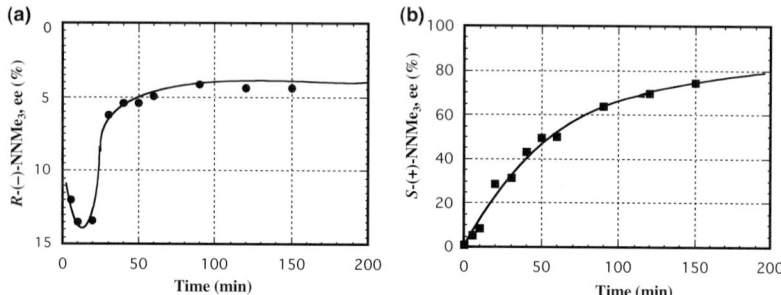

Figure 14. Time profile of the changes in the ee values (a) in the deposited crystals and (b) in the mother liquor during crystallization from the supersaturated EtOH solution of *S*-rich NNMe$_3$ (0.9% ee, 1.0 g/16 mL) at 25°C. (Reprinted with permission from ref 18. Copyright 2003 American Chemical Society.)

4 to 6% than those in the area distant from the crystals.   As will be described in the section 7.3, this dissolution behavior of one enantiomer seems to correspond to the polymorphic transition of the metastable γ-form to the stable δ-form.   Similarly, the in situ ATR-FTIR measurement indicates that three new absorption bands at 1246, 1225 and 1198 cm^{-1} corresponding to S-O stretching vibrations of the powder sample of the stable δ-form appear after crystallization begins and their intensity gradually increases for the first 90 min (Figure 16).[18] Since the crystal structure of the metastable γ-form is most likely

Figure 15.   Visual observation of polymorphic transition followed by redissolution of one enantiomer from the once formed crystals during crystallization from the supersaturated EtOH solution of nearly racemic NNMe$_3$ (0.1 mol/L); (a) 15 min and (b) 60 min after crystallization began. (Reprinted with permission from ref 18. Copyright 2003 American Chemical Society.)

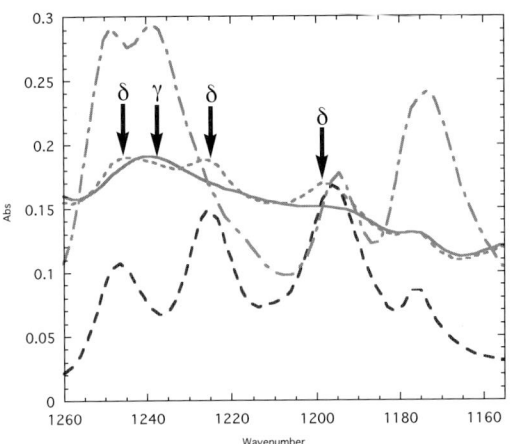

Figure 16. Comparison of the in situ ATR-FTIR spectra monitored by means of React IR spectroscopy at the beginning (gray solid line, —) and the end (gray dotted line, ⋯) of crystallization from the supersaturated EtOH solution of NNMe$_3$ (0.1 mol/L) with the spectra in the solid state of nearly racemic NNMe$_3$ (black dashed line, - - -, δ-form crystal structure) and racemic NNMe$_3$-OPr (gray dashed and doted line, - · - · -, γ-form crystal structure).   Arrows indicate the absorptions corresponding to the δ- or γ-form supramolecular structure. (Reprinted with permission from ref 18. Copyright 2003 American Chemical Society.)

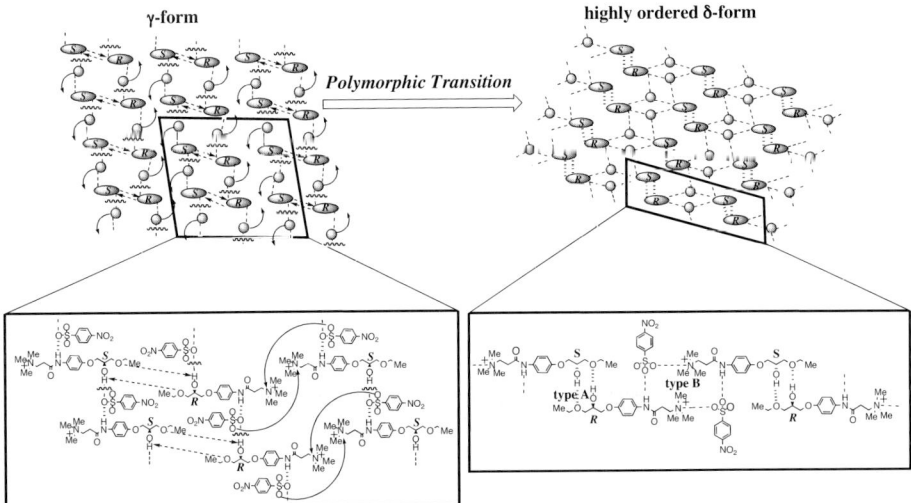

Figure 17.   Polymorphic transition with respect to NNMe$_3$ in a case where equal numbers of homochiral *R* and *S* chains are alternatingly aligned in the γ-form crystal, leading to the highly ordered δ-form crystal.

similar to the stable supramolecular structure in solution, it is reasonable that only single phase-change can be observed by the in situ ATR-FTIR measurement.

By comparing the crystal structures between the metastable γ-form composed of the homochiral 1D-chain and the stable δ-form comprised of the heterochiral 2D sheet, the mode of this single phase transition can be proposed (Figure 17). Polymorphic transition is initiated by rearrangement of the hydrogen bonds inside the crystal lattice. It is very possible for the nearest *R* and *S* molecules in the two adjacent chains in the γ-form to form new hydrogen bonds between the ethoxy oxygen atoms and the hydroxy groups by slight movement of the two molecules in the crystal. This rearrangement of hydrogen bonds accompanied by slight movement of the sulfonate ions so as to form the cyclic dimers of types A and B occurs one after another in the crystal lattice to lead to the heterochiral 1D-chains and then the 2D sheet structure.

7.2 SLOW PHASE TRANSITION

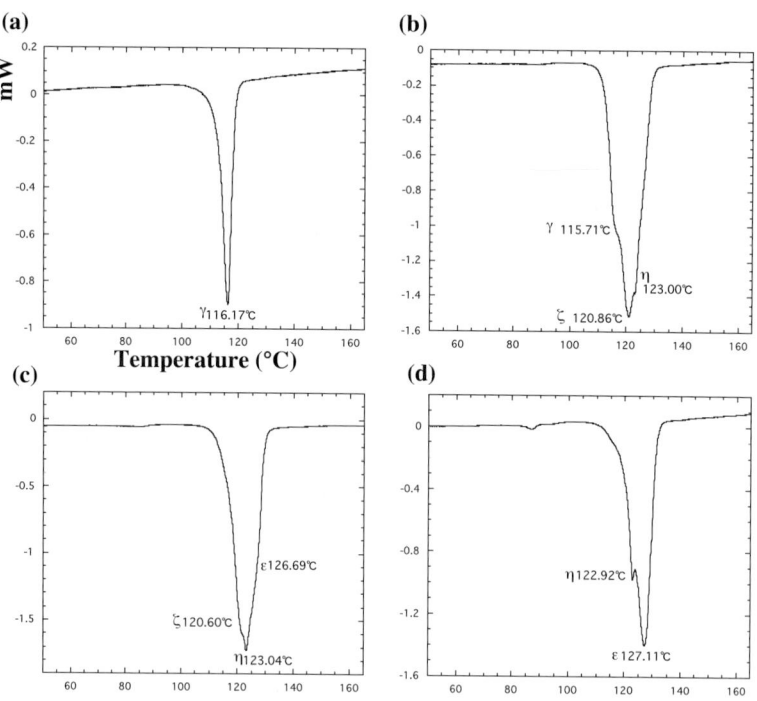

Figure 18. DSC aspects of gradual polymorphic transition of the deposited crystals observed during crystallization of racemic NPMe₃ from EtOH: (a) γ-form, (b) a mixture of γ, ζ and η-forms after 8 days, (c) a mixture of ζ, η and ε-forms after 13 days, and (d) a mixture of η and ε-forms after 19 days. (Reprinted with permission from ref 18. Copyright 2003 American Chemical Society.)

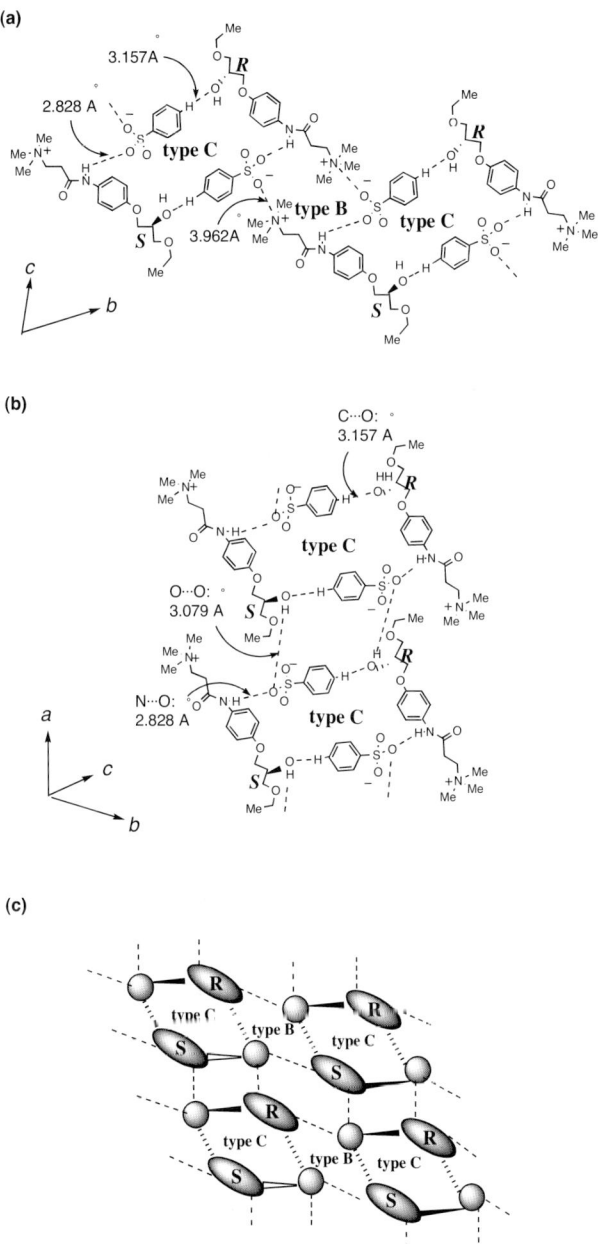

Figure 19. Intermolecular interactions in the ε-form crystal of racemic NPMe₃ viewed (a) down the *a* axis and (b) along the *a* axis, and (c) the schematic representation of the heterochiral step-like sheet structure, in which the ellipsoid and circle indicate the ammonium ion and sulfonate ion, respectively. The dashed lines show the intermolecular hydrogen bonds and electrostatic interactions. (Reprinted with permission from ref 26. Copyright 2003 American Chemical Society.)

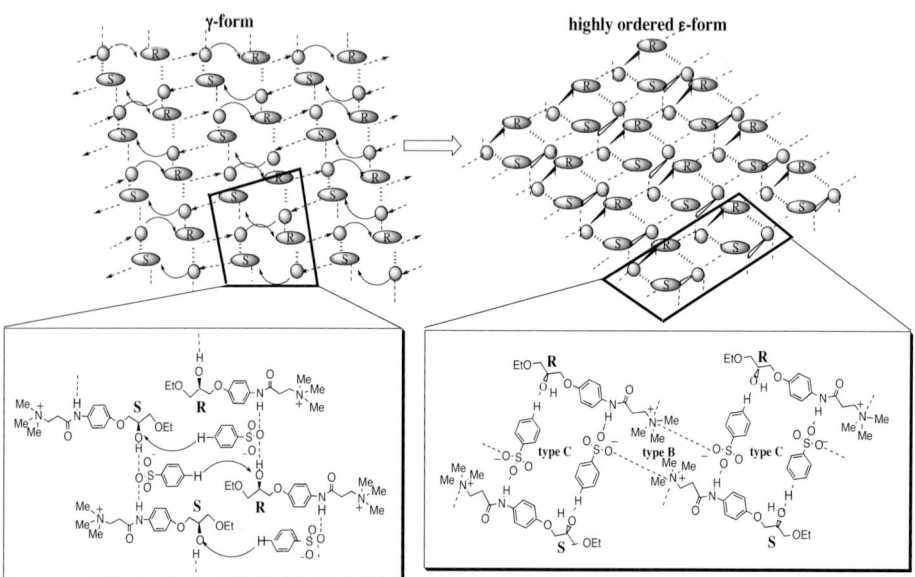

Figure 20. Polymorphic transition with respect to NPMe₃ in a case where equal numbers of homochiral *R* and *S* chains are alternatingly aligned in the γ-form crystal, leading to the highly ordered ε-form crystal.

DSC traces of the crystals deposited by crystallization from the supersaturated solution of racemic $NPMe_3$ (0.259 mol/L) in EtOH showed the occurrence of successive polymorphic transitions; the first-formed γ-form polymorph (an endothermal peak at 116°C) was gradually transformed into the most stable ε-form one (127°C) over 14 days via two different metastable ζ- and η-forms (121 and 123°C, respectively) (Figure 18).[18]   The overall rate of the successive polymorphic transitions of the γ-form crystal into the ε-form one, which corresponds to that of the enrichment of one enantiomer in solution, was greatly accelerated by increasing the solute concentration.

The metastable γ-form crystal structure prior to the slow phase transitions was determined by the X-ray crystallographic analysis, but the crystal structures of three other new polymorphs (ζ-, η- and ε-forms) were undetermined by the same X-ray analysis, because these polymorphs were obtained as powder samples. Since the nearly racemic ε-form crystals could be obtained as the monophasic powder sample by just leaving the crystallization mixture untouched, its crystal structure was solved from its X-ray diffraction (XRD) data by the direct-space approach using the Monte Carlo algorithm with the subsequent Rietveld refinement.[26,27]   Consequently, the R_{wp} value has been satisfactorily converged to 9.1%.

The crystal structure of the ε-form is similar to that of the γ-form, but there is a decisive difference between them; the former is comprised of heterochiral supramolecular 1D dimer chains (Figure 19), while the latter consists of homochiral supramolecular R and S 1D ones (Figure 13). That is, the ε-form crystal structure is characterized by a centrosymmetric cyclic dimer (type C) which is formed through the sandwich-type intermediary of the two phenylsulfonate ions between a pair of R and S long-chain cations by two kinds of intermolecular interactions; (i) the strong $C(sp^2)H\cdots O$ interaction between the para hydrogen atom of a phenylsulfonate ion and the hydroxy group in the neighboring long-chain cation and (ii) the strong hydrogen bond between one oxygen atom of the phenylsulfonate ion and the nearest amide NH (Figure 19a). Furthermore, the heterochiral cyclic dimers of type C interact with each other through the electrostatic interaction between another oxygen atom of the sulfonate ion and the nearest ammonium nitrogen atom, forming another head-to-head cyclic dimer of type B and thereby a heterochiral supramolecular 1D dimer chain structure on the bc plane (Figure 19a). The intermolecular O-H$\cdots$O-SO$_2$Ph hydrogen bond along the a axis observed in the γ-form becomes weaker but still remains in the ε-form, so that the heterochiral 1D dimer chains interact with each other along the a axis by this hydrogen bond, eventually giving rise to a heterochiral step-like sheet structure (Figure 19b and c).

The mode of the polymorphic transition of the γ-form crystal into the ε-form one can be easily rationalized in terms of the slight movement of the phenylsulfonate ions inside the crystal lattice so as to strengthen the $C(sp^2)H\cdots O$ interaction between the para hydrogen atom of the phenylsulfonate ion and the hydroxy group in the neighboring long-chain cation, to loosen one hydrogen bond (OH$\cdots$OSO$_2$Ph), and to retain another hydrogen bond (NH$\cdots$OSO$_2$Ph), eventually forming a heterochiral supramolecular step-like sheet structure (Figure 20). This strong $C(sp^2)H\cdots O$ interaction seems responsible for the inhibition of the expected polymorphic transition of the γ-form crystal to the δ-form one that is frequently observed for other compounds, leading to the isolation of the ε-form crystal. This is a very rare case that $C(sp^2)H\cdots O$ interaction exerts a primary influence on the control of the crystal structure[28] and thereby the mode of polymorphic transition. It is surprising that such a slight molecular movement inside the crystal lattice could induce the dynamic enantiomeric resolution phenomenon, i.e., Preferential Enrichment.

7.3 MODE OF REARRANGEMENT OF HYDROGEN BONDS IN THE CRYSTAL LATTICE

Before discussing the overall mode of the polymorphic transition, let's consider the association mode of the homochiral R and S chains in the metastable γ-form crystal formed kinetically from the highly supersaturated solution.

Supposed that the R enantiomer is slightly excess in the supersaturated solution, the resulting metastable γ-form crystal must become R-rich, too. This slightly R-rich crystal exists as *a racemic mixed crystal* which consists of three types of alignment modes of the R and S homochiral chains; the major regular alignment of the R and S chains, along with two minor irregular alignments in which an odd number of R chains and an even number of R ones

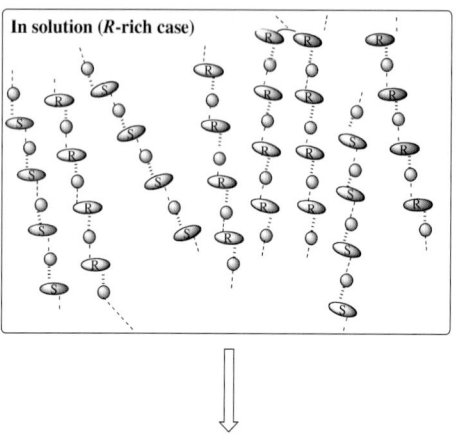

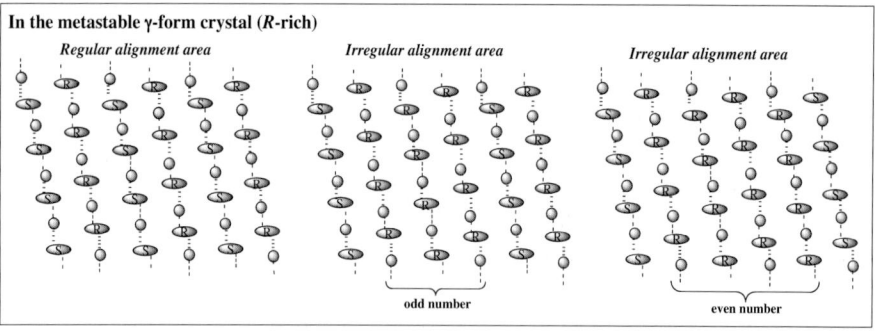

Figure 21.   Association mode of homochiral R and S chains in the metastable γ-form crystal formed kinetically from the highly supersaturated solution of slightly R-rich NNMe$_3$ or NPMe$_3$.

are respectively surrounded by two S chains (Figure 21).

Now let's move to the mode of the polymorphic transition.   The fashion of alignment of the R and S homochiral chains in the metastable γ-form crystal must define the type of the resulting δ- or ε-form crystal structure after the phase transition; (i) where equal numbers of R and S homochiral chains are alternatingly aligned in an antiparallel direction along one axis, a completely ordered δ- or ε-form crystal structure will be formed after the polymorphic transition (Figures 17 and 20), (ii) where an odd number of homochiral R chains are aligned between two S chains, after the polymorphic transition, the fairly ordered δ- or ε-form crystal structure will be formed without disintegration of the crystal (Figures 22 and 23), and (iii) where an even number of homochiral R chains are aligned between two S chains, after the polymorphic transition, partial disintegration in the resulting δ- or ε-form crystal occurs to release the R-rich area into solution, because the R-rich area was surrounded by the sites, where the OH···OEt or C(sp^2)H···O interaction cannot be formed, respectively, as well as the sites with weak electrostatic

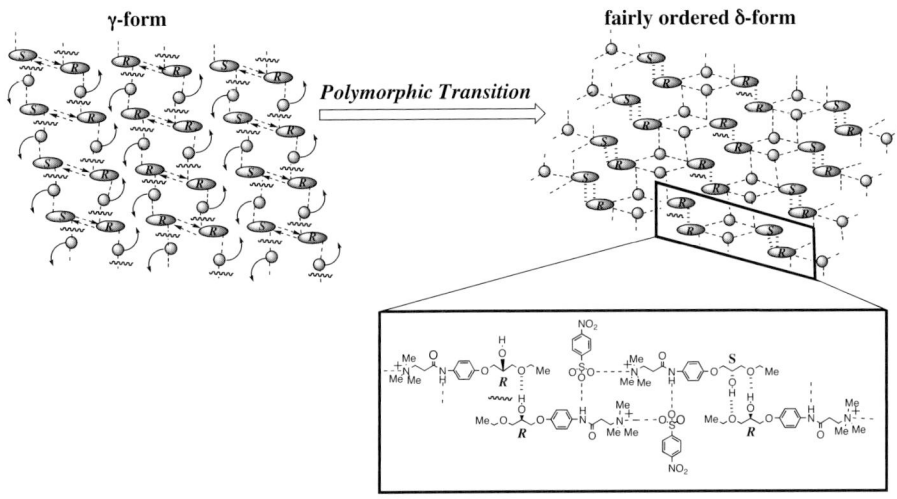

Figure 22. Polymorphic transition with respect to NNMe₃ in a case where an odd number (three in this case) of homochiral *R* chains are surrounded by two *S* chains in the γ-form crystal, leading to the fairly ordered δ-form crystal without crystal disintegration.

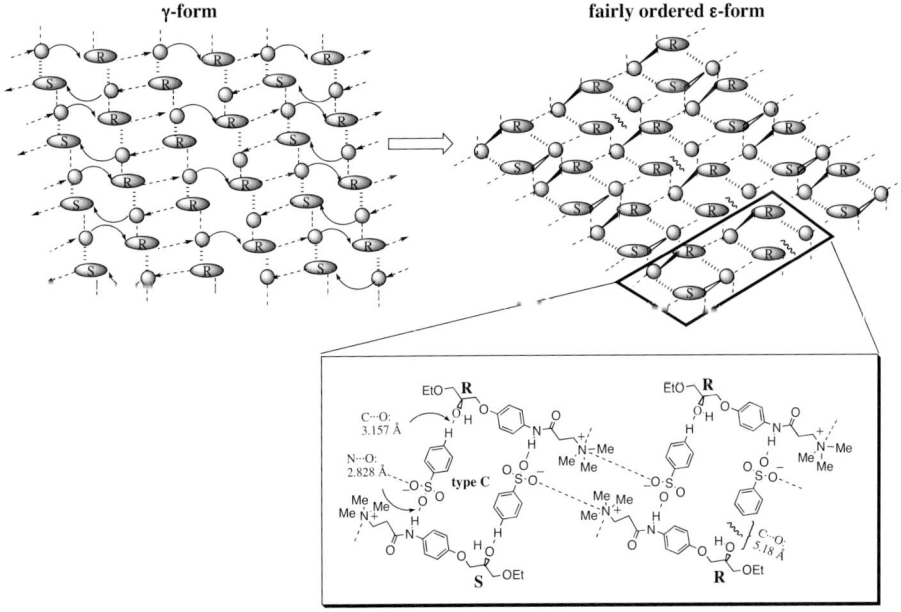

Figure 23.   Polymorphic transition with respect to NPMe₃ in a case where an odd number (three in this case) of homochiral *R* chains are surrounded by two *S* chains in the γ-form crystal, leading to the fairly ordered ε-form crystal without crystal disintegration. (Reprinted with permission from ref 26. Copyright 2003 American Chemical Society.)

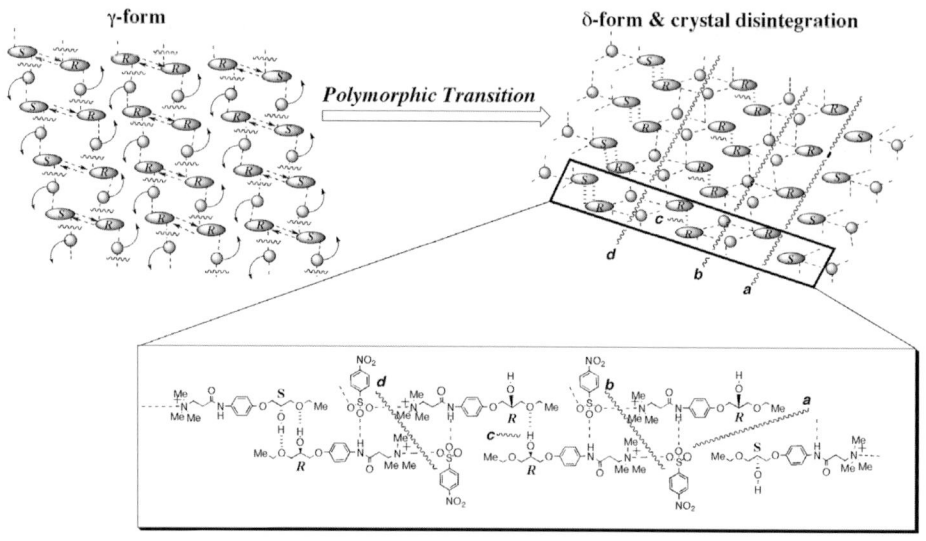

Figure 24. Polymorphic transition with respect to NNMe₃ in a case where an even number (four in this case) of homochiral *R* chains are surrounded by two *S* chains in the γ-form crystal, leading to the local crystal disintegration in the δ-form crystal that occurs at the sites a, b, c and d.

interactions (Figures 24 and 25). This third case corresponds to the visually observable redissolution phenomenon shown in Figure 15 and must be responsible for the two characteristic phenomena of Preferential Enrichment, *a considerable enrichment of one enantiomer in solution* and *a slight enrichment of the opposite enantiomer in the deposited crystals.* That is, if the metastable γ-form crystals contain the *R* enantiomer in a small excess (e.g., ca. 5% ee), probably the sum of the *R* chains is larger than that of the *S* ones in the crystal, and the probability for an even number of *R* chains to be aligned between two S chains should be higher than the opposite case. If so, after the polymorphic transition, a substantial amount of the *R* enantiomers must be liberated into solution eventually to give the fairly *R*-enriched solution and the slightly *S*-enriched crystals.

8. Mechanism of Preferential Enrichment

The mechanism of Preferential Enrichment can be interpreted in terms of an interplay between (i) a polymorphic transition at the beginning of crystallization resulting in *a considerable enantiomeric enrichment in the mother liquor* and *a slight enrichment of the opposite enantiomer in the deposited crystals* and (ii) the chiral discrimination by the deposited crystals in the subsequent crystal growth process leading to *a further enrichment of the same enantiomer in the mother liquor.*

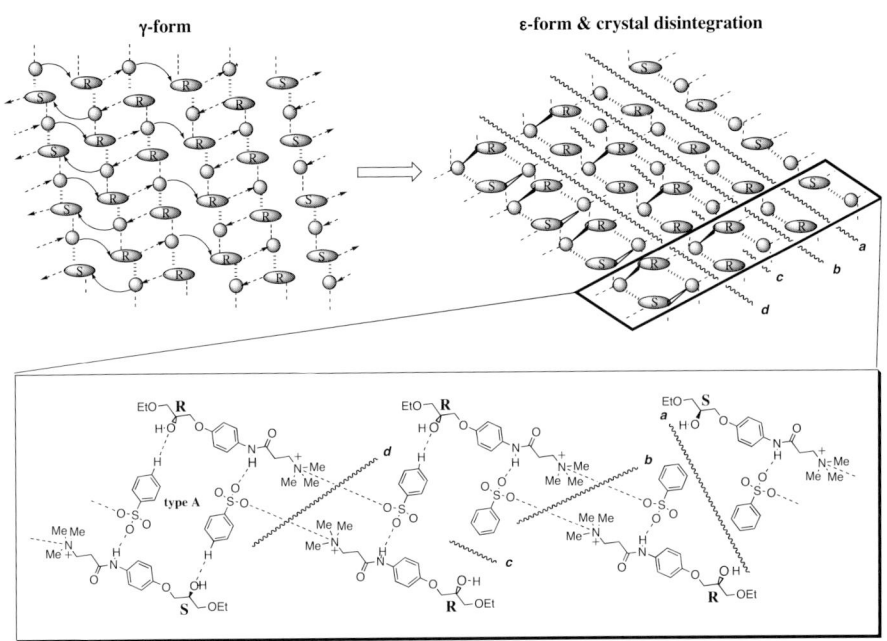

Figure 25. Polymorphic transition with respect to NPMe₃ in a case where an even number (four in this case) of homochiral *R* chains are surrounded by two *S* chains in the γ-form crystal, leading to the local crystal disintegration in the ε-form crystal that occurs at the sites a, b, c and d. (Reprinted with permission from ref 26. Copyright 2003 American Chemical Society.)

    The proposed overall mechanism of Preferential Enrichment in the case of NNMe₃ is illustrated in Figure 26; in this case, an enrichment of the *R* enantiomer occurs in the mother liquor by crystallization from the slightly *R*-rich (ca. 5% ee) supersaturated solution. In solution, homochiral oligomers are formed preferentially and aggregate to form a supramolecular cluster, which undergoes a phase transition to give the metastable γ-form crystal in a process of nucleation (Figure 21). Since the *R* enantiomer is slightly excess in solution, the resulting metastable crystal becomes slightly *R*-rich, too, and this crystal contains the large *R* chain-rich areas (represented by four consecutive *R* chains) and the small *S* chain-rich areas (represented by three consecutive *S* chains) besides the alternating *R* and *S* chains alignment. The next step is the polymorphic transition at the beginning of crystal growth arising from the rearrangement of the hydrogen bonds in the crystal lattice as described in the section 7.3. After the polymorphic transition, the large *R*-rich area can no longer stay inside the transformed crystal due to the partial disintegration of the crystal, dissolving into solution (see Figure 24). However, the small *S*-rich area can be maintained in the crystal lattice by newly formed hydrogen bonds to form an *S*-rich crystal with a fairly ordered δ-form. Eventually, together with a slight enrichment of the opposite S enantiomer in the deposited crystals, an enrichment of the R enantiomer in the mother liquor occurs to some extent at this stage.   Subsequently, the resulting slightly S-rich δ-form crystals induce chiral discrimination to promote the growth of the crystals with the same handedness and ee values.   Thus, the enantiomeric purity of the R enantiomer in the mother liquor is gradually raised until the crystal growth ceases.   In addition, the higher solubility of the materials with high ee values than those with low ee ones accelerates the R enantiomer to be enriched in the mother liquor.

Finally, it should be stressed that when the original supersaturated solution for crystallization was strictly racemic (0.0% ee), the probability for either the R or the S enantiomer to be enriched in solution after crystallization was 50%;[9a] this is because initial capricious formation of the very first nonracemic metastable crystal nucleus should doom which enantiomer is enriched in solution later.

9. Concluding Remarks

Preferential Enrichment is a secondary phenomenon caused by a polymorphic transition occurring during crystallization from a highly supersaturated solution. This unique dynamic enantiomeric resolution phenomenon has proved to be observable for *a fairly ordered racemic mixed crystal* showing a polymorphism; a solvent-assisted solid-to-solid type of polymorphic transition from the kinetically-formed metastable crystalline phase comprising homochiral R and S chains into the thermodynamically stable crystalline phase consisting of a heterochiral 2D sheet structure during crystallization is responsible for this phenomenon.   That is, it is essential that homochiral R and S 1D chain structures are stable in solution while a heterochiral 2D sheet structure is stable in the crystal.

At the same time, the fairly random aggregation of the R and S 1D chains induced by crystallization from a highly supersaturated solution seems to be one of the factors essential to the occurrence of Preferential Enrichment. Therefore, Preferential Enrichment is applicable to other chiral organic substances that exist as *a highly ordered racemic mixed crystal* which can hardly be distinguished from *a racemic compound*, as exemplified in the case of $NNMe_3$.   By analogy with this result, by carefully choosing the appropriate crystallization conditions, Preferential Enrichment might be applicable to even *a certain racemic compound* that satisfies the structural requirements for supramolecular association in solution and in the crystalline state.

Apart from the significance as a novel enantiomeric resolution phenomenon, the investigation on the mechanism of Preferential Enrichment has also shed light on the hitherto unknown mechanism of a polymorphic transition occurring during crystallization from a supersaturated solution.   A combination of several techniques employed here would also be useful to elucidate the unknown mechanism of another type of polymorphic transition occurring during crystallization.

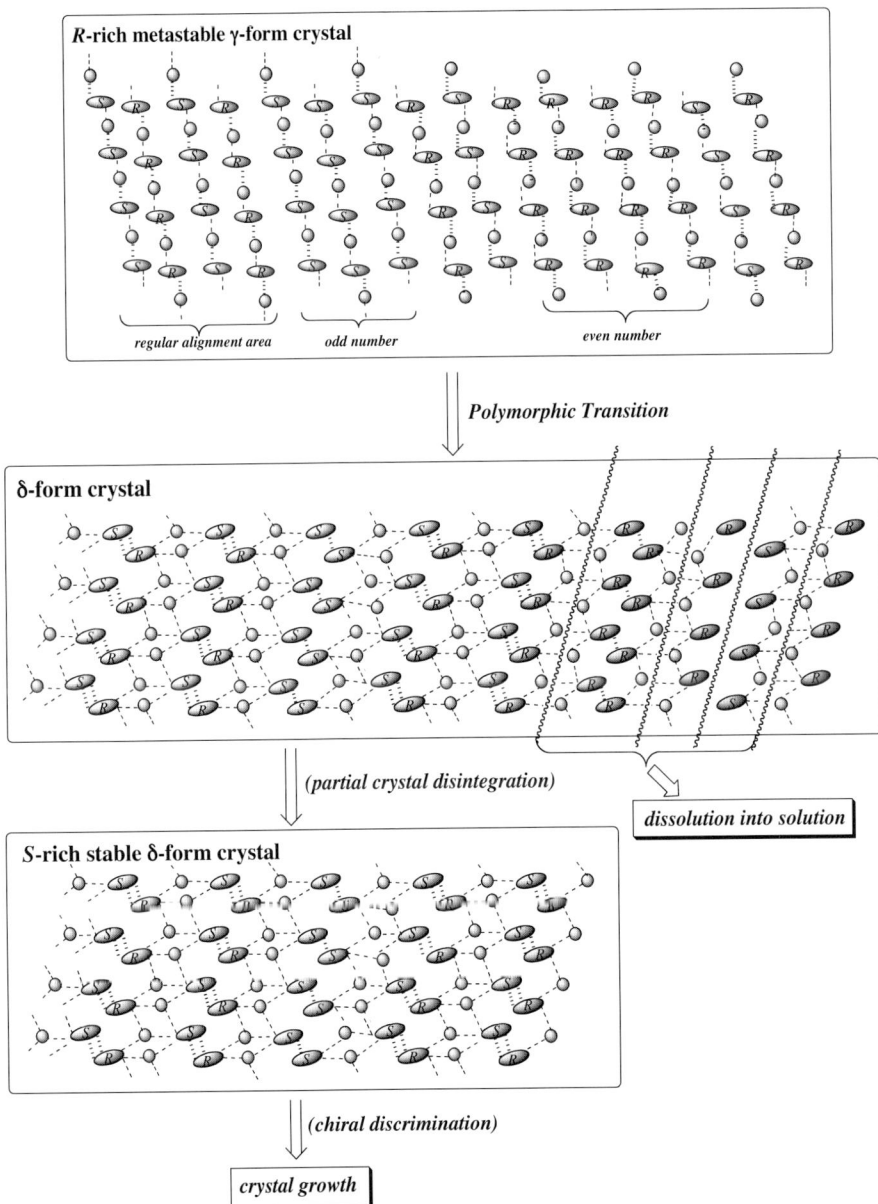

Figure 26. Mechanism of Preferential Enrichment characterized by homochiral molecular association, polymorphic transition, followed by crystal disintegration, and chiral discrimination in the case of NNMe$_3$. (Reprinted with permission from ref 18. Copyright 2003 American Chemical Society.)

10. References

1. (a) *Polymorphism in Pharmaceutical Solids*; (Drugs and the Pharmaceutical Sciences Vol. 95); Brittain, H. G., Ed.; Marcel Dekker: New York, 1999. (b) *Chirality In Industry*; Collins, A. N., Sheldrake, G. N., Crosby, J., Eds.; Wiley: New York, 1992. (c) *Chirality In Industry II*; Collins, A. N., Sheldrake, G. N., Crosby, J., Eds.; Wiley: New York, 1997.

2. (a) *Chromatographic Chiral Separations* (Chromatographic Science Series, Vol. 40); Zief, M.; Crane, L. J., Eds.; Marcel Dekker: New York, 1988. (b) *Chiral Separations, Applications and Technology* ; Ahuja, S., Ed.; American Chemical Society: Washington, DC, 1997. (c) *Chiral Separation Techniques*; Subramanian, G., Ed.; Wiley-VCH: Weinheim, 2001.

3. (a) Newman, P. *Optical Resolution Procedures for Chemical Compounds*; Optical Resolution Information Center: New York, 1978, 1981, and 1984; Vol. 1-3.

4. (a) *CRC Handbook of Optical Resolutions via Diastereomeric Salt Formation*; Kozma D., Ed.; CRC Press, Boca Raton, FL, 2001. (b) Kinbara, K.; Saigo, K. *Top. Stereochem.* **2003**, *23*, 207. (c) Vries, T.; Wynberg, H.; van Echten, E.; Koek, J.; ten Hoeve, W.; Kellog, R. M.; Broxterman, Q. B.; Minnaard, A.; Kaptein, B.; van der Sluis, S.; Hulshof, L.; Kooistra, J. *Angew. Chem. Int. Ed. Engl.* **1998**, *37*, 2349.

5. Toda, F. *Top. Curr. Chem.* **1987**, *140*, 43.

6. (a) Pasteur, L. *Ann. Chim. Phys.* **1848**, *24*, 442. (b) Gernez, M. *C. R. Hebd. Seances Acad. Sci.* **1866**, *63*, 843. (c) Collet, A.; Brienne, M. -J.; Jacques, J. *Chem. Rev.* **1980**, *80*, 215. (d) Collet, A. Optical Resolution In *Comprehensive Supramolecular Chemistry*; Reinhoudt, D. N., Ed.; Pergamon: Oxford, 1996; Vol. 10, pp 113-149. (e) Kinbara, K.; Hashimoto, Y.; Sukegawa, M.; Nohira, H.; Saigo, K. *J. Am. Chem. Soc.* **1996,** *118*, 3441.

7. Eliel, E.; Wilen, S. H.; Mander, L. N. *Stereochemistry of Organic Compounds*; Wiley: New York, 1994; pp 297-464.

8. Jacques, J.; Collet, A.; Wilen, S. H. *Enantiomers, Racemates and Resolutions*; Krieger Publishing Co.: Malabar, FL, 1994.

9. (a) Ushio, T.; Tamura, R.; Takahashi, H.; Yamamoto, K. *Angew. Chem. Int. Ed. Engl.* **1996**, *35*, 2372. (b) Ushio, T.; Tamura, R.; Azuma, N.; Nakamura, K.; Toda, F.; Kobayashi, K. *Mol. Cryst. Liq. Cryst.* **1996**, *276*, 245. (c) Ushio, T.; Yamamoto, K. *J. Chromatogr. A*, **1994**, *684*, 235. (d) Koda, A.; Yanagihara, Y.; Matsuura, N. *Agents Actions* **1991**, *34*, 369.

10. Tamura, R.; Ushio, T.; Nakamura, K.; Takahashi, H.; Azuma, N; Toda, F. *Enantiomer* **1997**, *2*, 277.

11. Tamura, R., Ushio, T.; Takahashi, H.; Nakamura, K.; Azuma, N.; Toda, F.; Endo, K. *Chirality* **1997,** *9*, 220.

12. (a) Takahashi, H.; Tamura, R.; Ushio, T.; Nakajima, Y.; Hirotsu, K. *Chirality* **1998**, *10*, 705. (b) Tamura, R.; Takahashi, H.; Hirotsu, K.; Nakajima, Y.; Ushio, T. *Mol. Cryst. Liq. Cryst.* **2001**, *356*, 185.

13. Tamura, R.; Takahashi, H.; Hirotsu, K.; Nakajima, Y.; Ushio, T.; Toda, F. *Angew. Chem. Int. Ed. Engl.* **1998**, *37*, 2876.

14, Tamura, R.; Takahashi, H.; Ushio, T.; Nakajima, Y.;Hirotsu, K.; Toda, F. *Enantiomer* **1998**, *3*, 149.

15. Tamura, R.; Takahashi, H.; Miura, H.; Lepp, Z.; Nakajima Y.; Hirotsu, K.; Ushio, T. *Supramol. Chem.* **2001**, *13*, 71.

16. Takahashi, H.; Tamura, R.; Lepp, Z.; Kobayashi, K.; Ushio, T. *Enantiomer* **2001**, *6*, 57.

17. Takahashi, H.; Tamura, R.; Fujimoto, D.; Lepp, Z.; Kobayashi, K.; Ushio, T. *Chirality* **2002**,

14, 541.

18. Tamura, R.; Fujimoto, D.; Lepp, Z.; Misaki, K.; Miura, H.; Takahashi, H.; Ushio, T.; Nakai, T.; Hirotsu, K. *J. Am. Chem. Soc.* **2002**, *124*, 13139.

19. Takahashi, H.; Tamura, R.; Yabunaka, S.; Ushio, T. *Mendeleev Commun.* **2003**, 119.

20. (a) Maruyama, S.; Ooshima, H. *J. Crystal Growth* **2000**, *212*, 239. (b) Kitamura, M.; Ueno, S.; Sato, K. Molecular Aspects of the Polymorphic Crystallization of Amino Acids and Lipids. In *Crystallization Processes*; Ohtaki, H., Ed.; Wiley: Chichester, 1998; pp 99-129. (c) Kitamura, M. *J. Crystal Growth* **1989**, *96*, 541.

21. (a) Dobashi, A.; Saito, N.; Motoyama, Y.; Hara, S. *J. Am. Chem. Soc.* **1986**, *108*, 307. (b) Jursic B. S.; Goldberg, S. I. *J. Org. Chem.* **1992**, *57*, 7172. (c) Cung, M. T.; Marraud, M.; Neel, J. *Biopolymers* **1978**, *17*, 149. (d) Harger, M. J. P. *J. Chem. Soc., Perkin Trans. 2* **1977**, 1882. (e) Harger, M. J. P. *J. Chem. Soc., Perkin Trans. 2* **1978**, 326.

22. Miura, H.; Ushio, T.; Nagai, K.; Fujimoto, D.; Lepp, Z.; Takahashi, H.; Tamura, R. *Cryst. Growth Des.* **2003**, *3*, 959.

23. (a) Bernstein, J. *Polymorphism in Molecular Crystals*, Oxford University Press, Oxford, 2002. (b) Bernstein, J.; Davey, R. J.; Henck, J-o. *Angew. Chem. Int. Ed. Engl.* **1999**, *38*, 3440. (c) Dunitz, J. D., Bernstein, J. *Acc. Chem. Res.* **1995**, *28*, 193. (d) McCrone, W. C. Polymorphism. In *Physics and Chemistry of the Organic Solid State*; Fox, D.; Labes, M. M.; Weissberger, A., Eds.; Interscience: New York, 1965; Vol. II, pp 726-767. Also see ref 1a.

24. Ostwald, W. *Grundriss der Allgemeinen Chemie*; Leipzig, 1899.

25. Parkinson, G. M.; Thomas, J. M.; Williams, J. O.; Goringe, M. J.; Hobbs, L. W. *J. Chem. Soc., Perkin 2* **1976**, 836.

26. Fujimoto, D.; Tamura, R.; Lepp, Z.; Takahashi, H.; Ushio, T. *Cryst. Growth Des.* **2003**, *3*, 973.

27. Harris, K. D. M.; Tremayne, M.; Kariuki, B. M. *Angew. Chem. Int. Ed. Engl.* **2001**, *40*, 1626.

28. (a) Desiraju, G. R.; Steiner, T. *The Weak Hydrogen Bond*; Oxford University Press: Oxford, 1999. (b) Desiraju, G. R. *Acc. Chem. Res.* **2002**, *35*, 565.

OPTICAL RESOLUTION BY MEANS OF CRYSTALLIZATION

H. NOHIRA, D. Sc., *Emeritus Professor of Saitama University*
Innovative Research Organization for New Century,
Saitama University
Shimo-ohkubo, Sakuraku, Saitama 338-8570, Japan

K. SAKAI, Ph. D.
R & D Division
Yamakawa Chemical Industry Co., Ltd.
Isohara, Kitaibaraki, Ibaraki 319-1541, Japan

The most desirable technology for obtaining optically active substances is a method of selective preparation of the desired optical isomer. Asymmetric syntheses or biological reactions are generally employed for this purpose. However, even today, an efficient and practical method for the separation of racemates via crystallization is still a useful key technology for preparing optically active compounds in an industrial-scale production.

In this chapter, we deal with resolutions involving crystallization techniques, i.e. preferential crystallization and diastereomeric crystallization.

1. Resolution by Direct Crystallization of Enantiomeric Mixtures[1-3]

A racemate that crystallizes as a racemic mixture, i.e. a conglomerate, can be resolved into two stereoisomers by a technique called *preferential crystallization* or *entrainment*.

1.1 CHARACTER OF RACEMIC MODIFICATION

Crystalline states of racemates are generally classified into the following three forms: racemic mixture (conglomerate), racemic compound, and racemic solid solution (Figure 1).

F. Toda (ed.), Enantiomer Separation, 165-191.

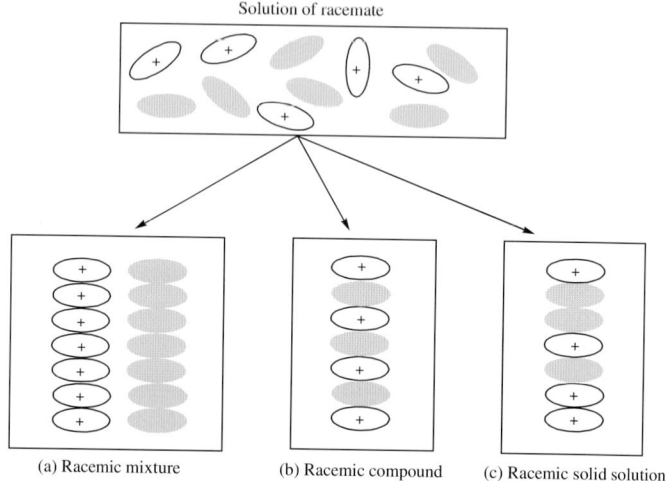

Figure 1. Crystalline states of racemate

(a) *Racemic mixture* (*conglomerate*)

In a racemic mixture, each crystal consists of molecules of the same chirality. Thus, a racemic mixture is an equimolar mixture of crystals of the (+) and (−)-forms. This occurs when each enantiomer has a greater affinity of molecules of the opposite chirality. This phenomenon is called a spontaneous resolution. The melting point of the ground mixture of these crystals is lower than that of the optically active component. In addition, the solubility of the racemic mixture is larger than that of the optically active forms (Figure 2).

(b) *Racemic compound*

In this case, molecules of different chirality are crystallized in a manner of pairing, and crystal structure is different from that of the optically active forms. Therefore, the melting point or solubility of the racemic compound would be lower or higher than those of the corresponding optically active forms (Figure 3). The IR spectrum or the X-ray diffraction pattern in crystal state is different from that of the optically active forms.

(c) *Racemic solid solution*

This state is produced by a random arrangement of the (+) and (−) molecules in the solid. Thus, melting point and solubility are different from the active form, depending on the composition ratio of the enantiomers.

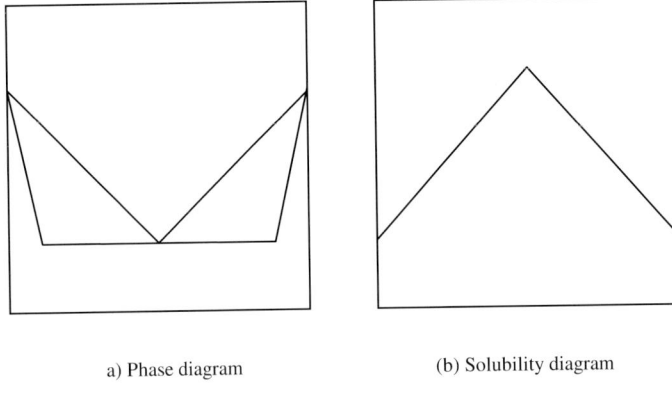

a) Phase diagram   (b) Solubility diagram

Figure 2   Binary phase diagrams and solubility diagram for racemic mixture

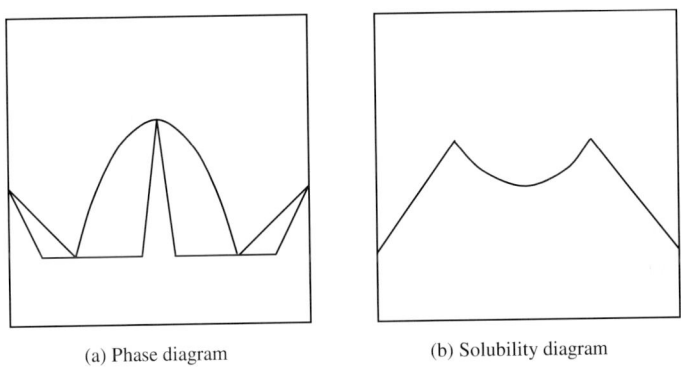

(a) Phase diagram   (b) Solubility diagram

Figure 3   Binary phase diagrams and solubility diagram for racemic compound

1.2 SPONTANEOUS CRYSTALLIZATION AND METHOD OF INOCULATION

In 1848, the French scientist Louis Pasteur prepared the sodium ammonium salt of racemic tartaric acid and allowed it to crystallize in large crystals which are visually distinctive from hemihedral forms.[4]   By discriminating the asymmetric faces of the crystals, he picked out the two kinds of crystals mechanically with a pair of tweezers and a loupe.   Finally he obtained two piles of crystals, one of (+) and one of (−)-sodium ammonium tartrate. This was the first separation of optically active compounds from their racemate.

In 2003, one hundred and fifty-five years after this historic achievement, it was selected as one of the most memorable discoveries in chemistry by a survey of Chemical & Engineering News (C & EN) readers.[5]

[Experimental-1] Example of spontaneous crystallization
1.2 g of (±)-*cis*-2-benzamindecyclohexanecarboxylic acid (**1**) benzylamine salt monohydrate (**2**)[6] was dissolved into 50 g of hot water and allowed to crystallize spontaneously to afford needle crystals, shown below (Figure 4). The mother liquor was carefully removed with a pipet and the separated crystals were washed with a small amount of water and dried in the air at ambient temperature. One of the needles was picked out with a pair of tweezers and accurately weighed. Each needle was dissolved with 2~5 mL of ethanol and optical rotation and optical purity were measured. Experimental results are summarized in Table 1.

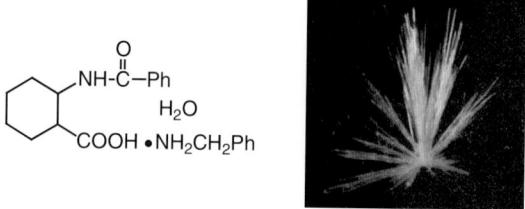

Figure 4   Needle crystals obtained by spontaneous crystallization

Table 1   Optical purity of needle crystals obtained by spontaneous crystallization of **2**

Weight of needle crystal (mg)	Optical rotation at 435 nm (°)	Specific optical rotation $[\alpha]_{435}^{25}$ (°)	Optical purity (% ee)
3.4	+0.016	+50	91
2.2	0.012	55	100
2.4	+0.010	+42	76
4.6	0.024	52	95
3.2	+0.012	+39	71

In general, however, the occurence of spontaneous resolution of racemate is rare. Moreover cases in which the crystals of the enantiomers have visually distinct hemihedral face are extremely rare.

A more useful variation of the mechanical separation applied by Pasteur is the method of inoculation, originally discovered by Gernez in 1866.[7] This is a method in which a supersaturated solution of a racemic mixture is inoculated with pure crystals of one of the enantiomers and let only the crystals of the same kind of enantiomer are allowed to grow selectively. This method is called preferential crystallization or entrainment.

[Experimental-2] Example of resolution by preferential crystallization
To a few 100 mL Erlenmyer Flask was added 0.6-0.8g of (±)-**2** and 30 mL of water, and the mixture was heated up to give a clear solution. The solution was then gradually cooled and inoculated with a needle crystal obtained from *Experimental-1*, and allowed to crystallize. When about 15-30 mg of crystals were deposited, they were filtered off and dried. Yield and optical rotation were measured. Crystals with the same chirality were combined and recrystallized from water (10~20 parts (v/w) versus crystals) to afford enantiomerically pure salt **2**; mp 104-105°C, $[\alpha]_D^{25}$ + and − 24.5°C (*c* 1, EtOH).

1.3 FINDING OF DERIVATIVES APPLICABLE TO PREFERENTIAL CRYSTALLIZATION

It is very fortunate if a target compound crystallizes as a racemic mixture. Even if it does not crystallize as a racemic mixture, it is better that it lead to suitable salt derivatives and the possibility of resolution can be examined.

When the target compound is an acid, various achiral organic or inorganic bases are reacted on, and the generated salt derivatives are examined. If the target compound is an amine, the salt derivatives with achiral acids are checked. The target optically active compound is prepared in advance by other methods, usually by the diastereomeric method described in Section 2. The melting point, solubility, IR spectrum, and X-ray diffraction pattern, etc., are examined for the derivatives that meet the requirements of the racemic mixture. These are: higher melting point, greater solubility and the same IR spectrum.

Table 2 shows the result of examinations for the derivatives of α-methylbenzylamine. As can be seen, the salt of α-methylbenzylamine with cinnamic acid forms the crystals of a racemic mixture (Figure 5), and optical resolution by preferential crystallization can be possible in the form of this salt.[8] Jacques and his co-authors have listed up 248 examples of organic conglomerates.[1]

Figure 5 Hemihedral crystals of the salt of α-methylbenzylamine with cinnamic acid, which were crystallized from water, and spontaneously resolved.
Each crystal weighed in a range of 1~2 mg.

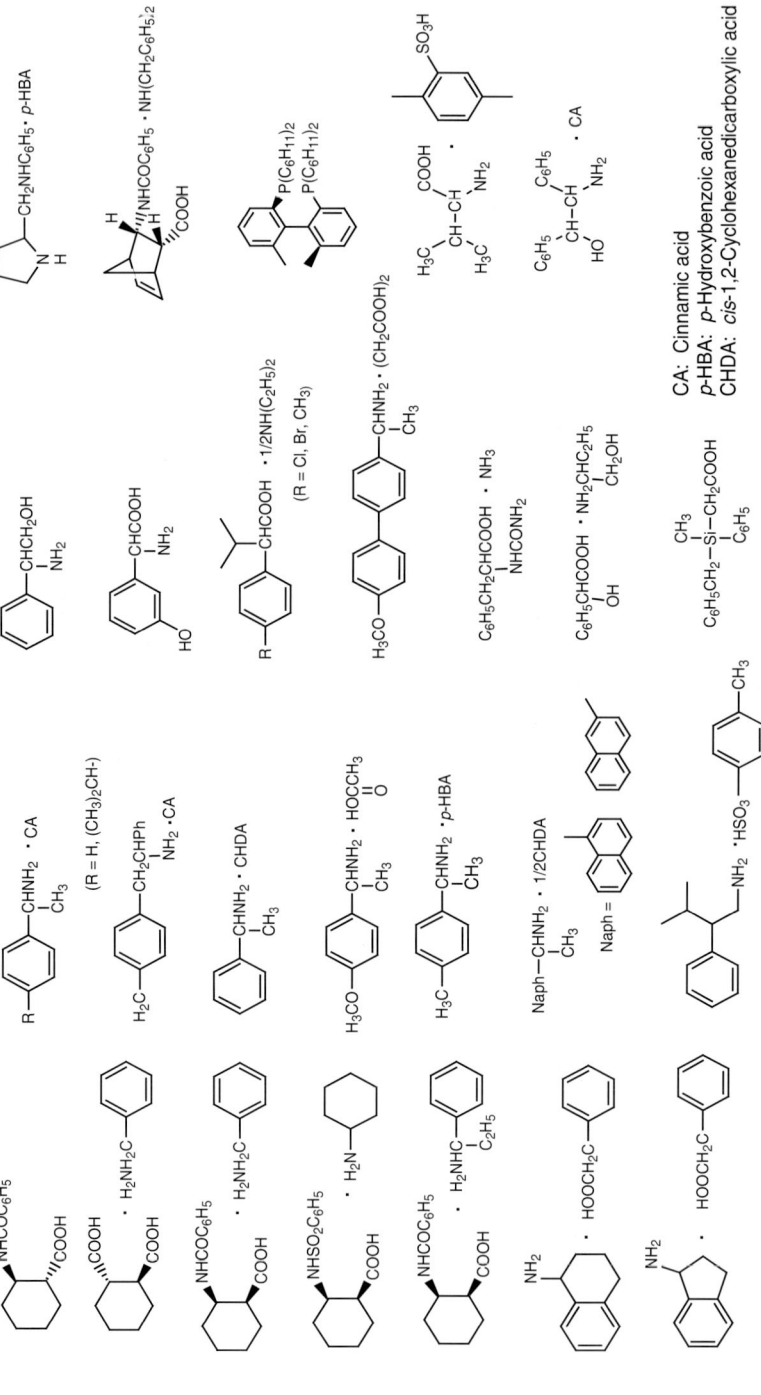

Figure 6　Resolvable compounds by preferential crystallization found in our group[9h]

CA: Cinnamic acid
p-HBA: *p*-Hydroxybenzoic acid
CHDA: *cis*-1,2-Cyclohexanedicarboxylic acid

Table 2   Comarison of the derivatives of α-methylbenzylamine as racemic mixtures

Acids	Melting point (°C)		Evaluation	Solubility	IR spectrum
	Racemic	Optically Active			
Benzoic acid	134	160	○	○	×
p-Toluic acid	160	172	○	△	N/A
p-Chlorobenzoic acid	151	179	○	△	△
p-Hydroxybenzoic acid	209	206	×	×	N/A
Phenylacetic acid	110	135	○	○	×
Cinnamic acid	146	166	○	○	○
Cyclohexanecarboxilic acid	132	133	△	△	N/A
Benzensulfonic acid	159	149	×	△	N/A
p-Toluenesulfonic acid	159	165	○	×	×
cis-1,2-Cyclohexanecarboxylic acid	146	155	○	○	○

Legend: ○=satisfactory, ×=unsatisfactory, △=cannot be assessed, N/A=not applicable

The resolvable compounds by preferential crystallization found in our group are listed in Figure 6.[9]

1.4 TYPICAL PROCESSES OF PREFERENTIAL CRYSTALLIZATION

If a conglomerate derivative is found, a moderate supersaturated solution of the racemic modification is prepared.   Then, appropriate quantity of the (+) or (−) seed crystal of the optically active compound is inoculated into the solution, and it is left standing or stirred gently to crystallize.   If a certain amount of the optically active compound (e.g. 1 g) was crystallized out by the first inoculation, then we can obtain about 2 g each of the optically active compounds after adding 2 g each of the racemic modification and repeating the operation.

This process is illustrated in a solubility phase diagram of three-component system as shown in Fig. 7.[2b]

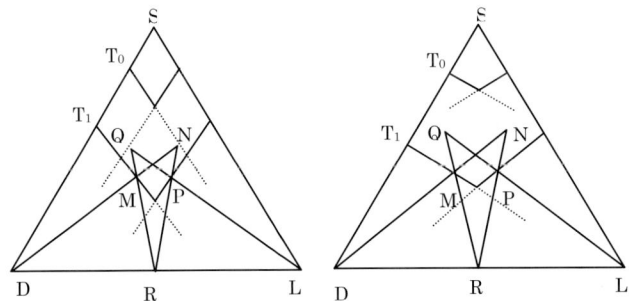

(a) Non-dissociated compounds   (b) Salt having the same achiral ionic component

Fig. 7   Triangular solubility phase diagrams of preferential crystallization

A mixture of composition M is dissolved at a temperature T_1. While cooling it to T_0, the solution is inoculated with pure (+)-seeds. The composition of the solution gradually changes from M to N as the (+)-enantiomer crystallizes. The crystals are filtered off and an equivalent weight of racemate is added to the solution. The mixture of composition P is dissolved by heating. During cooling, the solution is inoculated with pure (–)-seeds and crystallized out until the solution reaches the composition Q. The (–)-crystals are collected, and the corresponding amount of racemate is added to return the solution to the composition M. Thus, we can resolve the racemate at the cycle of $M \rightarrow N \rightarrow P \rightarrow Q \rightarrow M$, and obtain (+)- and (–)-isomers in turn.

If ionic component X^- is the same in the conglomerate salt A^+X^-, and the salt is dissociated into the ions, the solubility ratio of the racemate and active component S_R/S_A becomes equal to $\sqrt{2}$. The solubility phase diagram in this case is shown in Figure 4a, where the operation area becomes wider, and operations become easier.

In general, racemic salts with a common ion are more applicable to preferential crystallization than racemates having non-dissociated covalent bonds.

1.5   STABILIZATION OF SUPERSATURATED SOLUTION BY USING CO-EXISTING SALT

In a preferential procedure, the amount of crystals with high optical purity which is regularly obtainable for each inoculation is usually only about $5\sim15\%$ of all racemic modifications.

Even for such a quantity of crystallization, separating the crystals while at the same preventing the crystallization of the opposite enantiomer is often difficult because the solution is supersaturated with the opposite enantiomer. In this case, the stability of the supersaturated solution can be improved remarkably by dissolving a readily soluble salt derivative of the racemic modification (Fig. 8).

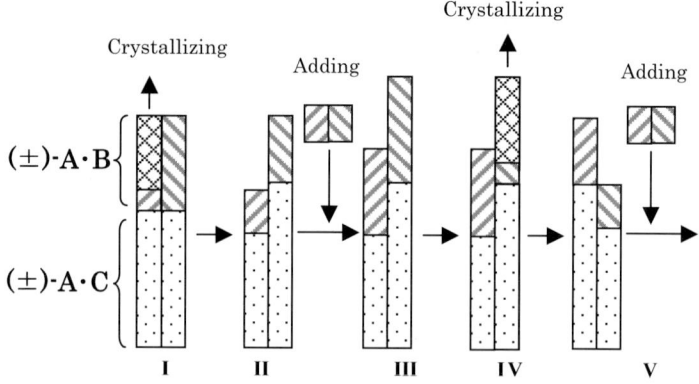

Figure 8   Effect of the co-existing salt

In Figure 8, the salt to be resolved by preferential crystallization is shown as A·B, and the readily soluble co-existing salt is shown as A·C. Starting with a mixture of composition I, the composition of the mother liquor becomes like composition II when the (+)-isomer is crystallized out from the solution. If the racemic salt (±)-A· B is then added, the composition of the mother liquor becomes III. When (−)-isomer is crystallized out from III, it is possible that almost twice as much of the (−)-isomer is crystallized out from the solution shown as IV. Thereafter, we can obtain the optically active salts alternatively by adding the racemic modification salt (±)-A·B and repeating the same operations.

In this process, the readily soluble co-existing salt is in equilibrium with the target salts, and acts as a buffer to the supersaturation of the opposite enantiomer. Thus, we can carry out the preferential crystallization procedure more easily and effectively. For instance, α-methylbenzylamine cinnamic acid salt can be efficiently resolved by adding hydrochloric acid salt of the racemic amine as a co-existing salt.

[*Experimental-3*] *Example of resolution by preferential crystallization in the presence of co-existing salt*[10]
To a 300 mL Erlenmyer Flask were added (±)-*cis*-2-benzamidecyclohexancarboxylic acid (CBA) (12.4g, 0.05 mol), (±)-1-amino-2-propanol (4.2 g, 0.055 mol) and water (60 mL), and dissolved by heat to give a clear solution. If an insoluble was found, it was filtrated to give a clear solution. To the clear solution was added a (±)-CBA· benzylamine (BA) salt (3.5 g, 0.01 mol). The solution was then stirred and heated to give a clear solution. Then the solution was gradually cooled to about 50°C and inoculate a with 5-10 mg of seed crystals of (+) or (−)-CBA·BA. About 15 minutes later, deposited crystals were filtered off. To the mother liquor was added (±)-CBA·BA (2.5 g) and the same procedures were repeated and the solution was inoculated with seed crystals of the opposite sign to the previous crystallization. Thus 1.0 g of salt was obtained at the first crystallization and about 2.5 g of the salt was obtained from the following alternating crystallizations ((+) or (−)-CBA·BA, 85-96% ee). A key point to obtaining stable resolution is important to maintain the volume of the solution. (±)-1-Amino-2-propanol acting as a counterpart of the co-existing salt can be added if necessary. Finally, the crystals with the same sign are combined and recrystallized from water (6-7 parts (v/w) versus salt) to give enantiopure (+) or (−)-CBA·BA.

[*Experimental-4*] *Typical experiment of preferential crystallization*[11]
To a 50 mL flask were added (±)-α-(*m*-hydroxyphenyl)glycine (HPG) (6.2g, 37.1 mmol), (+)-HPG (0.20 g, 1.2 mmol), methanesulfonic acid (2.07g, 21.6 mmol) and water (30 mL) which was heated and dissolved to give a clear solution. The solution was then cooled and inoculated with seed crystals (25 mg). Slurry was cooled over iced water for 2 hours and the deposited crystals were filtered to afford (+)-HPG with 98%ee (0.57g). To a mother liquor was supplementary added (±)-HPG and the same operations repeated to afford (+)-HPG with 98% ee (0.85 g).

In this method, HPG methanesulfonate is in equilibrium state with HPG and acts as a buffer against crystallization pressure from the HPG enantiomer. Therefore, the liquid state and solid state of the enantiomer crystallized by inoculation can be kept in a stable relationship during the resolution process. Thus, enantiomer can be obtained in the same quantity as the added racemate. Examples of resolution using co-existing salt are shown in Table 3.

Table 3   Effect of co-existing salt in preferential crystallization

Resolvable compounds	Co-existing salt	Ref.
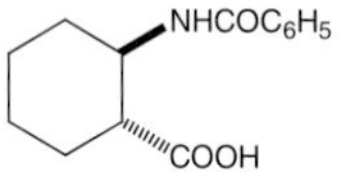 NHCOC$_6$H$_5$ / COOH (cyclohexane structure)	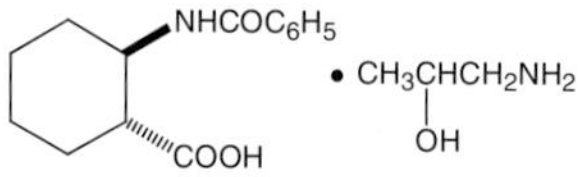 NHCOC$_6$H$_5$ / COOH (cyclohexane structure) • CH$_3$CHCH$_2$NH$_2$ / OH	12
C$_6$H$_5$CHCH$_3$ / NH$_2$ C$_6$H$_5$CH=CHCOOH	C$_6$H$_5$CHCH$_3$ / NH • HCl	8
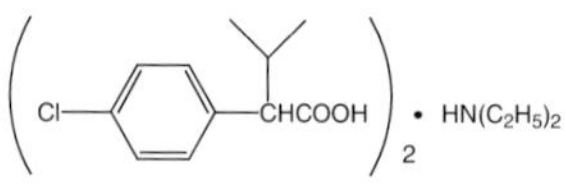 (Cl—⟨benzene⟩—CHCOOH • HN(C$_2$H$_5$)$_2$)$_2$	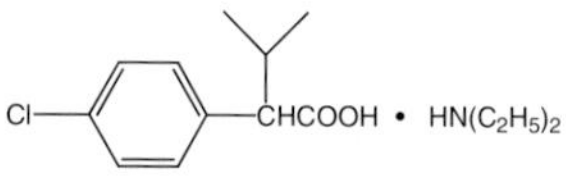 Cl—⟨benzene⟩—CHCOOH • HN(C$_2$H$_5$)$_2$	13
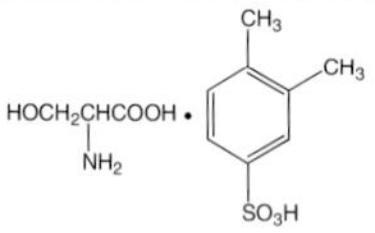 HOCH$_2$CHCOOH • / NH$_2$ (with CH$_3$, CH$_3$ benzene, SO$_3$H)	HOCH$_2$CHCOOH / NH$_2$ • HCl	14

2. Diastereomeric Crystallization Method[15-18]
This is the classical but most widely applied method for resolution at the first trial. It is still widely used in industrial-scale production for obtaining various valuable compounds from their racemates.

2.1 OUTLINE OF DIASTEREOMERIC CRYSTALLIZATION METHOD
Assume that (±)-B is a compound to be resolved.   When the compound is reacted with a certain resolving agent, for example (+)-A, a pair of diastereomeric derivatives are: (+)-B·(+)- A and (−)-B·(+)-A.   Thus, their chemical and physical properties such as melting points and solubility should be different from each others. Therefore, separating these diastereomers is considered to be the same as separating two kinds of compounds.   Most conveniently two diastereomeric salt crystals are separated by fractional crystallization.   Optically pure (+)- and (−)-B can be recovered by decomposing the separated and refined respective diastereomeric salts by means of a suitable process (Figure 9).

Acidic or basic racemates, such as carboxylic acids or amines, are most easily applicable to this method because of the following reason: a racemic carboxylic acid or amine is readily reacted with an optically active amine or carboxylic acid to afford a pair of diastereomeric salts.

Nowadays, a variety of optically active amines and carboxylic acids are available as resolving agents.[17,18]

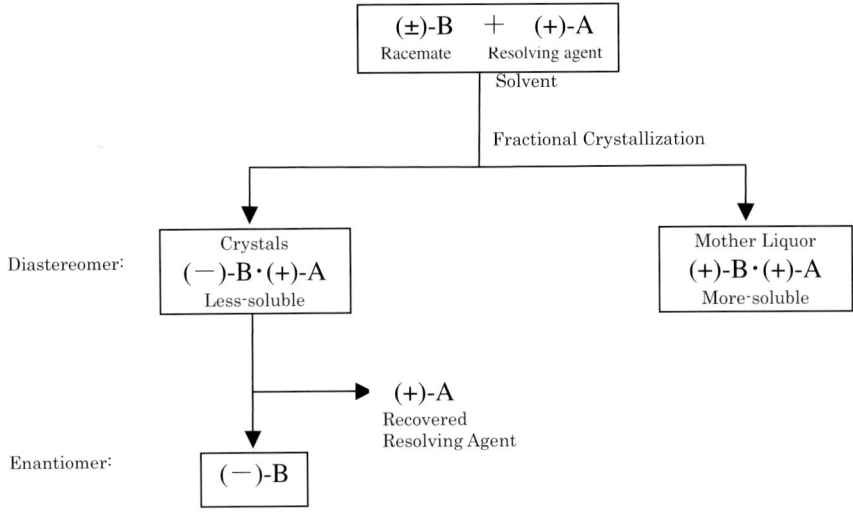

Figure 9   Typical process flow of diastereomeric method

In general, the organic salts generated by this procedure have a moderate solubility and crystallinity in common solvents.   Optically active acids or amines can be easily recovered by the double decomposition of the diastereomeric salts.

2.2 RESOLVING AGENT

The optically active compounds used to generate the diastereomers are called resolving agents.

It is preferable for resolving agent to have higher optically optical purity and to be abundant compounds.   The resolving agents are classified into three groups: basic, acidic, and other kinds.   They can also be classified into the same three groups according to their origin.

Typical resolving agents are summarized in Table 4.   Typical acidic resolving agents from natural products are malic acid, tartaric acid and its derivative, camphor sulfonic acids derived from camphor.   10-Camphor sulfonic acid is a strong acid and indicates high capability of making salts with weak bases.

Many resolving agents derived from natural products have been used since the first discovery of the process, and they are still effective.   Many examples are described in the literature.[17,18]   Synthetic resolving agents, such as α-methylbenzylamine, 1-phenyl-2-(p-tolyl)ethylamine, and mandelic acid are very important ones, and widely used for the production of various optically active compounds. (S)-3-Methyl-2-phenylbutylamine, a new versatile synthetic resolving agent we have proposed, is effective for the resolution of chrysanthemic acid, ibuprofen, ketoprofen, naproxen and others.[19]

Table 4   Typical resolving agents

	Name of compound	Molecular formula	Molecular weight	mp and bp [°C]	Physical properties [α]$_D$ (°)	Applicable compounds
	BASIC RESOLVING AGENTS					
Natural based	Quinine	C$_{20}$H$_{24}$N$_2$O$_2$	324.4	mp 177	−169 (c 2, EtOH)	Chrysanthemic acid
	Cinchonidine	C$_{19}$H$_{22}$N$_2$O	294.4	mp 210	−109.2 (EtOH)	α-Methylbutyric acid
	Brucine	C$_{23}$H$_{26}$N$_2$O$_4$	394.5	mp 178	−127 (CHCl$_3$) HCl salt	α-Hydroxybutyric acid
	Ephedrine	C$_{10}$H$_{15}$NO	165.2	mp 34	−35.5 (c 5, H$_2$O)	Pantoyllactone
	L-(+)-Lysine	C$_6$H$_{14}$N$_2$O$_2$	146.2	mp 224.5 dec.	+25.9 (c 2, 6N HCl)	Chrysanthemic acid
	(+)-Dehydroabietylamine	C$_{20}$H$_{31}$N	285.2	mp 44 - 45	+46 (c 4, MeOH)	Pantoyllactone
Synthetic	(+) and (−)-α-Methylbenzylamine	C$_8$H$_{11}$N	121.2	mp 144 - 146 HCl salt	+31.9 (c 6.4, MeOH) or −40.3 (neat)	Mandelic acid, Malic acid
	(+) and (−)-α-Ethylbenzylamine	C$_9$H$_{13}$N	135.2	bp 98/23 torr	+ or −6.6 (c 2, EtOH)	Ketoprofen
	(+) and (−)-1-(1-Naphthy)ethylamine	C$_{12}$H$_{13}$N	171.2	bp 153/11 torr	+ or −82.8 (neat)	α-Chloropropionic acid
	(+) and (−)-cis-2-(Benzylamino)cyclohexanemethanol	C$_{14}$H$_{21}$NO	219.3	mp 68	+ or −23.5 (c 1, MeOH)	cis-Permethrinic acid
	(+) and (−)-1-Phenyl-2-(p-tolyl)ethylamine	C$_{15}$H$_{17}$N	211.3	bp 148/1 torr	+ or −62.9 (c 1, MeOH)	Naproxen, Chrysanthemic acid
	(+) and (−)-3-Methyl-2-phenylbutylamine	C$_{11}$H$_{17}$N	163.3	bp 115-117/17 torr	+ or −2.32 (neat), or +19.2 (c 1, MeOH)	Ibuprofen, Ketoprofen
	ACIDIC RESOLVING AGENTS					
Natural based	(+)-Tartaric acid	C$_4$H$_6$O$_6$	150.1	mp 168 − 170	+12.0 (c 20, H$_2$O)	α-Amino-ε-caprolactam
	(−)-Malic acid	C$_4$H$_6$O$_5$	134.1	mp 100	−2.3 (c 2, H$_2$O); +28 (c 5, Pylidine)	α-Methylbenzylamine
	(−)-Dibenzoyltartaric acid	C$_{18}$H$_{14}$O$_8$	258.3	mp 138 − 140	−118.5 (c 4.9, EtOH)	Propranolol
	(+)-10-Campharsulfonic acid	C$_{10}$H$_{16}$O$_4$S	232.3	mp 193 − 195 dec.	+43.5 (c 4.3, EtOH)	α-Phenylglycine
Amino acid	(−)-Pyroglutamic acid	C$_5$H$_7$NO$_3$	129.1	mp 162 − 163	−23.6 (c 5, pH7)	α-Amino-ε-caprolactam
	(+)-Aspartic acid	C$_4$H$_7$NO$_4$	133.1	mp 270 − 271	+25.0 (c 2, 6N HCl)	1-phenyl-2-(p-tolyl)ethylamine
Synthetic	(+) and (−)-α-Phenylethanesulfonic acid	C$_8$H$_{10}$O$_3$S	186.2	---	+ or −6.2 (c 1, H$_2$O)	p-Hydroxyphenylglycine
	(+) and (−)-Mandelic acid	C$_8$H$_8$O$_3$	152.1	mp 133	+ or −156.6 (c 3, H$_2$O)	2-Amino-1-butanol
	(+) and (−)-α-Phenylpropionic acid	C$_9$H$_{10}$O$_2$	150.2	bp 143/12 torr	−94.2 (neat)	Noradrenalin
	(+) and (−)-cis-2-Benzylamidecyclohexanecarboxylic acid	C$_{14}$H$_{17}$NO$_3$	247.3	mp 209 − 210	+ or −41.0 (c 1, EtOH)	1-(1-Naphthy)ethylamine
	OTHER TYPES OF RESOLVING AGENTS					
Synthetic	(+) and (−)-Phenylglycine	C$_8$H$_9$NO$_2$	151.2	mp 305 - 310	+ or −157.8 (dil. HCl)	10-Camphorsulfonic acid
	(+) and (−)-α-Methylbenzyl isocyanate	C$_9$H$_9$NO	147.2	bp 76-9/7 torr	+ or −12.9 (c 1, CHCl$_3$)	Alcohols

2.3 FINDING OF GOOD RESOLVING AGENT

When we are resolving a racemate by the diastereomeric method, the most important task is to find a suitable resolving agent and recrystallizing solvent with a great solubility difference between two of the diastereomers.

Even today, however, it is quit difficult to predict what resolving agent is efficient for a target racemate. In fact, we have to search for the most efficient one using as many resolving agents as possible. This trial resolution is performed as follows.

An equimolar amount of a resolving agent is reacted with a several mmol amount of a target racemate. The crystallizing condition is adjusted so as to crystallize out about a half-molar amount of the diastereomeric salt. After the target compound is recovered by decomposing the salt, the optical purity of the obtained enantiomer is measured.

In order to easily compare the resolution results, we defined the resolution efficiency (E) as 1/100 value of the product of the theoretical yield (a half amount of the racemate is assumed to be 100 %) and the optical purity (O.P., % ee) of the resulted enantiomer; or:

$$E\ (\%) = \text{Yield}\ (\%) \times \text{O.P.}\ (\%)\ /\ 100$$

Thus, we can reason that the resolving agent with the greatest possible value of E is regarded as being a good resolving agent.

Table 5   Suitable resolving agents for ($\pm$)-2-hydroxy-4-phenylbutyric acid

Resolving agent	Solvent	Yield [%]	Optical purity [%ee]	Resolution efficiency E
(1)	2-Propanol	74	6.4	4.7
(2)	2-Propanol	66	52.8	34.8
(3)	Methanol	58	26.6	15.4
(4)	2-Propanol	61	13.3	8.1
(5)	2-Propanol	49	23.9	11.7
(6)	2-Propanol	65	20.9	13.6

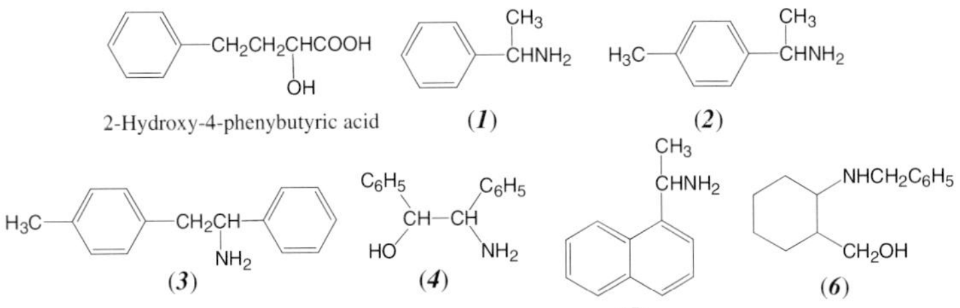

Table 5 shows the result of a trial resolution carried out on racemic 2-hydroxy-4-phenylbutanoic acid, one of the raw materials for the synthesis of

angiotensin converting enzyme (ACE) inhibitors such as Benazepril and Cilazapril.[20]

It can be seen from this table that α-(p-tolyl)ethylamine is judged as the most promising resolving agent for the acid.

Table 6   Suitable solvent for resolution of 2-hydroxy-4-phenylbutyric acid with enantiopure 1-(p-tolyl)ethylamine

Solvent	Yield [%]	Optical purity [% ee]	Resolution efficiency E
2-Propanol	66	52.8	34.8
Dioxane	57	80.6	45.9
Water:Methanol (4:1)	65	30.5	19.8
Acetone	52	71.8	37.3
4-Methy-2-pentanone(MIBK)	53	95.2	50.5

Table 6 shows the efficiency of resolution E obtained by the experiment using various solvents.   From this experiment, it can be seen that dioxane and 4-methyl-2-pentanone (MIBK) are suitable solvents.

In general, there is often no significant influence on relative solubility of two diastereomeric salts even if the crystallization solvent is changed.   However, in special cases, both of the diastereomeric salts are obtained by changing the solvent.

For example, in the optical resolution of diltiazem, a benzothiazepin derivative, with optically active mandelic acid, one of the diastereomeric salts with the same stereochemical sign (+)·(+) or (−)·(−) is crystallized from ethyl acetate whereas another with the opposite stereochemical sign (+)·(−) or (−)·(+) is obtained from a mixed solvent of ethyl acetate and benzene (1:1).[21]   In the resolution of 1-phenyl-2-(p-tolyl)ethylamine with the same chiral acid, the same sign salt crystallizes from 50 % aqueous methanol and the opposite one from 2-propanol.[22]

A racemic alcohol is resolved in the form of a monoester derived from phthalic anhydride of succinic anhydride.   Thus, the diastereomeric salt of the monoester with a resolving amine is purified by crystallization.   Also, optically active monoesters are recovered from the salts, and hydrolysed to give optically active alcohols (Fig. 10).[23]

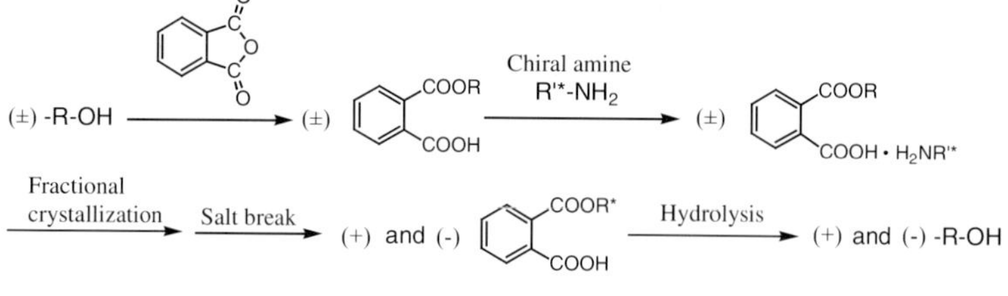

Figure 10   Resolution of alcohols via phthalic acid mono ester

Amphoteric compounds such as amino acids can be resolved as acid or amine forms after deriving corresponding esters or N-acyl compounds.   Racemic alcohols and amines are also resolved by use of optically active isocyanates, where the alcohols and amines are derived the corresponding diastereomeric urethanes or ureas.

Moreover, it is possible to lead these compounds into the diastereomeric esters or amides using of optically active acid anhydrides. In this case, the diastereomers with a covalent bond need not always necessarily be a crystalline compound because there are still some other methods left to separate these diastereomers, such as thin layer chromatography, high-performance liquid chromatography, etc.

Such racemic lactones as substituted -butyrolactones and -valerolactones are resolved by means of the diastereomeric salt formation or the diastereomeric amide formation method.[24,25] For example, -decalactone is successfully resolved in the form of its diastereomeric amides derived from α-methylbenzylamine (Figure 11).[24]

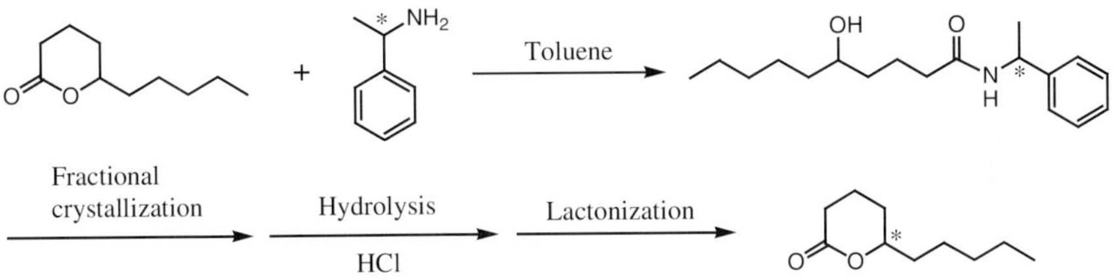

Figure 11   Resolution of -decalactone

2.4 ISOLATION OF PURE ENANTIOMERS FROM PARTIALLY RESOLVED COMPOUNDS BY USE OF CRYSTALLIZING CHARACTERISTICS OF RACEMIC MIXTURES

In the diastereomeric method, an optically active compound with intended high purity might not be easily obtained by the recrystallization of the diastereomeric salt because it forms a solid solution or a double salt.

In certain cases, the enantiomer of high purity is usually obtained by recrystallizing the enantiomer mixture which is recovered by decomposing the diastereomeric salt.   In such a case, it is favorable to know the crystallizing characteristic of the target compound.   For instance, the crystallizing characteristic of an enantiomer mixture can be outlined by its binary phase diagram.

It is known that the melting characteristic of a crystal is correlated with its solubility to solvents.   That is, the solubility of a compound showing a high melting point and nicely crystallized solid is relatively small whereas it is the largest at the eutectic point where the melting point is lowest.

The following is an example of the method that is applied to obtain an optically active compound with high purity.   Racemic *trans*-1,2,-cyclohexanedicarboxylic acid can be resolved by the diastereomeric salt formation method using an optically active α-methylbenzylamine as a resolving agent.   However, it is necessary to repeat the recrystallization of the diastereomeric salt to obtain an enantiomer with high optical purity.   The binary phase diagram of this dicarboxylic acid is shown in Figure 12a.   Since the eutectic point is in 88% ee, it is necessary to obtain the diastereomeric salt with 88% ee or more to obtain the enantiomer with high optical purity by a simple recrystallization.

On the other hand, the binary phase diagram of *trans*-1,2-cyclohexanedicarboxylic acid anhydride which is obtained from the diacarboxylic acid and acetyl chloride is shown in Figure 12b. If the compound with optical purity of 34% ee or more is recrystallized, the acid anhydride of high optical purity can be obtained.

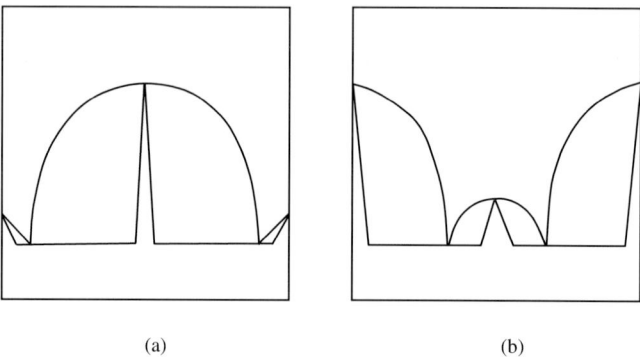

(a)   (b)

Figure 12   Phase diagrams of (a) *trans*-1,2,-cyclohexanedicarboxylic acid and
(b) *trans*-1,2,-cyclohexanedicarboxylic acid anhydride.

In an actual procedure, the carboxylic acid is reacted with an optically active α-methylbenzylamine to crystallize out the less-soluble salt in the quantities of about 50% of the whole diastereomeric salts. The double decompositions of both salts existing in the precipitate and mother liquor give the dicarboxylic acids of (+) and (−) 60% ee, respectively. When these partially resolved carboxylic acids are recrystallized from water, the precipitated crystals are almost racemic, and the carboxylic acids of 88% ee remain in water. They can be converted into the corresponding acid anhydrides by the action of acetyl chloride. Acid anhydrides of almost 100% ee can be obtained by the recrystallization from acetone, after recovering the active acids from the mother liquor. Optically pure (+)- and (−)-*trans*-1,2-cyclohexanedicarboxylic acids can be obtained by the hydrolysis of these anhydrides.

3. Preferential Crystallization Accompanying Asymmetric Transformation

If one enantiomer obtained by resolution is the desired one, the other recovered from the mother liquor of the resolution should be racemized to use for the next resolution. In general, whether the racemization method for unwanted enantiomer is known or not will be key to producing optically active compounds on an industrial scale. In most cases, optical resolution and racemization are non-divisable for the realization of a practical and economical process. For instance, nickel chloride complex of DL-α-amino-ε-caprolactam has been successfully resolved. The complex, dissolved in ethanol, was heated and inoculated with enantiopure complex, then racemization and crystallization of the complex occurred at the same time. As a result, enantiopure caprolactam with 97% ee has been obtained in 92% yield from racemate.[26]

DL-Acylamino acid, such as DL-acetylleucine and DL-benzoylphenylglycine, were resolved in the same manner using a mixture of acetylanhydride and acetic acid as a solvent while the unwanted enantiomer was transformed.[27]

4. Alternative Preferential Crystallization of Less-soluble Diastereomeric Salts

Generally, four kinds of diastereomeric salts are derived from a combination of racemic acid [(±)-A] and racemic base [(±)-B] as shown below.   Surprisingly, it has been found that one pair of less-soluble salts is spontaneously resolved from the four salts and four stereoisomers are resolved simultaneously by preferential crystallization.

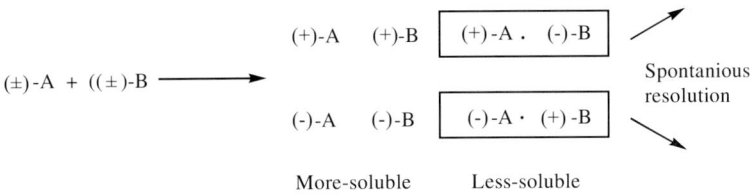

Examples of spontaneously resolved compounds as less-soluble salts
(1) Z-DL-Alanine, Z-DL-Valine, Z-DL-Methionin, Z-DL-Phenylalanin·(±)-Ephedrine[28]
(2) (±)- α -Phenylglycine·(±)-1-camphorsulfonic acid[29]
(3) N-Carbamoyl-DL-valine·(±)-α-aminocaprolactam[30]
(4)   DL-Alanine·DL-mandelic acid[31]
(5) (±)- α-Methylbenzylamine·(±)-malic acid[32]
(6) (±)-Mandelic acid·(±)-2-aminobutanol[33]

A combination of (±)-N-benzyloxycarbonylamino acid (Z-amino acid) and (±)-ephedrine is a good example of one method of affording four diastereomers efficiently if seed crystals of less-soluble salt are obtained in advance (Figure 13).

As shown above, a combination of (±)-N-benzyloxycarbonylamino acid (Z-amino acid) and (±)-ephedrine is a successful model resulting in four diastereomer in high yield and optical purity from the racemic acid and base.   However, in general, the yield of crystals obtained per crystallization is usually small.

If the racemic base (±)-B can be resolved with enantiopure acid (+)- or (−)-A, or if the racemic (±)-A can be resolved with enantiopure base (+)- or (−)-B, reciprocal resolution can be applied as shown below (Figure 14).

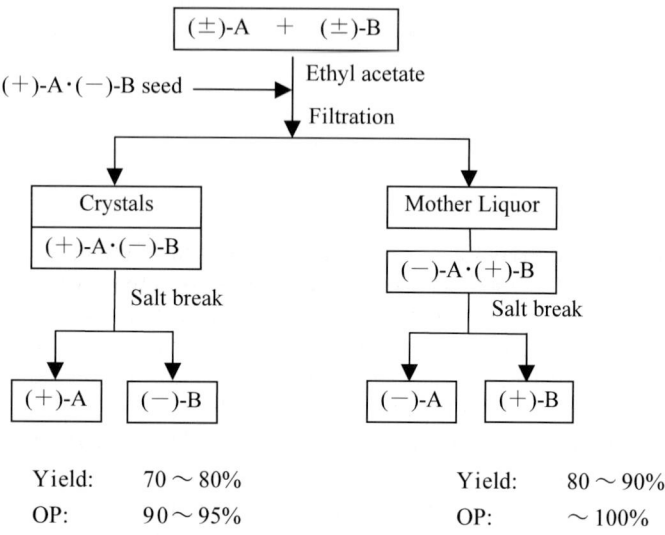

Figure 13   Optical resolution of *N*-benzyloxycarbonylamino acid (A) and ephedrine (B)
             by preferential crystallization

Theoretically, the procedure shown in Figure 14 can be effective for all of the combinations of preferentially resolvable less-soluble diastereomeric salts.   However, practical "activity" of optically active material gradually decreases since resolution efficiency is not 100% in every process.   In fact, yields of reciprocal resolution of (±)-phenylglycine (PG) and (±)-10-camphorsulfonic acid (CSA) gradually decreases as shown in Table 7.

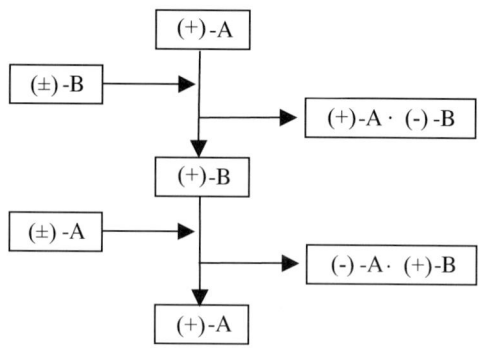

Figure 14   Reciprocal resolution of (±)-A and (±)-B

Table 7   Reciprocal resolution of (±)-phenylglycine (PG) and (±)-10-camphorsulfonic acid (CSA)

Cycle	(−)-PG·(+)-CSA		(+)-PG·(−)-CSA	
	Yield (%)	$[\alpha]_{435}$ (°)	Yield (%)	$[\alpha]_{435}$ (°)
1	85	− 78.8	84	+ 77.8
2	48	− 81.4	34	+ 82.8
3	28	− 88.9	21	+ 82.2

Standard optical rotations of (−)-PG·(+)-CSA and (+)-PG·(−)-CSA: $[\alpha]_{435}$ − and + 89.0
(*c* 2.0, 1N HCl)

As can be seen in Table 7, activity preservation of the optically active material is calculated as about 60% and therefore the rate of activity reduction is calculated as about 40%. To overcome this defect, a new method using a mixture of optically active and racemic compounds as a resolving agent has been devised, based on the consideration that the process will work steadily if a decreased amount of optically active component is supplemented systematically.[34]

To a mixture of (a + b) mol of (−)-substrate (A), b mol of (+)-resolving agent (B) and a mol of (±)-B are added 2(a + b) mol of (±)-A, 2a mol of (+)-B and 2b mol of (±)-B, and the mixture is stirred and heated up to obtain clear solution. If 2(a + b) mol of (−)-A·(+)-B salt is crystallized, (a + b) mol of (+)-A, b mol of (−)-B and a mol of (±)-B is retained in a mother liquor. Thus if 2(a + b) mol of (±)-A, 2a mol of (−)-B, and 2b mol of (±)-A are supplemented to the mother liquor, 2(a + b) mol of (+)-A·(−)-B salt should be crystallized and component in mother liquor will become the same as the previous one (Figure 15).

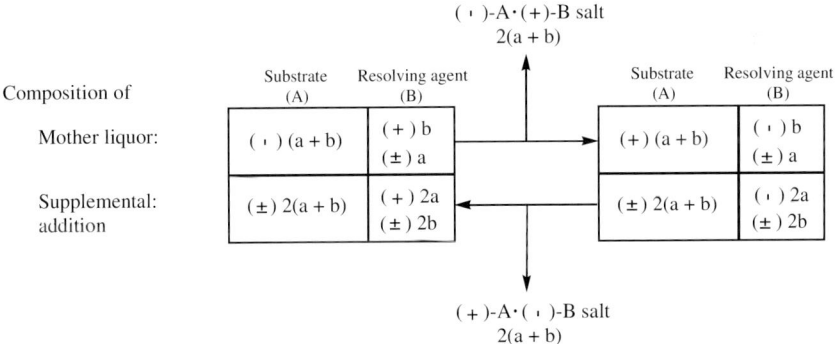

Figure 15   Alternative preferential crystallization of less-soluble diastereomeric salts using a mixture of optically active and racemic material as a resolving agent

By using this method, both substance (A) and resolving agent (B) can be resolved using a resolving agent composed of optically active and racemic form with a : b ratio. Experimental results of the resolution of (±)-phenylglycine (PG) with 10-camphorsulfonic acid (CSA) (a : b = 2 : 1) are shown below (Table 8).

Table 8   Resolution of (±)-phenylglycine (PG) with 10-camphorsulfonic acid (CSA) (a : b = 2 : 1)

184

Cycle	(−)-PG · (+)-CSA		(+)-PG · (−)-CSA	
	Yield (%)	$[\alpha]_{435}$ (°)	Yield (%)	$[\alpha]_{435}$ (°)
1	106	− 74.7	86	+ 82.2
2	84	− 88.5	84	+ 88.8
3	89	− 85.9	97	+ 89.5
4	98	− 89.3	95	+ 88.8

As can be seen from the table, salt with high optical purity can be obtained as expected. This process and method can be applied to various types of (±)-A and (±)-B systems if they are preferentially crystallized.

5. New Industrial-scale Resolution by Diastereomeric Salt Formation Utilizing Molecular Recognition Mechanism

Upon the resolution process development, finding the optimum resolving agent is the first concern for the chemist. Although various attempts have been made by numerous researchers, a useful concrete theory to find a satisfactory resolving agent has not yet been proposed.

The following are our current research results indicating that not only the molecular structures and hydrogen-bonding abilities of resolving agents and racemic substrates but also those of solvent are very important in realizing successful resolution.

5.1 CHIRALITY IMPROVEMENT WITH CRYSTAL MORPHOLOGY CONTROL
35-39

It is widely known that 1-phenylethylamine (PEA) can be resolved with enantiopure mandelic acid (MA) via diastereomeric salt formation in industrial-scale production. However, the optical purity of PEA obtained was not stable (97-99% ee) whereas more than 99% ee has been steadily obtained in a laboratory.

In order to clarify this strange fact, the actual manufacturing conditions have been carefully investigated. As a result, it was found that salt crystal shape was variant (thin long hexagonal and thick hexagonal), causing a decrease in optical purity. The thin long crystal shape caused difficulties in separating mother liquor containing undesired enantiomer (Figure 16). To avoid lowering optical purity by contamination with the mother liquor, optimum resolution conditions for obtaining thick hexagonal crystals have been investigated. As a result, it has been found that secondary amine *bis*(1-phenylethyl)amines (bisPEA), having at least one of the same absolute configurations of optically active PEA composed of less-soluble diastereomeric salt, were quite effective in producing thick crystals. If the less-soluble salt is (R)-PEA · (R)-MA, the effective concentrations of (R,R)- and (R,S)-bisPEA were 0.007 and 0.29 mol%, respectively, in the racemic substrate (RS)-PEA existing in the resolution system (Figure 16). In other words, crystal habit modification with chiral additive is structure- and stereo-specific.

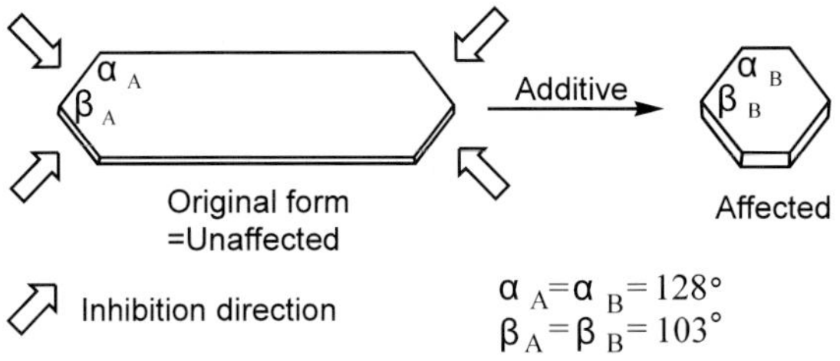

Figure 16   Schematic crystal morphology of the less-soluble diastereomeric salt
in the resolution of (*RS*)-PEA with enantiopure MA.

In actual production, effective additive bisPEA is a by-product of the racemization step, and racemized PEA containing enough bisPEA to produce crystal habit modification is recycled to the next resolution step as part of the raw material (*RS*)-PEA (Figure 17).

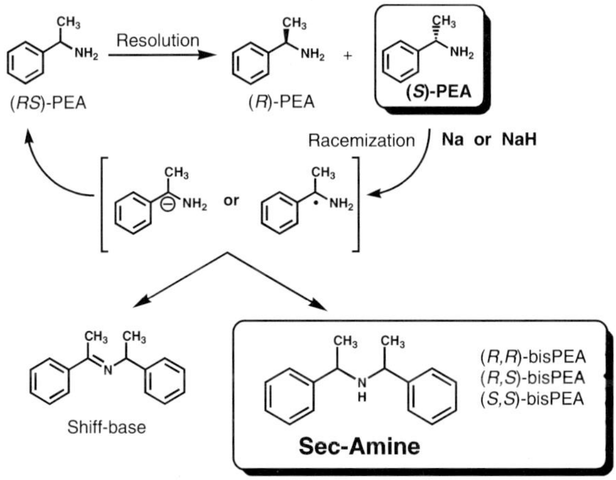

Figure 17   Resolution of PEA with enantiopure MA

The above indicated example suggests that the hydrogen-bonding network among chiral molecules is very important for controlling chiral discrimination. It can be concluded that understanding the crystal structure of the less-soluble salt and hydrogen-bonding network is key to controlling their interaction and realizing successful resolution.

Since this new methodology has been applied in an actual production process, optical purity of enantiopure PEA has been well stabilized. Detailed crystal habit modification mechanisms of the phenomenon[35-38] and its application to other compounds[39] are reported in the literature.

5.2 NEW APPROACH FOR FINDING OPTIMUM CRYSTALLIZATION CONDITION: TAYLOR MADE CRYSTALLIZATION WITH PROTIC ADDITIVE [40-43]

(S)-3-(Dimethylamino)-1-(2-thienyl)propan-1-ol (DMT) was known as a key intermediate for duloxetine production and it can be resolved with (S)-MA. To minimize production steps and avoid contamination by impurities during demethylation, (S)-3-(methylamino)-1-(2-thienyl)propan-1-ol (MMT) was selected as a new intermediate for production (Figure 18). However, resolution of (RS)-MMT had not been found. Thus, at first, the same resolution conditions that apply for (RS)-DMT were tested but no crystals were deposited.

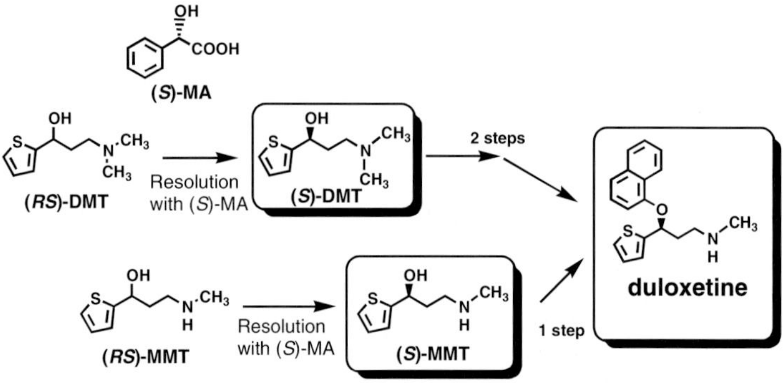

Figure 18   Duloxetine production process

To solve this problem, namely to realize deposition of diastereomeric salt, our working hypothesis[40,41] for obtaining diastereomeric salt crystals called "Space Filler" was applied. That is, as shown in Figure 19, the count numbers from α-carbon concerning the functional groups, such as NH$_2$ or COOH, are attached to the heavy atoms in a distant position of molecule. The distance to that point is called the "molecular length" of the molecule, as shown in Figure 19. If the molecular length of the resolving agent is identical to that of the racemic substrate, resolution is expected to be successful. If the difference in molecular lengths between the resolving agent and racemic substrate is one or more, this difference can be compensated with a protic solvent such as water (molecular length=1) or methanol (molecular length=2) or other protic solvent such as alcohols (molecular length>3). This working hypothesis is based on an idea of the crystallographic closest packing and hydrogen-bonding abilities of the molecules.

In the case of resolution of (RS)-MMT (molecular length=6) with enantiopure MA (molecular length=5), the difference in molecular lengths is one (=1). Therefore water was chosen as a protic additive to make a new hydrogen-bonding network. Thus (S)-MMT was deposited as a less-soluble diastereomeric salt with MA and water molecules (1:1:1 molar ratio).[42,43]

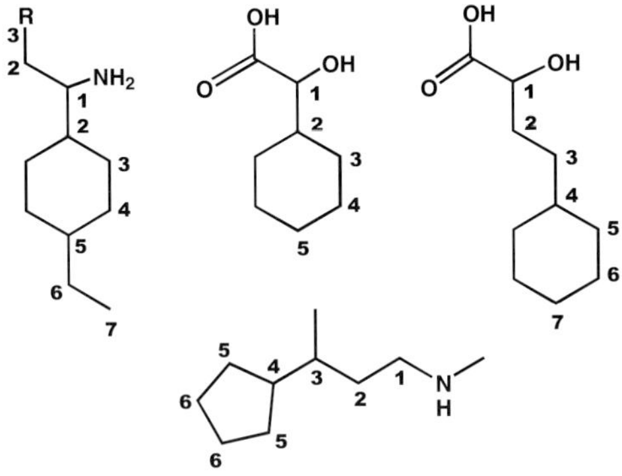

Figure 19   Concept of Space Filler: Determination of molecular length

Success or failure in resolution by the diastereomeric salt formation method is not determined only by the molecular structures of compounds used and physical properties of the salt crystals but also by the resolution environment such as solvent. Therefore, the proposed working hypothesis may not always be effective in all combinations of the resolution system.   However, this idea will be helpful in minimizing tedious trial & error experimental efforts in the laboratory.

5.3 CHIRALITY CONTROL BY THE SOLVENT DIELECTRIC CONSTANT: NEW DECISIVE FACTOR FOR CHIRAL DISCRIMINATION[44-48]

During the development of the resolution process of (RS)-α-amino-ε-caprolactam (ACL), it was found that N-tosyl-(S)-phenylalanine (TPA) was an effective resolving agent and the chirality of the target substrate deposited as the less-soluble diastereomeric salt was variant depending on the dielectric constant of the solvent used as the resolution solvent.[44,45]   Experimental results are shown in Figure 20.[46]   As can be seen from Figure 20, the (S)-salt mainly containing (S)-ACL was deposited as the less-soluble diastereomeric salt from the solvents with a relatively medium range of dielectric constant ($29 < \varepsilon < 58$) such as 45-100% MeOH, 70-90% EtOH, DMSO, DMF etc.   On the other hand, the (R)-salt mainly containing (R)-ACL was deposited as a less-soluble diastereomeric salt from the solvents with outer ε range for (S)-ACL; $\varepsilon < 27$ such as EtOH, 85-100% 2-PrOH, EDC or $\varepsilon > 62$ such as 30% EtOH, 10-35%MeOH or water.   From the X-ray crystallographic analyses of those crystals, it was found that they were (S)-ACL·(S)-TPA·H_2O and (R)-ACL·(S)-TPA, respectively (Figure 21). That is, water molecules has played a very important role in changing the molecular recognition system.

Utilizing this interesting phenomenon, a practical continuous resolution procedure was devised with a solvent switch method.[47,48]   At first, (S)-salt was resolved with (S)-TPA from MeOH, and the mother liquor containing (R)-enriched ACL and equimolar resolving agent (S)-TPA was concentrated to dryness.   To the residue was added 2-PrOH to give enantiopure (*R*)-salt.   In this case, the most desired enantiomer should be obtained from the mother liquor of the first resolution step because one enantiomer-enriched sample usually gives a much better result than that of the racemic sample as a starting material.

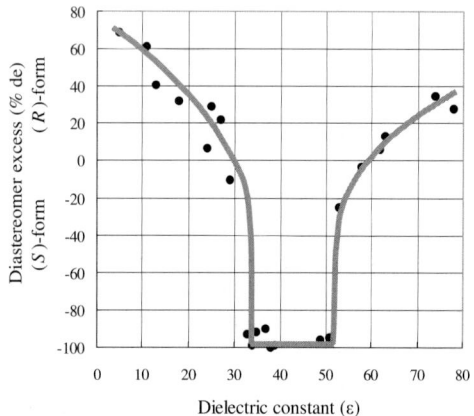

Figure 20   Relation between diastereomer excess (De%) and dielectric constant of solvent.

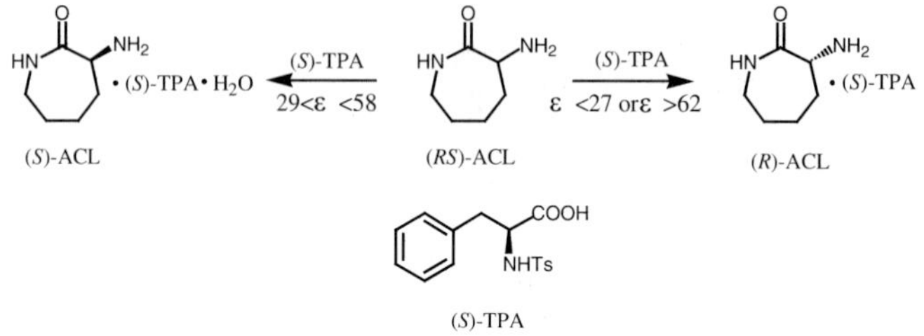

Figure 21   Resolution of ACL with (*S*)-TPA only by changing solvent

5.4   PROSPECT OF INDUSTRIAL RESOLUTION BY DIASTEREOMERIC SALT FORMATION

The three examples shown in this section clearly indicate that attention should be paid to solvent effects in addition to the usual concern for the chiral discrimination, such as molecular structures of resolving agents and racemic substrates and their hydrogen-bonding abilities.   From these research results, we can conclude that solvent, as a reaction environment, is very important for quickly fixing the optimum resolution conditions.   In other words, if we could illuminate mechanisms of solvent inclusion and chiral discrimination among molecules participating in the resolution reaction, we may quickly form reliable strategies and realize a cost effective resolution process.

References

1. Jacques, J. Collet, A. and Wilen, S. H. (1981) *Enantiomers, Racemates, and Resolutions,* Wiley, New York.
2. (a) Wilen, S.H., Collet, A. and Jacques, J. (1977) Strategies in Optical Resolutions, *Tetrahedron* **33**, 2725-2736. (b) Collet, A., Brienne, M. J., and Jacques, J. (1980) Optical Resolution by Direct Crystallization of Enantiomer Mixture, *Chem. Review* **80**, 215-230.
3. Sheldon, R. A. (1993) *Chirotechnology,* Marcel Dekker, New York.
4. (a) Pasteur, L. M. (1848) Memoire sur la relation qui peut exister entre la forme crystalline et la composition chimique, et sur la cause de la polarization rotatoire, *Compt. Rend.* **26**, 535-538, (b) Pasteur, L. (1848) Sur les relation qui penvent exister entre la forme crystalline, la composition chimique et le sens de la polarization rotatoire, *Ann. Chim. et Phys.* [3] **24**, 442-459.
5. Freemantle, M. (Aug. 15, 2003) *Chemcal & Engineering News,* p 27.
6. Nohira, H., Watanabe, K. and Kurokawa, M. (1976) Optical resolution of *N*-benzyl-*cis*-2-aminocyclohexanecarboxylic acid by preferential crystallization, *Chem. Lett,* **1976**, 299-300.
7. Gernez, M. (1866) Separation des tartrates gauches et des tartrates droits, a l'aide des solutions sursaturees, *Compt. Rend.* **63**, 843.
8. Nohira, H., Kai, M., Nohira, M., Nishikawa, J., Hoshiko, T. and Saigo, K. (1981) Optical resolution of α-methylbenzylamine and 1-phenyl-2-(*p*-tolyl)ethylamine by preferential crystallization, *Chem. Lett.,* **1981**, 951-952.
9. (a) Nohira, H. (1989) Optical resolution of organic compounds by means of crystallization, *Nippon Kagaku Kaishi*, 903-914. (b) Nohira, H. (1992) Optical resolution of organic compounds by means of crystallization and their application, *J. Synth. Org. Chem. Jpn.*, **50**, 14-23.
10. Nohira, H. (Oct. 1991) Evolution and multipilication of chirality by means of crystallization,, *Chemistry Today (Gendai Kagaku) No.* **247**, 51-54.
11. Nohira, H. and Yumino, Y. (1982) Optical resolving method of DL-*m*-phenylglycine, Jpn. Kokai Tokkyo Koho JP57-146744; *Chem. Abstr.* **98**: 126619 (1983).
12. (a) Nohira, H., Ehara, K. and Miyashita, A. (1970) Resolution and rotation of *trans*-2-aminocyclohexanecarboxylic acids and derivatives, *Bull. Chem. Soc. Jpn.* **43**, 2230-2233. (b) Nohira, H. and Miura, H. (1975) Optical resolution of (+/-)-menthol, *Nippon Kagaku Kaishi*, 1122-1123.
13. Nohira, H., Terunuma, D., Kobe, S., Asakura, I. and Miyashita, A. (1986) Optical resolution of 2-(4-substituted phenyl)-3-methylbutanoic acids with diethylamine by preferential crystallization, *Agric. Biol. Chem.* **50**, 675-680.
14. Nohira, H. and Kurogami, S. (1983) Optical resolution of DL-serine, Jpn. Kokai Tokkyo Koho JP58-180464; *Chem. Abstr.* **100**: 121616 (1984).
15. Bayley C. R. and Vaidya, N.A. (1992) Resolution of Racemates by Diastereomeric Salt Formation, A.N. Collins, , G.N. Sheldrake, and J. Crosby (eds.): *Chirality in Industry,* pp 69-86, Wiley, New York.
16. Ager, D.J. Ed., (1999) *Handbook of Chiral Chemicals,* Marcel Dekker, New York.
17. Newman, P. *Optical Resolution Procedures for Chemical Compounds,* (1978) *Vol. 1, Amines and Related Compounds,* (1981) *Vol. 2, Acid,* (1984) *Vol. 3, Alcohols, Phenols, Thiols, Aldehydes and Ketones,* Optical Resolution Information Center, New York.
18. Kozma, D. Ed., (2002) CRC handbook of Optical Resolution via Diastereomeric salt Formation, CRC Press, Boca Raton Florida.
19. (a) Nohira, H., Endo, K. and Nishiyama, T. (1986) Optical resolution of (+/-)-trans-chrysanthemumic acid, Jpn. Kokai Tokkyo Koho JP61-172,853; *Chem. Abstr.* **106**: 84047 (1987). (b) Chikusa, Y., Fujimoto, T., Ikunaka, M., Inoue, T., Kamiyama, S., Maruo, K., Matsumoto, J., Matsuyama, K., Moriwaki, M., Nohira, H., Saijo, S., Yamanishi, M., and Yoshida, K. (2002) (*S*)-3-Methyl-2-phenylbutylamine, a versatile agent to resolve chiral, racemic carboxylic acids, *Organic Process Research & Development*, **6**, 291-296.

20.   Sakai, K., Yoshida, S., Hashimoto, Y., Kinbara, K., Saigo, K. and Nohira, H. (1998) Reciprocal resolution of 1-(4-methylphenyl)ethylamine and 2-hydroxy-4-phenylbutyric acid, and habit modification of a less-soluble diastereomeric salt with a chiral additive, *Enantiomer* **3**, 23-35.

21.   Nohira, H. and Nohira, M. (1983) Optical resolution of benzothiazepine derivative, Jpn Kokai Tokkyo Koho JP58-32872; *Chem. Abstr.* **99**: 70782 (1983).

22.   Nohira H., Murata, H., Asakura, I., and Terunuma, D. (1984) Optical resolution of 1-phenyl-2-)p-tolyl)ethylamine, Jpn. Kokai Tokkyo Koho JP59-110656; *Chem. Abstr.* **102**: 5896 (1984).

23.   Ingersoll, A. W. (1944) The Resolution of Alcohols, *Organic Reactions* **2**, 376-414.

24.   Nohira, H., Mizuguchi, K., Murata, T., Yazaki, Y., Kanazawa, M., Aoki, Y. and Nohira, M. (2000) Optical resolution of fragrant lactones, *Heterocycles* **52**, 1359-1370.

25.   Riswoko, A., Aoki, Y., Hirose, T. and Nohira, H. (2002) Optical resolution of 5-alkyl- -valerolactones and synthesis of optically active 5-fluoroalkanols, *Enantiomer* **7**, 33-39.

26.   Sifniades, S., Boyle, W. J. Jr. and Van Peppen, J. F. (1976) Synthesis of L-lysine. Simultaneous resolution / racemization of α-amino-ε-caprolactam, *J. Am. Chem. Soc.*, **98**, 3738-3739.

27.   Yamada, S., Hongo, C. and Chibata, I. (1980) Asymmetric transformation of N-acyl-DL-amino acids, *Chem. Ind. (London)*, 539-540.

28.   Wong, C. H. and Wang K. T. (1978) Mutual resolution of (+/-)ephedrine and Z-DL-amino acid induced by seeding chiral salt, *Tetrahedron Lett.* **40**, 3813-3816.

29.   Watanabe, T., Hayashi, S., Ouchi, S., Senoo, S. (1973) Optical resolution of phenylglycine and camphorsulfonic acid, Jpn. Kokai Tokkyo Koho JP48-78137; *Chem. Abstr.* **80**: 71099 (1974).

30.   Ohonogi, J., Shibata, K., Hongo, C., Shibazaki, M. (1971) Optical division of α-aminocaprolactam and N-carbamoylvalanine, Jpn. Tokkyo Koho JP46-12131; *Chem. Abstr.* **75**: 36689 (1971).

31.   Tashiro, Y., Nagashima, K., Aoki, S., Yasuboshi, Y. (1980) Preparation of optically active alanine-mandelic acid complex, Jpn. Kokai Tokkyo Koho JP55-57545; *Chem. Abstr.* **93**: 132809 (1980).

32.   Nohira, H., Yajima, M., Fujimura, R. (1981) Method of optical resolution of (+/-)-2-amino-1-butanol and/or (+/-)-mandelic acid, Eur. Pat. 36265; *Chem. Abstr.* **96**: 122427 (1982).

33.   Nohira, H., Nohira, M. and Yoshida, S., Okada, A. and Terunuma, D. (1988) Optical resolution of α-ethylbenzylamine and its application as a resolving agent, *Bull. Chem. Soc. Jpn.* **61**, 1395-1396.

34.   Nohira, H., Fujii, H., Yajima, M., and Fujimura, R. (1980) Method of optical resolution of (+/-)-phenyl glycin and/or (+/-)-camphor sulfonic acid, Eur. Pat. 30871; *Chem. Abstr.* **96**: 34875 (1982).

35.   Sakai, K., Murakami, H., Saigo, K., Nohira, H. (1994) Production of optically active α-methylbenzylamine, Jpn. Patent Appl. 1994-1757; *Chem. Abstr.* **120**: 298229 (1994).

36.   Murakami, H., Sakai, K., Tobiyama, T. (2000) Production of optically active α-methylbenzylamine, Jpn. Patent Appl. 2000-297066; *Chem. Abstr.* **133**: 296269 (2000).

37.   Sakai, K., Maekawa, Y., Saigo, K., Sukegawa, M., Murakami, H. and Nohira, H. (1992) Habit modification of a diastereomeric salt with an additive in optical resolution, *Bull. Chem. Soc. Jpn.* **65**, 1747-1750.

38.   Sakai, K. (1999) Application of habit modification of diastereomeric salt crystals obtained from optical resolution via crystallization: manufacture of enantiomerically pure 1-phenylethylamine, *J. Org. Synth. Chem. Jpn*, **57**, 458-465.

39.   Sakai, K., Yoshida, S., Hashimoto, Y., Kinbara, K., Saigo, K. and Nohira, H. (1998) Reciprocal resolution of 1-(4-methylphenyl)ethylamine and 2-hydroxy-4-phenylbutyric acid, and habit modification of a less-soluble diastereomeric salt with a chiral additive, *Enantiomer* **3**, 23-35.

40.   Sakai, K., Saigo, K., Murakami, H., Nohira, H. (1993) Relation between molecular structure and resolvability on optical resolution via diastereomeric salt formation, *Symposium on Chiral Compounds* (Tokyo), Oct 22.

41.   K. Sakai (1994) Molecular recognition on optical resolution process, Ph.D. Dissertation (Saitama University) (1994).

42. Sakai, K., Sakurai, R., Yuzawa, A., Hatahira, K. (2002) Preparation method of optically active 3-(methylamino)-1-(2-thienyl)propan-1-ol and its intermediate for preparation, Jpn. Patent Appl. 2002-289068.
43. Sakai, K., Sakurai, R., Yuzawa, A., Kobayashi, Y., Saigo, K. (2003) Resolution of 3-(methylamino)-1-(2-thienyl)propan-1-ol, a key intermediate for duloxetine, with (S)-mandelic acid, *Tetrahedron Asymmetry,* **14**, 1631-1636.
44. Sakai, K. (2003) Industrial-scale resolution via diastereomeric salt formation, *Symposium Molecular Chirality 2003* (Shizuoka, Japan), IL-8, Oct. 19.
45. Sakurai, R., Sakai, K., Yuzawa, A., and Hirayama, N. (2003) Chirality control by solvent in resolution process, *Symposium Molecular Chirality 2003* (Shizuoka, Japan), PA-11, Oct 19.
46. Sakai, K., Sakurai, R. and Hirayama, N. (2004) Chiral discrimination controlled by the solvent dielectric constant, *Tetrahedron Asymmetry* **15**, *in press.*
47. Sakai, K., Sakurai, R., Yuzawa, A. and Hirayama, N. (2003) Practical continuous resolution of α-amino-ε-caprolactam by diastereomeric salt formation using a single resolving agent with a solvent switch method, *Tetrahedron Asymmetry* **14**, 3713-3718.
48. Sakai, K., Sakurai, R., Yuzawa, A. and Hatahira, K. (2003) Preparation method of optically active α-amino-ε-caprolactam and its intermediate for preparation, Jpn. Patent Appl. 2003-338118.

LIPASE–CATALYZED KINETIC RESOLUTION OF RACEMATES: A VERSATILE METHOD FOR THE SEPARATION OF ENANTIOMERS

ASHRAF GHANEM

Department of Organic Chemistry, University of Geneva,
Quai Ernest Ansermet 30, 1211 Geneva 4, Switzerland.

Corresponding author e-mail: ashraf.ghanem@chiorg.unige.ch

Abstract: This review article focuses on some of the recent developments in the rapidly growing field of lipase-catalyzed kinetic resolution of racemates. Special emphasis is given on the practical utility of the transesterification mode in organic solvents.

1. Introduction

1.1. ENZYMES IN ORGANIC SOLVENTS

Enzymatic catalysis in nonaqueous media significantly extents conventional aqueous-based biocatalysis.[1-18] Water is a poor solvent for nearly all applications in industrial chemistry since most organic compounds of commercial interest are very sparingly soluble and are sometimes unstable in aqueous solutions. Furthermore, the removal of water is tedious and expensive due to its high boiling point and high heat of vaporization. In contrast, biocatalysis in organic solvents offer several advantages. Among these advantages are: the use of low boiling point organic solvents facilates the recovery of the product with better overall yield, non-polar substrates are converted at a faster rate due to their increased solubility in the organic solvent,[15] deactivation and/or substrate or product inhibition is minimized, side reactions are largely suppressed, immobilization of enzymes is not required, denaturation of enzymes (loss of the native structure and thus catalytic activity) is minimized and shifting thermodynamic equilibria to favor synthesis over hydrolysis.

1.2. LIPASES IN ORGANIC SYNTHESIS

Lipases (triacylglycerol acyl hydrolases, EC 3.1.1.3), have been well established as a valuable catalyst in organic synthesis.[12] They are usually distinguished from carboxyl esterases (EC 3.1.1.1) by their substrate spectra, i.e. esterases prefer water soluble substrates and lipases show significantly higher activity towards their natural substrates,

F. Toda (ed.), Enantiomer Separation, 193-230.

triglycerides. Since the hydrolytic reaction is reversible in nonaqueous systems, these biocatalysts can also catalyze the formation of esters from acyl donors and alcohols.

Lipases do not require cofactors. A range of enzymes are commercially available in free and immobilized form. Most lipases accept a broad range of non-natural substrates and are thus very versatile for applications in organic synthesis. In many cases they exhibit good to excellent stereoselectivity.

Lipases have been widely used in three main types of reactions yielding optically pure compounds. These are kinetic resolution of racemic carboxylic acids or alcohols, enantioselective group differentiations of *meso* dicarboxylic acids or *meso* diols and enantiotopic group differentiation of prochiral dicarboxylic acid and diol derivatives.[3]

In kinetic resolutions of racemic alcohols via transesterification, the acyl donors of choice are enol esters such as vinyl acetate or isopropenyl acetate. The vinyl alcohol formed as a byproduct when using vinyl acetate undergoes keto-enol tautomerization yielding the corresponding carbonyl compound (acetaldehyde), while the isopropenyl alcohol released when using isopropenyl acetate tautomerizes to acetone making the reaction practically irreversible in both cases. Thus, these transesterification are much faster compared to reactions using free carboxylic acids or simple esters such as ethyl acetate.

In contrast to asymmetric synthesis, a kinetic resolution yields at maximum 50 % of the desired enantiomer.[19] To achieve higher yields, the non-wanted enantiomer can be separated and re-racemized in a second step. Alternatively, this can be achieved by a dynamic kinetic resolution (DKR).[20] Several methods have been described,[21] and are reviewed below.

2.   Methods to obtain optically pure compounds

The methods used to access enantiomeric compounds can be divided into three categories depending on the type of starting material used.[22]

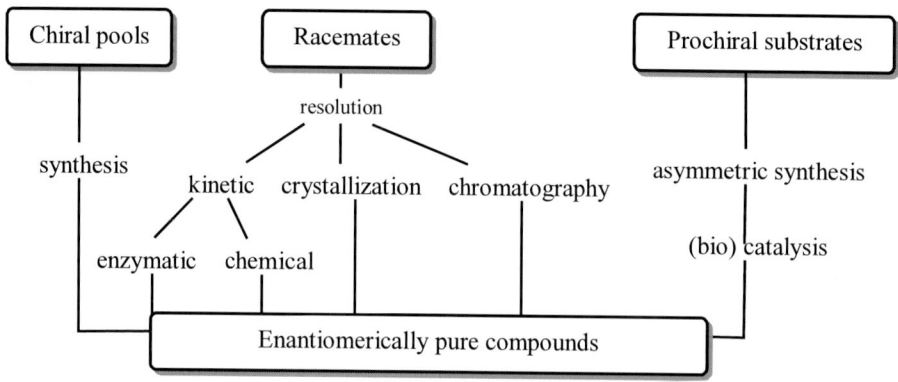

Figure 1: Methods to obtain enantiomerically pure compounds.
Two ways are used for the preparation of enantiomers using enzymes[23,24] (a) Stereoselective synthesis
(b) Resolution of the racemate.

2.1. STEREOSELECTIVE SYNTHESIS

(a) Stereoselective synthesis   (b) Resolution of the racemate

Figure 2: Stereoselective synthesis versus resolution of a racemate

In asymmetric synthesis, a chiral compound is synthesized from an achiral precursor in such a way that the formation of one enantiomer predominates over the other.[23] The asymmetry of the reaction is induced by the presence of a diastereomeric complex and is a result of the formation of two distinct diastereomeric transition states separated in energy by the amount $\Delta\Delta G^{++} > 0$. The ratio of the rate constants for the formation of the two enantiomers, k_R and k_S, is related to $\Delta\Delta G^{++}$ according to equation (2.1.1).[25] Assuming a kinetically controlled reaction, the k_R/k_S ratio will be reflected in the relative amount of each enantiomer formed.

$$\Delta\Delta G^{++} = \left| RT \ln (k_R/k_S) \right|$$ 2.1.1

The enantiomeric complex may be required either in a stoichiometric amount with respect to the reactant, or in a catalytic amount. However, complicated preparation procedures or expensive reagents can sometimes limit the availability of certain enantiomeric complexes.

2.1.1. Desymmetrisation of prochiral substrates

The desymmetrization of prochiral substrates including meso and P-stereogenic substrates has been become a powerful method in the asymmetric synthesis.[26] The advantage of desymmetrization over conventional kinetic resolution is the potential ability to achieve high enantiomeric excess even at high conversion with a theoretical yiled of 100%.[22] Among the widely used substrates in this approach, prochiral ketones and alcohols received much attention. For ketones, the known Baeyer Villiger oxidation or deprotonation using chiral lithium amide bases serves to differentiate the two-prochiral groups attached to the carbonyl of the ketone. A part from the chiral amide approach, Baeyer Villiger mono-oxygenase enzymes were used successfully in the desymmetrization of prochiral and meso cyclohexanones.[27] In a complementary method, lipases have been used in the desymmetrization of enol esters derived from two synthetically important class of cyclic and bicyclic prochiral ketones[26] and in the desymmetrization of prochiral alcohols or acetate.[28]

2.2. RESOLUTION OF RACEMATES

Despite the impressive new progress in asymmetric synthesis, the dominant production method to obtain single enantiomer in industrial synthesis consists of the resolution of racemates.[29] The resolution of enantiomers can be divided into four categories consisting of direct preferential crystallization, crystallization of diastereomeric salts, chromatography and kinetic resolution.

2.2.1. Preferential crystallization

(Also referred to as resolution by entrainment) is widely used on the industrial scale, e.g. in the manufacture of chloramphenicol[30] and α-methy-L-dopa.[31]
Haarmann & Reimer, the market leader in synthetic (–)-menthol, utilizes the preferential crystallization of menthyl benzoate enantiomers. This can be induced by seeding the bulk with one of the pure enantiomers and is used in the production of (–)-menthol.[32]
This process is technically feasible only with racemates that form conglomerates (ones that consist of mechanical mixtures of crystals of the two enantiomers in equal amounts). Unfortunately, less than 20% of all racemates are conglomerates, the rest comprising true racemic compounds that cannot be separated by preferential crystallization. The success of the preferential crystallization is depending on the fact that for a conglomerate, the racemic mixture is more soluble than either of the enantiomers.[22]

2.2.2. Diastereomer crystallization

In case the racemate is a true racemic mixture, this cannot be separated by preferential crystallization, but can be resolved using the diastereomer crystallization developed by Pasteur in 1848. A solution of the racemic mixture in water or methanol is allowed to react with a pure enantiomer (resolving agent), thereby forming a mixture of diastereomers that can be separated by crystallization.

2.2.3. Kinetic resolution

The third method used in the resolution of racemates is the kinetic resolution. The success of this method is depending on the fact that the two enantiomers react at different rates with a chiral entity. The chiral entity should be present in catalytic amounts; it may be a biocatalyst (enzyme or a microorganism) or a chemocatalyst (chiral acid or base or even a chiral metal complex). Kinetic resolution of racemic compounds is by far the most common transformation catalyzed by lipases, in which, the enzyme discriminate between the two enantiomers of racemic mixture, so that one enantiomer is readily transferred to the product faster than the other.[1-18] (cf. fig 3)

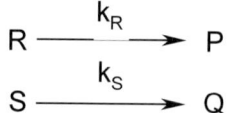

Figure 3: Catalytic kinetic resolution

The kinetic resolution occurs when $k_R \neq k_S$ and the reaction is stopped somewhere between 0 and 100% conversion. Ideally one enantiomer reacts much faster than the other, for example, if the reactant (R) is the only reacting enantiomer ($k_s = 0$). In this case, and at 50% conversion of the initial 50/50 mixture leads to a final mixture of 50%

reactant (S) and 50% product (P). This route has the advantage of easy separation of both enantiomers by using a single enzyme.

2.2.4. Dynamic kinetic resolution

Such conventional kinetic resolution reported above often provide an effective route to access to the enantiomerically pure/enriched compounds. However, the limitation of such process is that the resolution of two enantiomers will provide a maximum 50% yield of the enantiomerically pure materials. Such limitation can be overcome in several ways. Among these ways are: the use of meso compounds or prochiral substrates,[33] inversion of the stereochemistry (stereoinversion) of the unwanted enantiomer (the remaining unreacted substrate),[34] racemization and recycling of the unwanted enantiomer and dynamic kinetic resolution (DKR).[21]

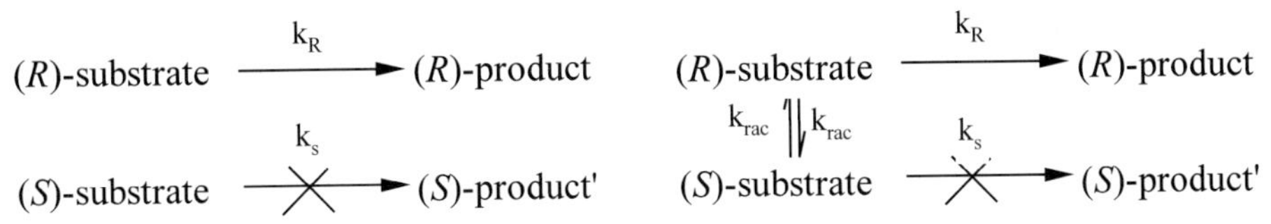

(R)-substrate $\xrightarrow{k_R}$ (R)-product   (R)-substrate $\xrightarrow{k_R}$ (R)-product

                                                         $k_{rac} \upharpoonleft\downharpoonright k_{rac}$

(S)-substrate $\xrightarrow{k_s}\!\!\!\times$ (S)-product'   (S)-substrate $\xrightarrow{k_s}\!\!\!\times$ (S)-product'

Figure 4: a) Convential kinetic resolution (max. 50% conv.) b) Dynamic kinetic resolution with theoretical 100% yield.

Conventional and dynamic kinetic resolution, the enantiomer (R)-substrate is transformed to (R)-product faster than the enantiomer (S)-substrate ($k_R > k_S$).

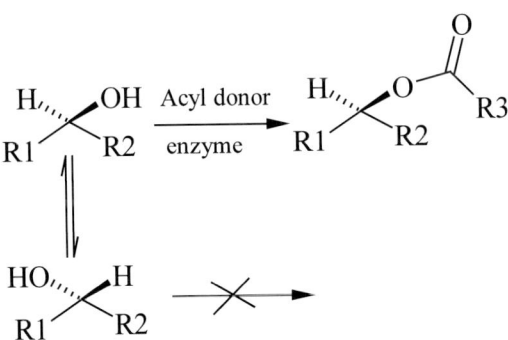

Figure 5: Dynamic kinetic resolution of secondary alcohols

The only difference is that in conventional kinetic resolution the enantiomer (S)-substrate is left behind as unreacted starting material while in case of dynamic kinetic resolution the substrate is continuously isomerised during the resolution process, thus (R) and (S)-substrates are in equilibrium, which allows for the possibility of converting all starting materials of (R)-substrate into (R)-product. Several conditions should be applied and are reviewed in literature.[21] For instance, Bäckvall et al[20] used a combination of enzyme and transition metal complex (Ru-catalyst) to perform the DKR of a set of secondary alcohols. Depending on the substrate, the chemical yield was ranging from 60 to 88 % with more than 99% ee.[20,35]

3. Enantioselectivity

Two important concepts should be understood in the enzyme-catalyzed reactions; the enantiomeric excess (ee) and the enantiomeric ratio E.

The enantiomeric purity of any compound is expressed in terms of its enantiomeric excess (ee) value defined as:

$$\% \ ee_R = \frac{R - S}{R + S} \times 100 \quad For \ R > S$$

Where R is the concentration of the (R)- enantiomer and S is the concentration of (S)-enantiomer. Thus, for a racemic compound the ee value is zero where as for an enantiomerically pure compound the ee value is 1 (or 100 %ee).

Since lipases are chiral, they possess the ability to distinguish between the two enantiomers of a racemic mixture. The parameter of choice to describe the stereoselectivity or the enantioselectivity of lipase-catalyzed reaction is the enantioselectivity, which is also called the enantiomeric ratio E. The E-value is defined as the ratio of specificity constant for the two enantiomers.

$$E_{RS} = \frac{(k_{cat}/k_M)_R}{(k_{cat}/k_M)_S}$$

Where k_{cat} is the rate constant or the turnover number and k_M is the Michaelis-Menten constant. Sih et al[36,37] developed this equation in terms of the enantiomeric excess of the product (ee_p), the unreacted substrate (ee_s), and the conversion (c). Thus, for reversible enzymatic rection, E value is expressed by the following equation:

$$E = \frac{Ln[1-(1+K)c(1+ee_p)]}{Ln[1-(1+K)c(1-ee_p)]} = \frac{Ln[1-(1+K)(c+ee_s\{1-c\})]}{Ln[1-(1+K)(c-ee_s\{1-c\})]}$$

Where K is the equilibrium constant. When the reaction is irreversible or the reverse reaction is negligible (K=0), this equation is reduced to the following equation:

$$E = \frac{Ln[1-c(1+ee_p)]}{Ln[1-c(1-ee_p)]} = \frac{Ln[(1-c)(1-ee_s)]}{Ln[(1-c)(1+ee_s)]}$$

Where (c) is expressed by the following equation:

$$c = \frac{ee_s}{ee_s + ee_p}$$

E can also be expressed in terms of the ee_s and ee_p only by the following equation:[12]

$$E = \frac{Ln\left[\dfrac{1 - ee_s}{1 + (ee_s/ee_p)}\right]}{Ln\left[\dfrac{1 + ee_s}{1 + (ee_s/ee_p)}\right]}$$

Thus, to calculate E value, one can measure two of the three variables: ee_s, ee_p and the extent of conversion (c). A non-selective reaction has an E-value of 1, while an E-value above 20 is the minimum for an acceptable resolution.[12]

3.1. CHIRAL RECOGNITION BY LIPASES

An enzyme model always describes the mechanism of the enantioselectivity in enzymatic reaction. The simplest models, more accurately referred to as rules, do not attempt to predict the degree of enantioselectivity, but only predict which enantiomer reacts faster. The earliest example of such model is the Prelog's rule,[38] which predicts the enantioselectivity of the reduction of ketones by yeast alcohol dehydrogenases based on the size of the two substituents on the carbonyl group. Other models are based on pockets, which give an indication of the size and shape of the molecules tolerated in the active site. Examples of such models are the model of Jones for pig liver esterase (PLE),[39] Subtilisin[40] and several lipases.[41-43] One example is the empirical rule of Kazlauskas for chiral recognition by lipases.[44] This rule predicts, as exemplified for lipase from *Pseudomonas cepacia*, enantiopreference towards a certain substrate, but can not predict the degree of enantioselectivity. It is translated into an active site model for lipases consisting of two pockets of different size, a large one and a small one (cf fig. 6).

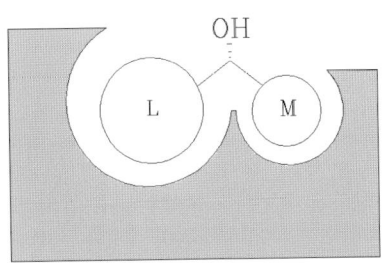 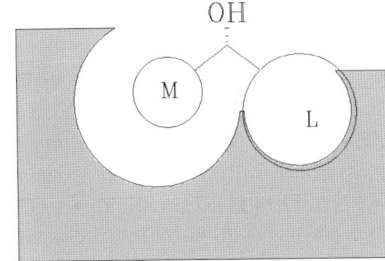

Figure 6: The fast reacting enantiomer (a) and the slow reacting one (b) in the active side model for lipases derived from Kazlauskas' rule.

The stereoselectivity for substrates bearing a small and a large substituent (e.g. a secondary alcohol as shown in fig.6) is explained by assuming that when the secondary alcohol is subjected to resolution by a lipase, the fast reacting enantiomer binds to the active side in the manner shown in fig. 6a, however, when the other enantiomer reacts with the lipase, it is forced to accommodate its large substituent into the smallest pocket (fig. 6b). This rule works well for secondary alcohols. However for primary alcohols, the rule is only applicable if an oxygen atom is attached to the stereocenter. A similar rue was also proposed for the resolution of carboxylic acids.

In addition, a range of lipases structures have been solved by X-ray crystallography or are available by homology modelling. This information together with sequence data in public databases (e.g. www.led.uni-stuttgart.de) allows further insights into the structure-function relationships of lipases. In addition, rational protein design allowed the alteration of the enantioselectivity of lipases. Thus, the directed evolution method combining random mutagenesis and high-throughput screening has been used as a versatil tool for tuning or engineering of the enantioselectivity of lipases. In this particular point of research, Reetz et al reported that the combination of error-prone PCR and DNA shuffling gave a lipase variant of *Pseudomonas aeruginosa* having completely inverted enantioselectivity.[45,46] Recently, Koga et al reported the inversion of the enantioselectivity of another thermostable lipase from *Burkhorderia cepacia* KWI-56 using a novel *in vitro* technique for construction and screening of a protein library by single-molecule DNA amplification by PCR followed by *in vitro* coupled transcription/translation system termed single-molecule-PCR-linked *in vitro* expression (SIMPLEX).[47]

4.   Analytical methods: determination of the enantiomeric excesses (ee)

The development of accurate methods for the determination of enantiomeric purity, which began in the late 1960´s, has been critical for the assessment of enantioselective synthesis. Thus a prerequisite in the enzyme-catalyzed kinetic resolution of racemates is a precise and reliable assessment of the degree of enantioselectivity (E), enantiomeric excess (ee) and conversion (c). Among these methods are: 1) polarimetric methods, 2) gas chromatographic methods, 3) liquid chromatographic methods and 4) NMR spectroscopy. The most convenient and sensitive methods used are chiral GC and HPLC.

4.1. GAS CHROMATOGRAPHIC METHODS

An attractive method for the determination of the enantiomeric excess of substrates and products resulting from the enzyme-catalyzed kinetic resolution of secondary alcohols is chiral gas chromatography (GC).[48,49] This sensitive method is quick, simple to carry out and unaffected by the presence of impurities in the analyzed sample, therefore, isolation and purification of the analyzed sample is not required. Very small sample size is required for the analysis; hence, reactions can be done on small scale.
This method is based on the fact that molecular association may lead to an efficient chiral recognition leading to enantiomeric separation when a chiral stationary phase (e.g. cyclodextrins) is used in GC. The gas (mobile phase, e.g. hydrogen, helium, nitrogen) is carrying the chiral analyte through the stationary phase. The enantiomers to be analyzed

undergo rapid and reversible diastereomeric interactions with the chiral stationary phase and hence may be eluted at different times. One of the limitations associated with this method is that the sample should be sufficiently volatile, thermally stable and resolvable on the chiral stationary phase used. The measurement of the enantiomeric excess using GC is linked with a high degree of precision (+/-0.05%) so that reliable data may be obtained.[50] It is to be noted that high enantiomeric excesses (ee) up to 99% may be detected.[51-58]

4.2. HPLC METHODS

HPLC methods follow the same principles and advantages as GC-analysis. The major difference is that more polar and also non-volatile compounds can be analyzed.

5.   Application of lipases in racemates resolution

Resolution of racemates via lipase-catalyzed kinetic resolution is one of the most attractive methods used to access to enantiomerically pure compounds. Of all methods used in kinetic resolution, transesterification in organic solvents catalyzed by lipase is the most dominant one. Thus, in the presence of a suitable acyl donor, an enzyme as well as the appropriate solvent, and at the optimum temperature, one enantiomer of the racemic mixture is selectively transferred to the corresponding ester leaving the second unreacted enantiomer in enantiomerically pure form. [51-58]
If a good leaving group is present on the acyl donor, as in case of trichloroethyl or trifloroethyl esters (Fig. 7a), the reaction of the halogenated alcohol with the formed ester (the backward reaction) is minimized allowing the shift of the equilibrium to the product formation. Oxime esters have been proposed as acyl tranfer agents (Fig. 7b) for irreversible process; however, this approach is limited due to some disadvantages incorporated to cosubstrate inhibition and reversibility of the reaction. The best method for irreversible transesterification procedure is achieved when using enol esters (Fig. 7c) where the back reaction is suppressed due to the tautomerization of the resulting enol alcohol (to acetaldehyde or acetone depending on wether a vinyl or an isopropenyl ester serve as acyl donors), thereby shifting the equilibrium to the required product. However, acetaldehyde may have some detrimental effects on some enzymes.[8] Isopropenyl acetate was proposed as an innocuous and more suitable acyl donor in lipase-catalyzed irreversible transesterification in organic solvents (Fig. 7d). The use of different reagents in irreversible acylation catalyzed by lipase has been recently reviewed.[59]

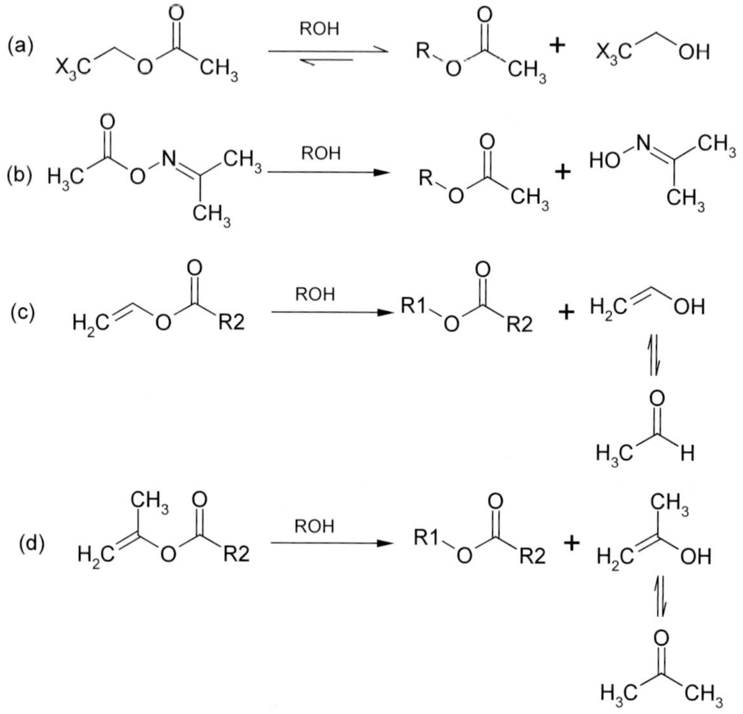

Figure 7: Lipase-catalyzed irreversible transesterification.

5.1. KINETIC RESOLUTION OF PRIMARY ALCOHOLS

Primary alcohols have been successfully used as substrates for lipases. Monterde et. Al[60] reported the resolution of the chiral auxiliary 2-methoxy-2-phenylethanol **1** via *Candida antarctica* lipase B (CAL-B)-catalyzed acylation using either vinyl acetate (R=H) or isopropenyl acetate (R= CH$_3$) as acyl donor (cf. fig. 8) and the alkoxycarbonylation using diallyl carbonate as the alkoxycarbonylation agent in THF at 30 °C (cf. fig. 9).

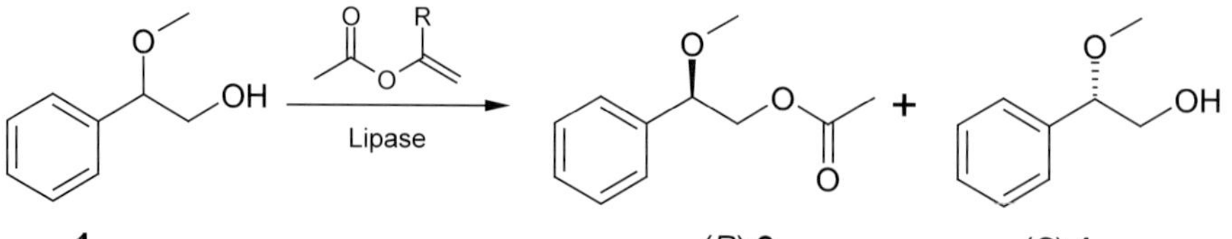

Figure 8: Lipase-catalyzed enantioselective acylation of 2-methoxy-2-phenylethanol (rac-**1**) using either vinyl acetate or isopropenyl acetate as acyl donor.[60]

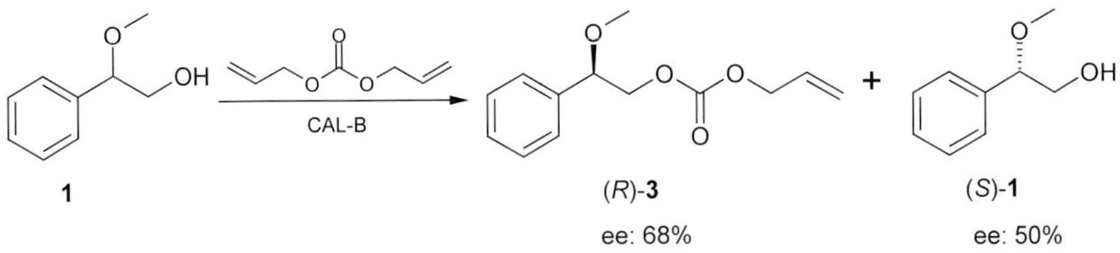

Figure 9: Lipase-catalysed enantioselective alkoxycarbonylation of 2-methoxy-2-phenylethanol (rac-**1**) using diallyl carbonate.[60]

The reverse reaction consisting of the PCL-catalyzed hydrolysis of the esters of three primary alcohols 2-methyl-3-phenyl-1-propanol **4**, 2-phenoxy-1-propanol **5** and solketal **6** was studied by Kazlauskas et al.[61] An enhancement of the enantioselectivity E was observed when changing the solvent from ethyl ether/phosphate buffer to 30% *n*-propanol in phosphate buffer and when changing the substrate from 1-acetate to 1-heptanoate. The same enhancement in E value was observed with **5** but there was no significant increase towards the alcohol **9**.

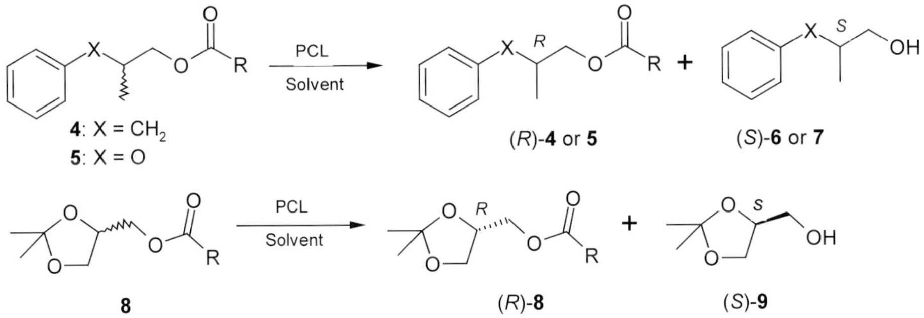

Figure 10: PCL-catalyzed hydrolysis of the esters of three primary alcohols.[61]

The improvement of the enantioselectivity E in kinetic resolution of a primary alcohol (**10**) through lipase-catalyzed transesterification was studied using a chiral acyl donor **11**. The combination of the lipase, solvent and acyl donor was effective for the enantioselectivity.[62]

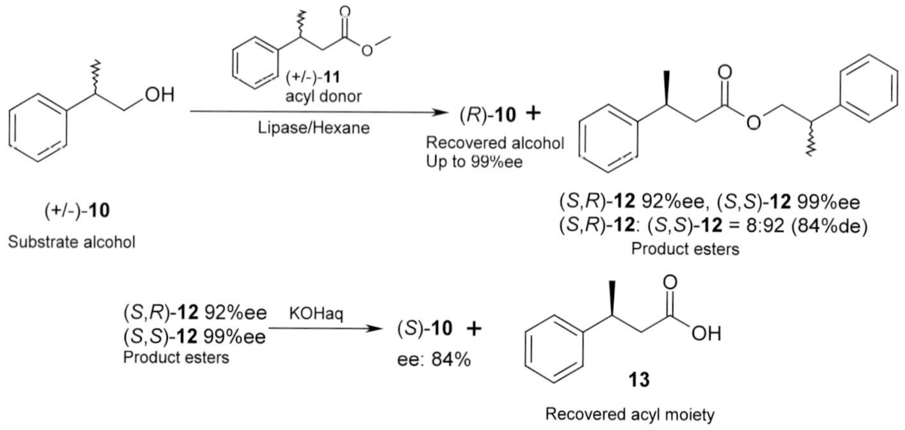

Figure 11: Representative kinetic resolution of primary alcohol (±)-**10** using chiral acyl donor (±)-**11**.[62]

The effect of a series of vinyl esters having bulky substituents such as vinyl 3-(aryl)propanoates bearing a variety of substituents on the phenyl ring on the enantioselectivity of lipase-catalyzed transesterification of 2-phenyl-1-propanol (**10**) was studied by Kawasaki et al.[63] The most effective transesterification of the alcohol **10** was achieved when using vinyl 3-(*p*-iodophenyl)- or 3-(*p*-trifluoromethylphenyl)propanoates with enantiomeric ratio E ranging from 116 to 138 affording (*S*)-**14** with 97% to 98%ee and (*R*)-**10** with 56% to 34% ee, respectively. Vinyl 3-phenylpropanoate showed also efficiency in conducting the lipase-catalyzed resolution of several 2-phenyl alkanols.

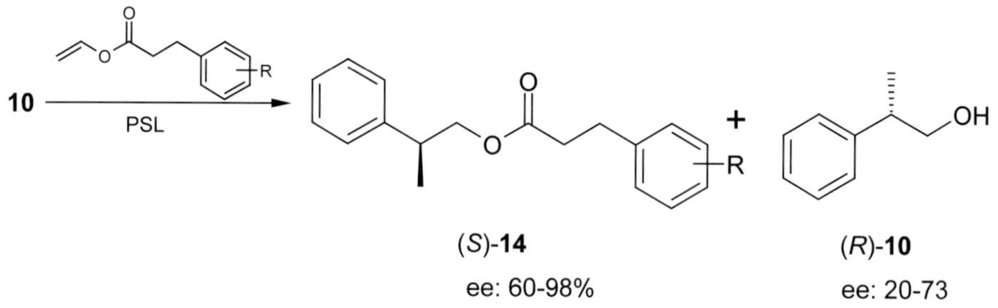

Figure 12: Lipase-catalyzed transesetrification of rac-**10** with vinyl 3-(substituted phenyl)propanoates.[63]

A practical resolution of 3-phenyl-2H-azirine-2-methanol **15** at a very low temperature (-40 °C) was reported by Sakai et al[64] to enhance the enantioselectivity in immobilized lipase-catalyzed resolution of **15** using vinyl butanoate as acyl donor in ether as organic solvent. The method was found to be effective in enhancing the enantioselectivity E and affords the primary alcohol (*S*)-**15** with 99% ee and the ester (*R*)-**16** with 91% ee.

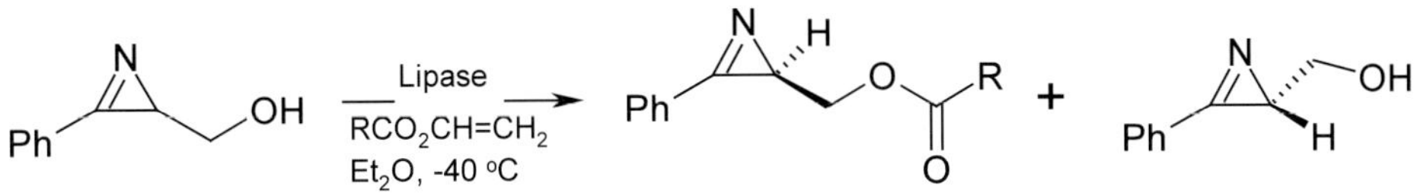

Figure 13: Toyonit-immobilized lipase-catalyzed resolution of **1** at -40 °C.[64]

5.2. KINETIC RESOLUTION OF SECONDARY ALCOHOLS

Secondary alcohols are by far the most frequently used targets in lipase-catalyzed resolutions. This is due to their importance in organic synthesis but also that lipases usually show much higher enantioselectivity in resolutions compared to primary and tertiary alcohols.

Numerous examples can be found in literature and only a few selected examples are included in this survey. Schurig et al reported a series of reports about the utility of isopropenyl acetate as an innocuous acyl donor in the lipase-catalyzed transesterification of secondary alcohols. The non-reacting alcohol enantiomers were obtained in > 99% ee.[51-57]

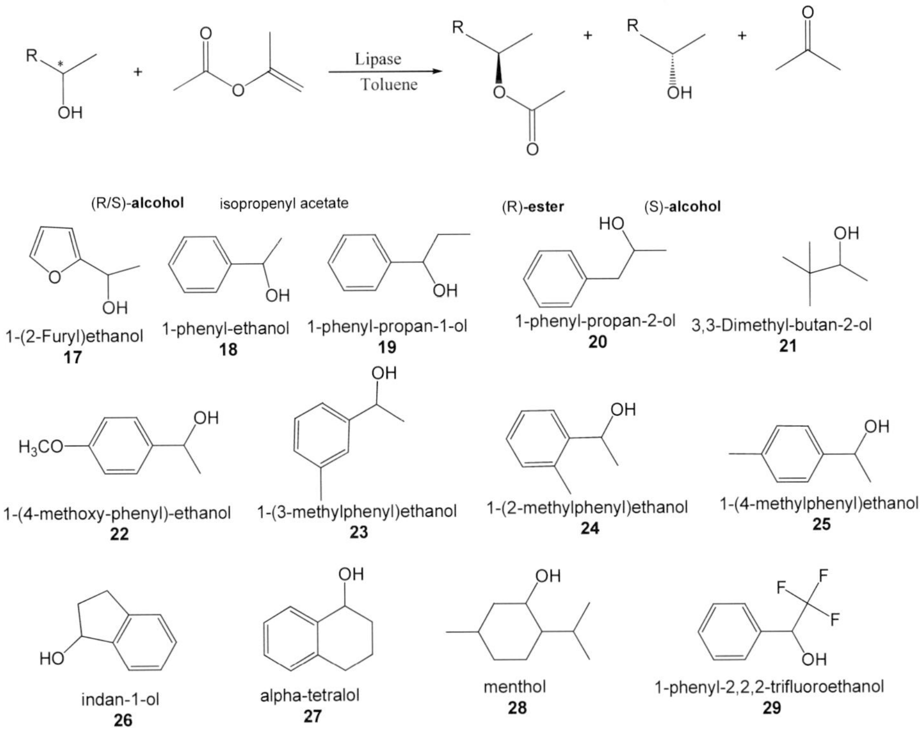

Figure 14: Lipase-catalyzed transesterification of secondary alcohols using isopropenyl acetate as acyl donor in toluene.[51-57]

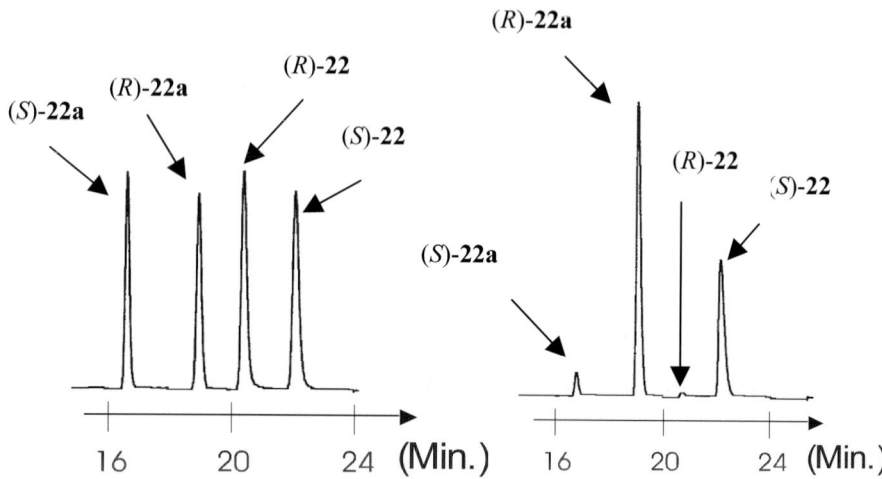

Figure 15: Gas chromatographic chiral separation of : (left) racemic 1-(4-methoxy-phenyl)ethanol **22** and its corresponding acetate **22a** (reference) and (right) lipase-catalyzed transesterification of 1-(4-methoxy-phenyl)ethanol **22** (4 hrs) using isopropenyl acetate as acyl donor in toluene as organic solvent: ee_s= 99.9 ee_p= 87 conv. =53.4, E=141.

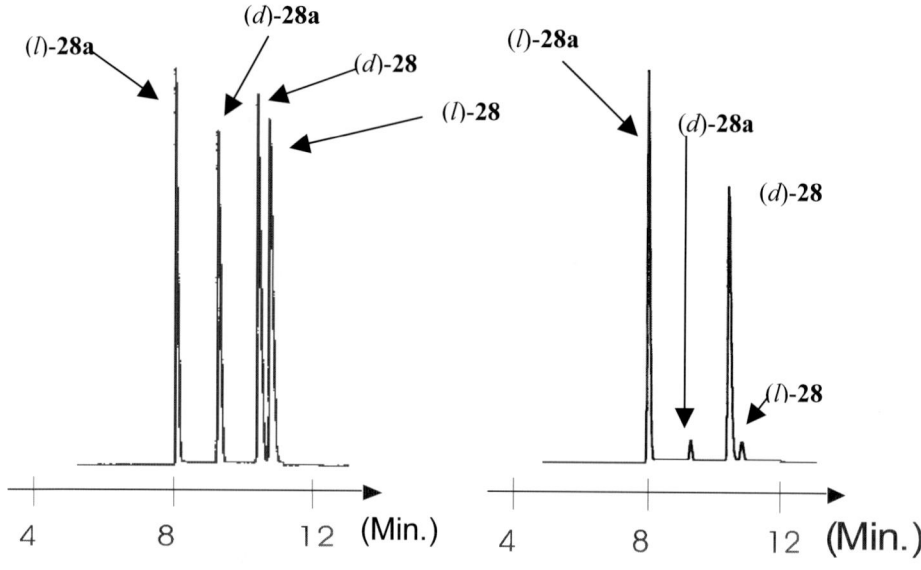

Figure 16: Gas chromatographic chiral separation of: (left) racemic (*dl*)-menthol **28** and its corresponding acetate **28a** (reference) and (right) lipase-catalyzed transesterification of (*dl*)-menthol (15 hrs) using isopropenyl acetate as acyl donor in toluene as organic solvent: ee_s= 85.2 ee_p= 88 conv. =49 E=42.

In the transesterification of (*R,S*)-secondary alcohols, the (*R*)-alcohol was the faster reacting enantiomer yielding the (*R*)-acetate in high *ee* and leaving (*S*)-alcohol as enantiomerically pure unreacted enantiomer. *Trans*-4-phenyl-3-butene-2-ol **29**, another substrate pocessing an allylic strain has been successfully resolved on gram-scale via lipase-catalyzed enantioselective acylation of the alcohol and hydrolysis of its acetate.[57]

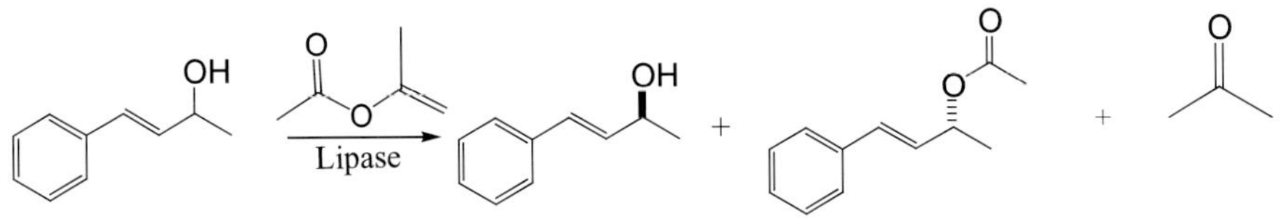

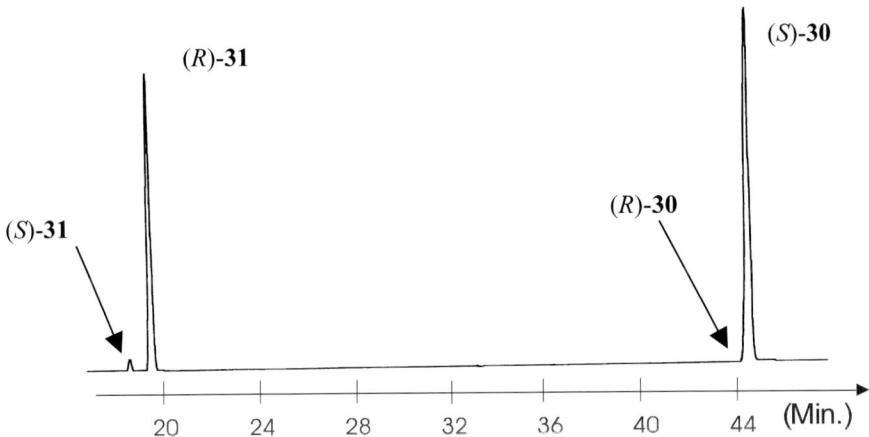

Figure 17: Lipase-catalyzed kinetic resolution of racemic **30** using isopropenylacetate as acyl donor in toluene as organic solvent.[57]

The lipase-catalyzed asymmetric transesterification was performed using isopropenyl acetate in organic media affording the (*S*)-alcohol in high enantiomeric excess (>99% ee).

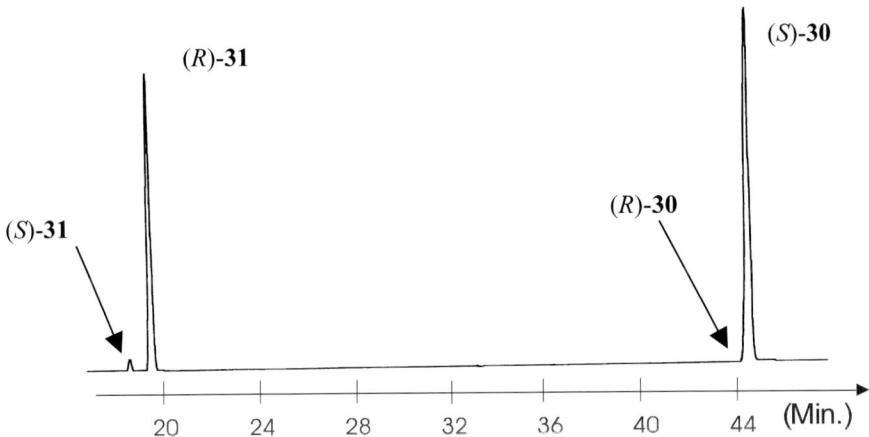

Figure 18: Gas-chromatographic separation of the enantiomer of both substrate (**30**) (as carbamate) and product (**31**) on heptakis-(2,3-di-*O*-methyl-6-*O*-tert-butyldimethylsilyl)-β-cyclodextrin of the *Pseudomonas fluorescens* lipase (PFL) catalyzed transesterification of (**30**) in toluene at t=9 hrs, ee$_s$ =99.9%, ee$_p$=92.2%, conv. =52%, E =284.

The reverse reaction was consisting of lipase-catalyzed hydrolysis of the racemic acetate afforded the (*R*)-alcohol in high enantiomeric excess (>99% ee).

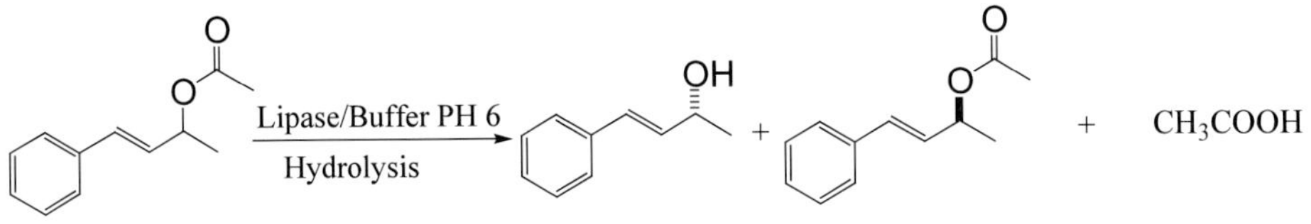

Figure 19: Lipase-catalyzed enantioselective hydrolysis of racemic (**31**) using phosphate buffer (pH =6) and toluene as organic solvent.[57]

In order to reduce the time needed to perform a complete kinetic resolution; Lindner et al[53] reported the use of the allylic alcohol **30** in enantiomerically enriched form rather than a racemic mixture in kinetic resolution. Thus, the kinetic resolution of **30** was performed starting from the enantiomerically enriched alcohol (R) or (S)-**30** (45%) ee obtained by the ruthenium-catalyzed asymmetric reduction of **32** with the aim to reach ~ 100 % ee in a consecutive approach. Several lipases were screened in resolving the enantiomerically enriched **30** either in the enantioselective transesterification of (S)-**30** (45% ee) using isopropenyl acetate as an acyl donor in toluene in non-aqueous medium or in the enantioselective hydrolysis of the corresponding acetate (R)-**31**, (45% ee) using a phosphate buffer (pH = 6) in aqueous medium. An E value of 300 was observed and the reaction was terminated after 3 h yielding (S)-**30** > 99% ee and the ester (R)-**31** was recovered with 86% ee determined by capillary GC after 50 % conversion.

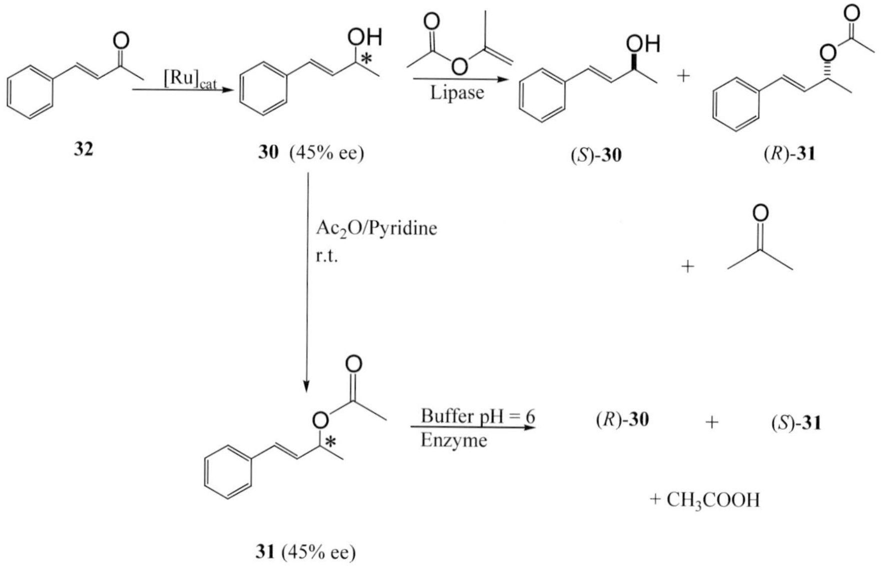

Figure 20: Ruthenium/ lipase-catalyzed separation of enriched (R,S)-**30**.[53]

Benzofuran-based structures are used as important chiral building blocks in the synthesis of biologically active compounds. Several 2-substitued benzofuran drugs are available nowadays such as Amiodarone (cardiac and anti-arrythmic) and Benziodarone (coronary vasodilator), adding to that they can be used in inhibiting HIV-1 reverse transcriptase or acting as antiaging compounds.

The kinetic resolution of racemic 1-(benzofuran-2-yl)ethanol rac-**33** having different substituents on the benzene was reported ring using lipase-catalyzed transesterification with vinyl acetate as acyl donor. The reaction afforded (1R)-1-acetoxy-1-(benzofuran-2-yl)ethanes (R)-**34** and (1S)-1-benzofuran-2-yl)ethanols (S)-**33** in highly enantipure form.[65]

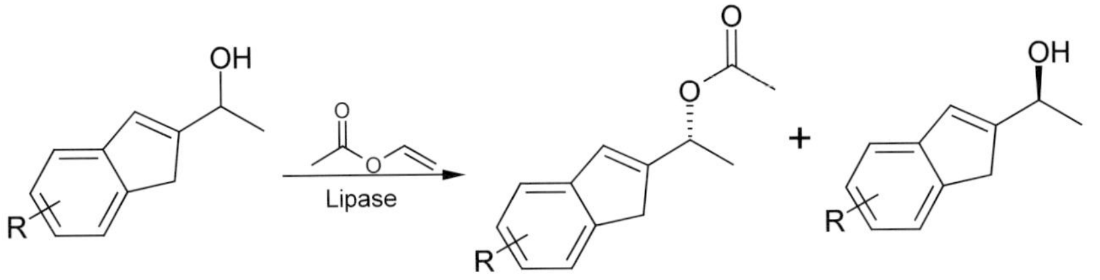

33   (R)-34   (S)-33

Yield: 43-48%   Yield: 48-51%
ee: >99%   ee: 98.6%

Figure 21: Lipase-mediated enantiomer separation of racemic 1-(benzofuran-2-yl)ethanol rac-**33**.[65]

Racemic 1-azido-3-aryloxy-2-propanols **35** was resolved by the lipase-catalyzed kinetic resolution using different acyl donors to access to the enantiomers in optically pure form.[66] The reduction of the azide group can afford the 1-amino-3-aryloxy-2-propanols, which is present in numerous biologically active compounds such as β-adrenolytic drugs (β-blockers) used in the treatment of angina pectoris, hypertension and other cardiac diseases.

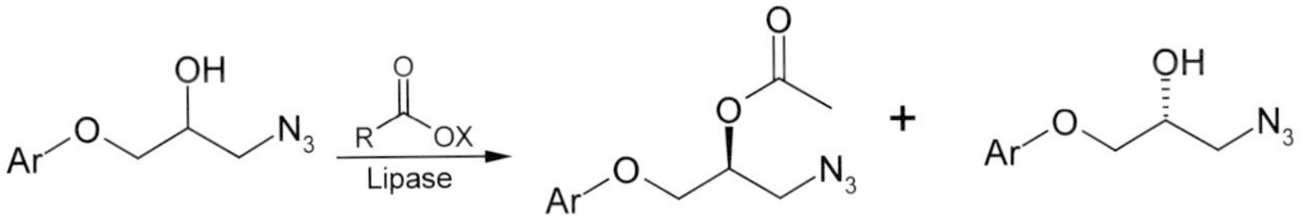

35   (R)-36   (S)-35

ee: 76-88%   ee: 62-96%

Figure 22: Lipase-catalyzed enantioselective acetylation of 1-azido-3-aryloxy-2-propanols **35**.[66]

Wielechowska et al[67] reported the lipase-catalysed transesterification of 1-alkylthio-3-aryloxypropan-2-ols **37** having various aromatic substituents using either vinyl or isopropenyl acetate as acyl donors in various organic solvents. The resulting product (*S*)-**38** was obtained with ee up to 91% while the remaining unreacted substrate (*R*)-**37** was recovered with an ee up to 85% depending on the substituents.

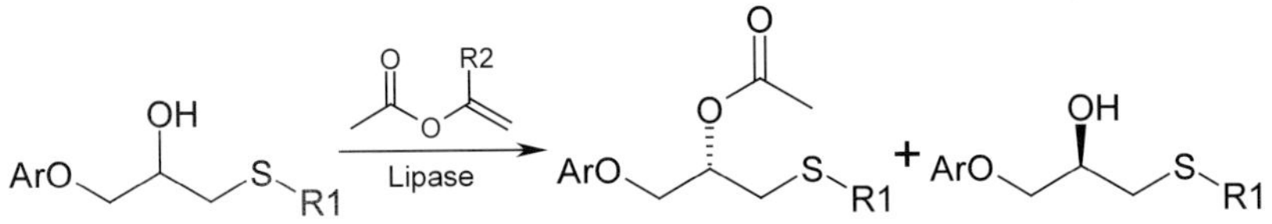

37   (S)-38   (R)-37

Figure 23: Lipase-catalysed transesterification of 1-alkylthio-3-aryloxypropan-2-ols **37**.[67]

4-hydroxy-2-methyl-2-*p*-tolyl-cyclopentane-carboxylic acid ethyl ester (±)-**39**, a key intermediate in the synthesis of sesquiterpenes containing quaternary centers (cuparene family) was resolved by the CAL-B catalyzed enantioselective transesterification using vinyl acetate as acyl donor. Both enantiomers were obtained in high enantiomeric excess (>98% ee).[68]

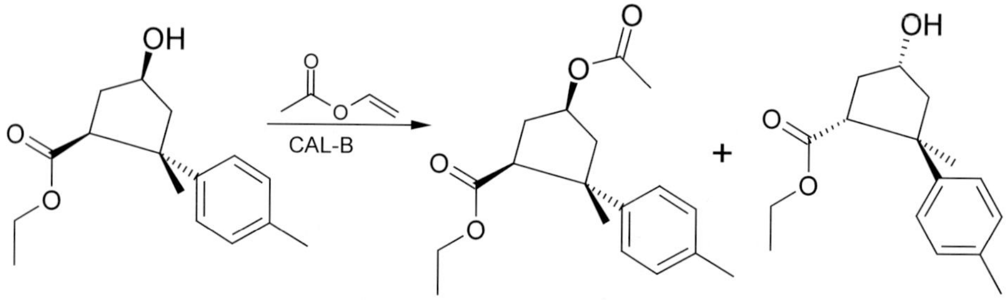

Figure 24: Lipase-catalyzed enantioselective transesterification of 4-hydroxy-2-methyl-2-*p*-tolyl-cyclopentane-carboxylic acid ethyl ester (±)-**39**.[68]

The enantioselective hydrolysis of the key industrial starting compound *dl*-methyl benzoate **41** was recently reported using a *Candida rugosa* isoenzyme LIP1 expressed in the methylotrophic yeast *Pichia pastoris* affording *l*-menthol **28** with > 99% ee.[69]

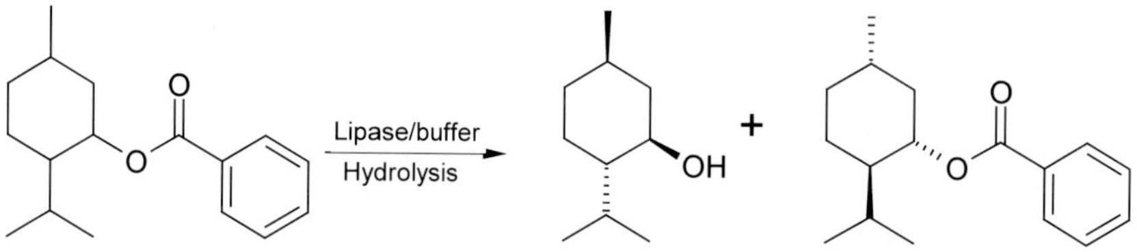

Figure 25: Lipase-catalyzed enantioselective hydrolysis of *dl*-menthyl benzoate (*dl*-**41**).[69]

A set of α–methylene-β-hydroxy esters **42** were resolved via enzymatic enantioselective transesterification with *Pseudomonas sp.* lipase (PCL), free and immobilized one using either vinyl or isopropenyl acetate as acyl donors under different conditions. The corresponding (*R*)-(+)-acetates (*R*)-**43** and the unreacted (*S*)-(-)-substrates (*S*)-**42** were obtained with an ee up to >99%.[70]

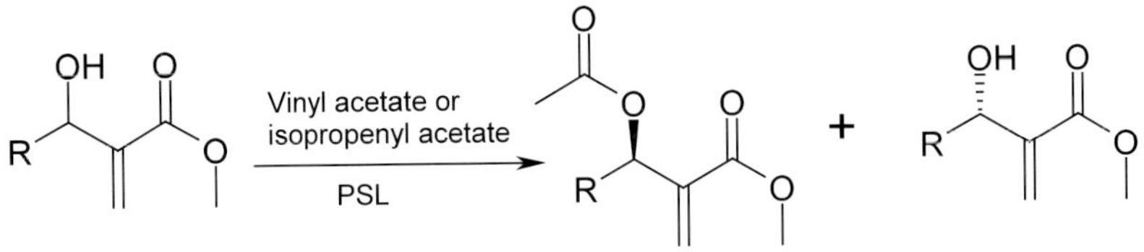

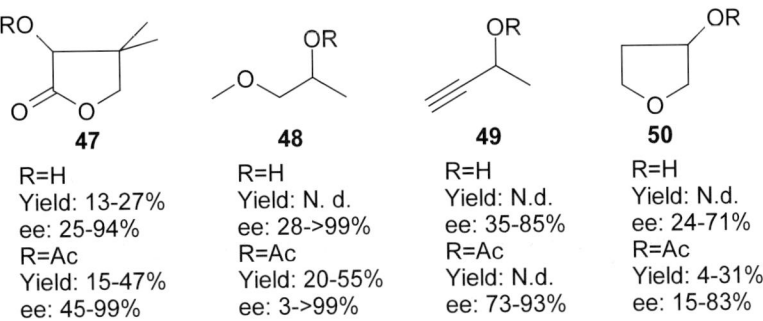

Figure 26: Lipase-catalysed enantioselective transesterification of α–methylene-β-hydroxy esters **42**.[70]

Joly et al[71] reported the use of *Candida cylindracea* lipase in the kinetic resolution of aryl-substituted β-hydroxy ketones **44** using either the transeterification of the free β-hydroxy ketone with vinyl acetate or the hydrolysis of its acetate. The transesterification mode afforded the alcohol with 30-70 % ee and the acetate with > 96% ee. That of hydrolysis afforded the alcohol with 64-93 % ee and the acetate with > 96% ee.

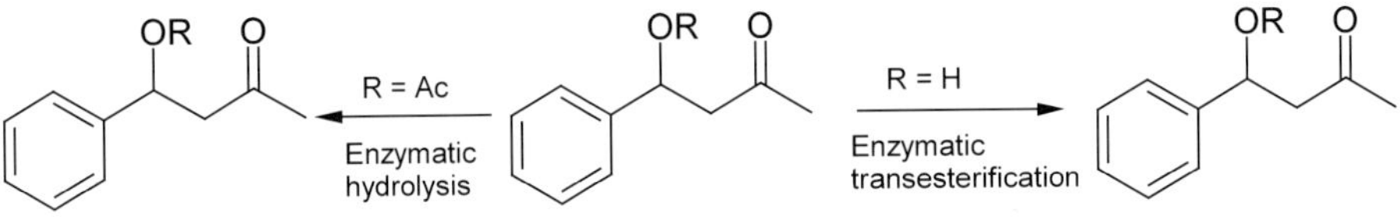

Figure 27: Lipase-catalyzed kinetic resolution of aryl-substituted β-hydroxy ketones **46**.[71]

The rapid screening of different hydrolases for the enantioselective hydrolysis of esters of the difficult to resolve substrates such as pentalactone **47**, 1-methoxy-2-propanol **48**, 3-butyn-2-ol **49** and 3-hydroxy-tetrahydrofuarn **50** was studied by Baumann et al. The screening was performed in a pH-indicator-based format in microtiter plates.[72]

	47	**48**	**49**	**50**
R=H	Yield: 13-27% ee: 25-94%	Yield: N. d. ee: 28->99%	Yield: N.d. ee: 35-85%	Yield: N.d. ee: 24-71%
R=Ac	Yield: 15-47% ee: 45-99%	Yield: 20-55% ee: 3->99%	Yield: N.d. ee: 73-93%	Yield: 4-31% ee: 15-83%

Figure 28: Difficult to resolve substrates used in the screening system.[72]

5.3. KINETIC RESOLUTION OF TERTIARY ALCOHOLS

Krishna et al[73] reported the enantioselective transesterification of a tertiary alcohol **51** using lipase A from *Candida antarctica* (CAL-A) and vinyl acetate as acyl donor in organic solvent.

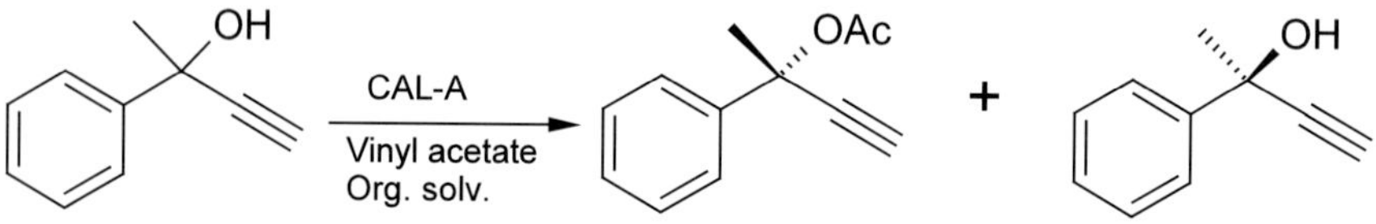

ee: 91-100%   ee: 9-26%

Figure 29: Lipase-catalyzed enantioselective transesterification of 2-phenylbut-3-yn-2-ol **51**.[73]

Attempts to resolve other tertiary alcohols (**44**,[74] **45**,[75] **46**[76]) are documented in literature.

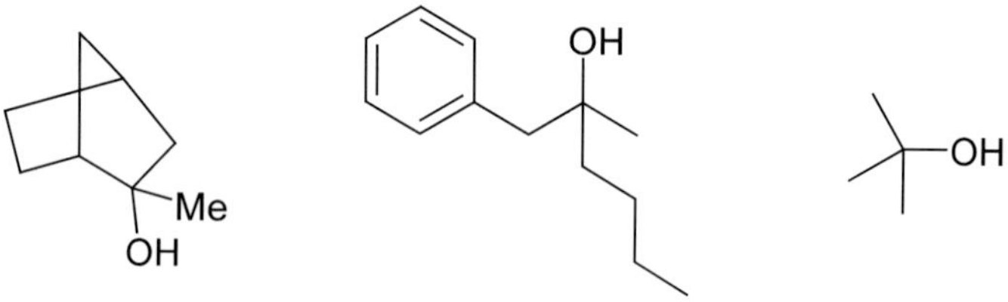

53   **54**   **55**

5.4. KINETIC RESOLUTION OF CHIRAL CARBOXYLIC ACIDS

2-Arylpropionic acids are important class of non-steroidal anti-inflammatory drugs (NSAID). Their pharmacological activity is mainly in one of both enantiomers. Thus, efforts had been made to access to the enantiomerically pure substance. The kinetic resolution of racemic 2-(2-fluoro-4-biphenyl) propanoic acid **56** and 2(4-isobutylphenyl) propanoic acid **59** (Ibuprofen) was performed via enzymatic esterification and transesterification using an alcohol and vinyl acetate, respectively in a membrane reactor. The unreacted acid is obtained in highly enantiomerically enriched form. A consecutive approach consisting of the enzymatic hydrolysis of the resulted esters is needed to achieve the alcohol in optically pure form.[77]

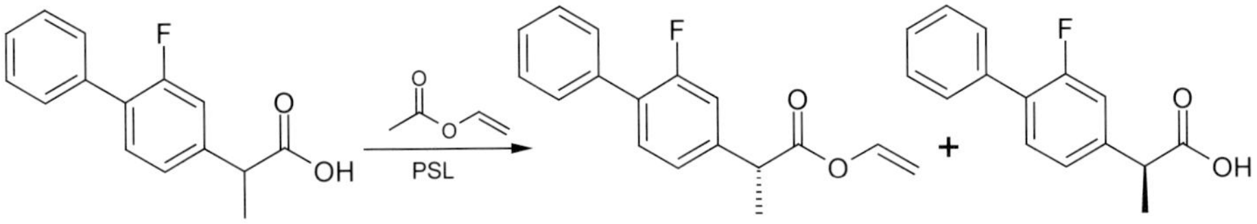

56   PSL   **57** + **56**

87% ee   84% ee

Figure 30: Lipase-catalysed transesterification of 2-(2-fluoro-4-biphenyl)propanoic acid **56** using vinyl acetate as acyl donor.[77]

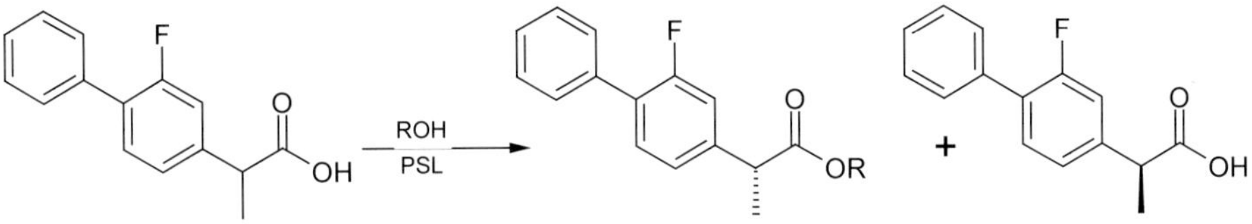

56   **58**   **56**

97% ee   92% ee

Figure 31: Lipase-catalyzed esterification of 2-(2-fluoro-4-biphenyl)propanoic acid **56** with various alcohols.[77]

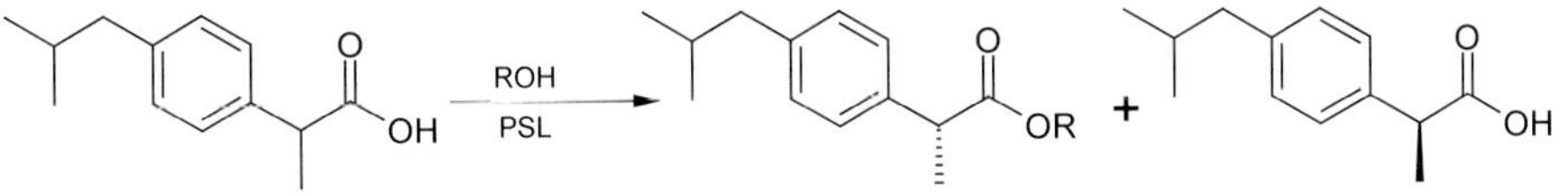

59   **60**   **59**

Up to 95% ee   Up to 92% ee

Figure 32: Lipase-catalyzed esterification of 2(4-isobutylphenyl)propanoic acid **59** with various alcohols.[77]

Henke et al[78] reported the resolution of 2(4-isobutylphenyl)propanoic acid **59** (ibuprofen) by transesterification of its corresponding vinylester **61** using lipase from *Candida antarctica*. Depending on the nucleophile, the vinylester **61** was recovered with 8-99% ee while the alkyl ester **62** or the free acid **59** is recovered with 16-75 % ee.

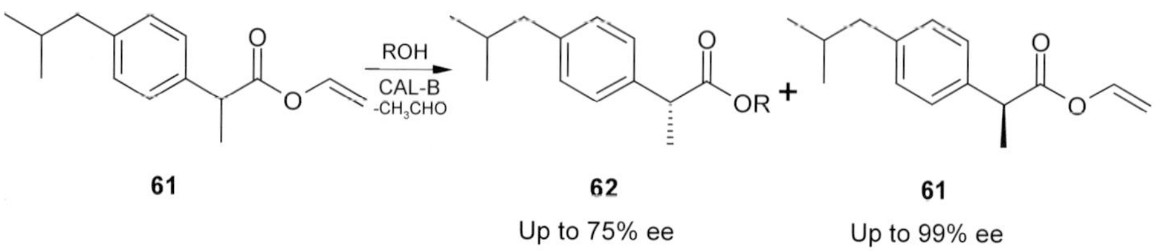

Figure 33: Lipase-catalysed transesterification of Ibuprofen vinyl ester **61**.[78]

The resolution of (R,S)-naproxen **54** using lipase-catalyzed esterification of the free acid with different diols and organic solvents was reported. The (S)-naproxen ester **55** was obtained in > 99 % ee when using 1,4-butandiol.[79]

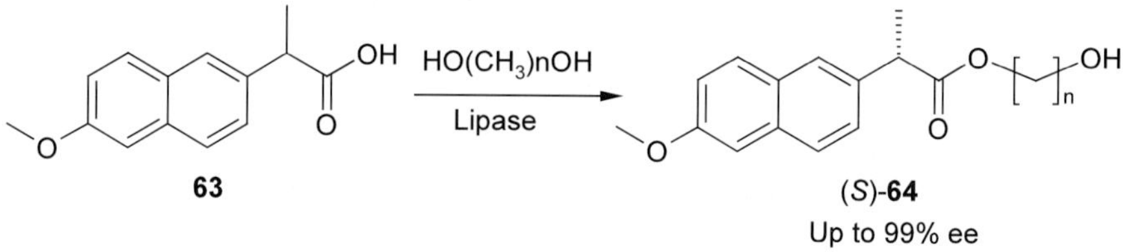

Figure 34: Lipase-catalysed kinetic resolution of (R,S)-naproxen **63**.[79]

A broad spectrum of racemic 2-hydroxy acids **65** were resolved via lipase-catalyzed enantioselective acetylation to their corresponding 2-acetoxy acids **66** using vinyl acetate as acyl donor in methyl *tert*-butyl ether as organic solvent. The unreacted enantiomer (R)-**65** was recovered with ee values up to >99%.[80]

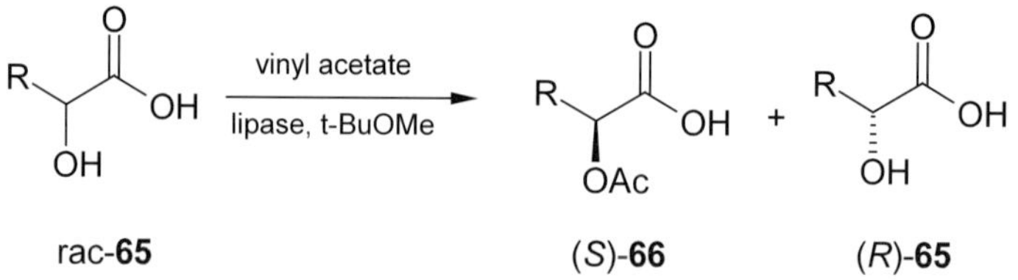

Figure 35: Lipase-catalyzed enantioselective acetylation of racemic 2-hydroxy acids **65** to their 2-acetoxy acids **66** using vinyl acetate as acyl donor in methyl *tert*-butyl ether as organic solvent.[80]

Kurokawa et al[81] reported the enzyme-catalyzed kinetic resolution of racemic N-carbamoyl, N-Boc, N-Cbz proline esters and prolinols using protease and *Candida antarctica* lipase B. The latter was efficient in the enantioselctive hydrolysis of both N-Boc and N-Cbz proline derivatives with E > 100.

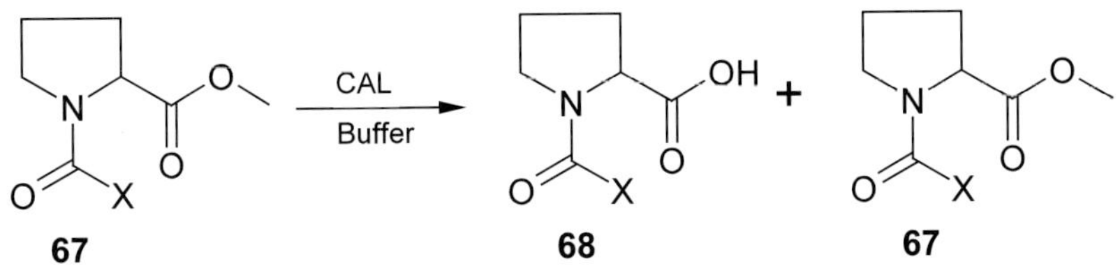

Figure 36: Lipase-catalyzed enantiomeric resolution of N-Boc-proline **67**.[81]

Based on the very different behaviors of lipases A (CAL-A) and B (CAL-B) from *Candida antarctica* towards polyfunctional compounds in non-aqueous media, Liljeblad et al[82] reported a novel lipase-catalyzed method for the resolution of N-heterocyclic amino esters using methyl pipecolinate **69** as a model compound. For this purpose, the chemo- and enantioselective alcoholysis and transesterification reaction of **69** in the presence of CAL-B and the N-acylations using CAL-A were studied. (cf. fig. 37 and 38).

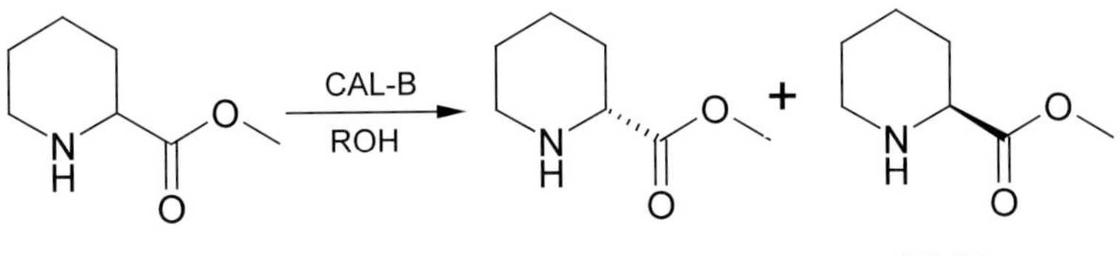

Figure 37: The chemo-and enantioselective alcoholysis of **69** using CAL-B.[82]

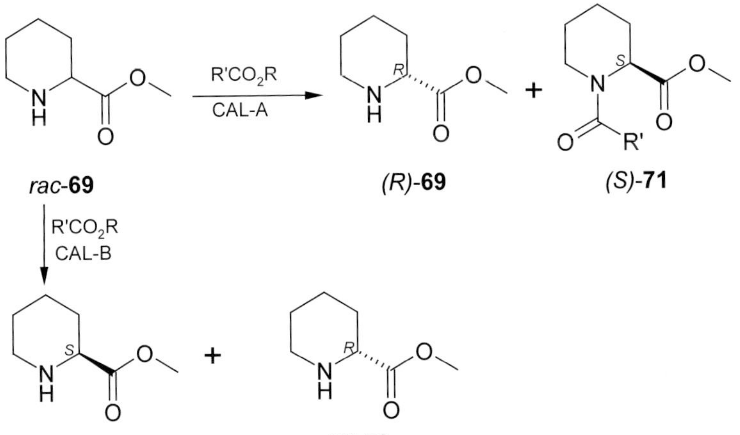

Figure 38: The transesterification reaction of **69** in the presence of CAL-B and the N-acylations using CAL-A.[82]

The study revealed that CAL-A is an efficient biocatalyst in the highly (S)-selective acylation at the secondary ring nitrogen of methyl pipecolinate **69**. The unreacted (R)-**69** was recovered with an ee up to 95% and the product (S)-**69** with >99 % ee with an E value >>100. While the catalysis by lipase B (CAL-B) leads to reactions at the methyl ester function of the substrate in an almost non-enantioselective manner.

5.5. KINETIC RESOLUTION OF DIOLS

5.5.1. Desymmetrization of racemic diols

Diols such as the optically active 1,1′-binaphthyl-2-2′-diol (BINOL) have been used as versatile templates and chiral auxiliaries in catalysts employed successfully in asymmetric synthesis. The application of enzymes in the enantioselective access to axially dissymmetric compounds was first reported by Fujimoto and coworkers.[83] In aqueous media, the asymmetric hydrolysis of the racemic binaphthyl dibutyrate (the ester) using whole cells from bacteria species afforded the (R)-diol with 96%ee and the unreacted substrate (S)-ester with 94% ee at 50 % conversion. Recently, in non-aqueous media, lipases from *Pseudomonas cepacia* and *Ps. fluorescens* have been employed in the enantioselective resolution and desymmetrization of racemic 6,6′-disubstituted BINOL derivatives using vinyl acetate.[84] The monoacetate (R)-73 (product) was obtained in 32-44 % chemical yields and 78-96% ee depending on the derivatives used. The unreacted BINOL (S)-72 was obtained in 30-52 % chemical yield and 55-80% ee.

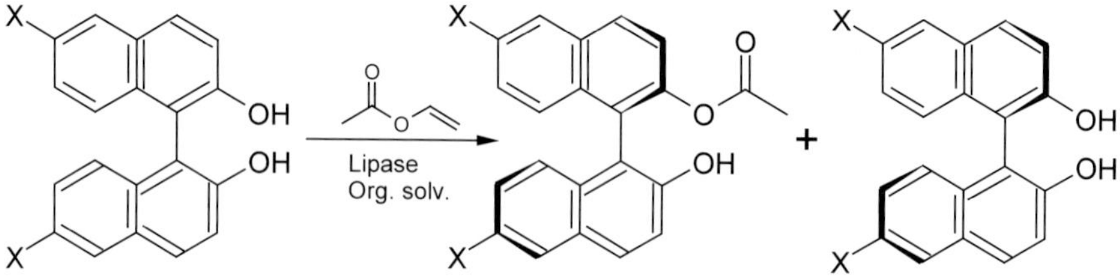

72

(R)-73
Yield: 32-44%
ee: 78-96%

(S)-72
Yield: 30-52
ee: 55-80%

Figure 39: Lipase-catalyzed stereoselective resolution and desymmetrization of binaphthols **72**.[84]

Biphenyls are recognised as stable analogues of BINOL. They are found in numerous natural products. Sanfilippo et al[85] reported the *Pseudomonas cepacia* lipase-catalysed kinetic resolution of 2,2′-dihydroxy-6,6′-dimethoxy-1,1′-biphenyl **74** using vinyl acetate as acyl donor in *tert*-butyl methyl ether as organic solvent. (R)-75 is obtained with an ee up to 98% while (S)-74 is recovered with an ee up to 96%.

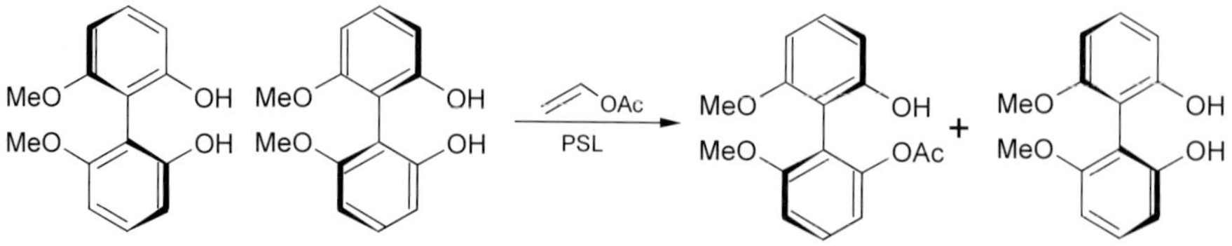

rac-**74**

(R)-**75**

(S)-**74**

Figure 40: Lipase-catalysed kinetic resolution of 2,2′-dihydroxy-6,6′-dimethoxy-1,1′-biphenyl **74**.[85]

Diols of different structures such as the meso-diol **76** (Fig. 41), the C2-symmetric diol rac-**79** (Fig. 42), the diol rac-**82** in which the primary hydroxy group is protected (Fig. 43) and the unprotected diol rac-**84** with a primary and secondary hydroxy group (Fig. 44) were used as substrates in the lipase-catalyzed transesterification using vinyl acetate as acyl donor in organic solvents with the aim to prepare chiral buildings blocks of high enantiomeric purity.[86]

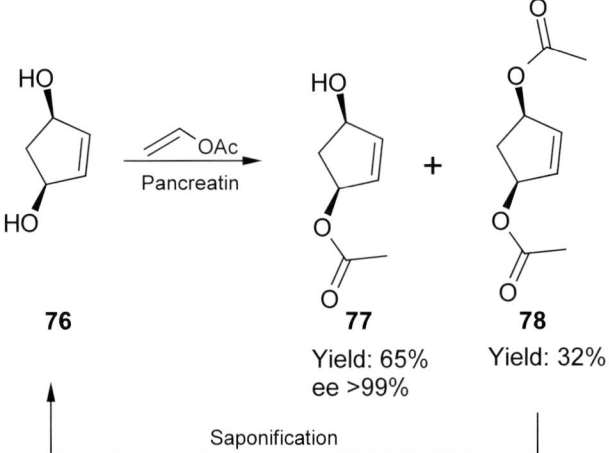

HO
HO
OAc
Pancreatin
HO
O
O
+
O
O
O
O
76
77
78
Yield: 65%
ee >99%
Yield: 32%
Saponification

Figure: 41: Lipase-catalysed desymmetrization of cis-2-cyclopeptene-1,4-diol **76**.[86]

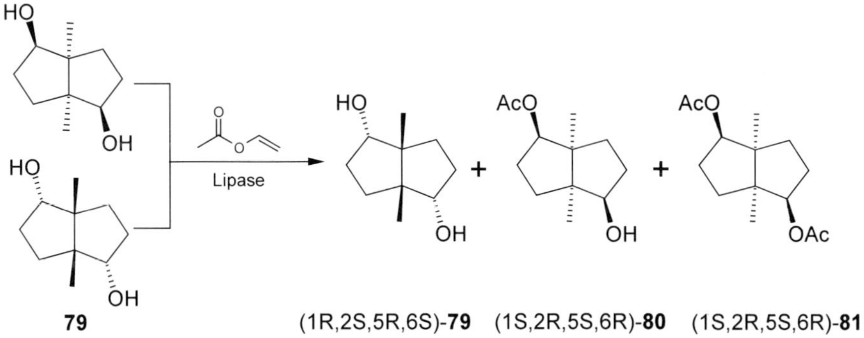

HO
OH
HO
OH
O
O
Lipase
HO
OH
+
AcO
OH
+
AcO
OAc
79
(1R,2S,5R,6S)-79
(1S,2R,5S,6R)-80
(1S,2R,5S,6R)-81

Figure 42: Lipase-catalysed kinetic resolution of endo-endo-cis-bicylo[3.3.0]octane-2,6-diol (rac-**79**).[86]

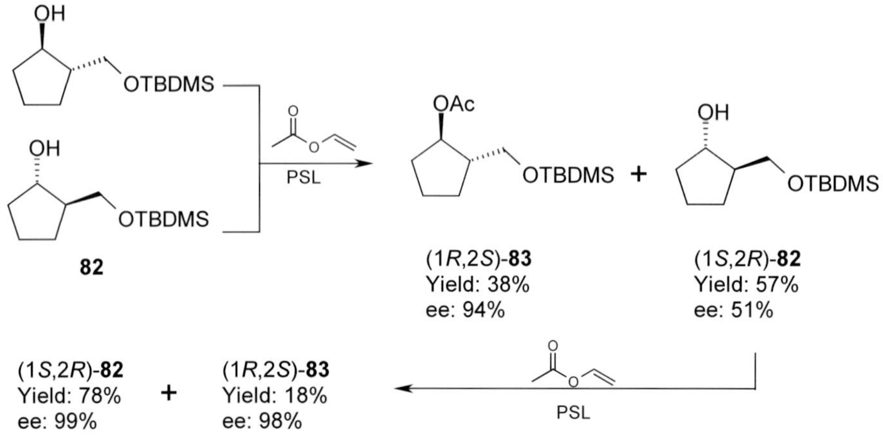

Figure 43: Lipase-catalyzed kinetic resolution of *trans*-2-(*tert*-butyldimethylsiloxymethyl) cyclopentanol (rac-**82**).[86]

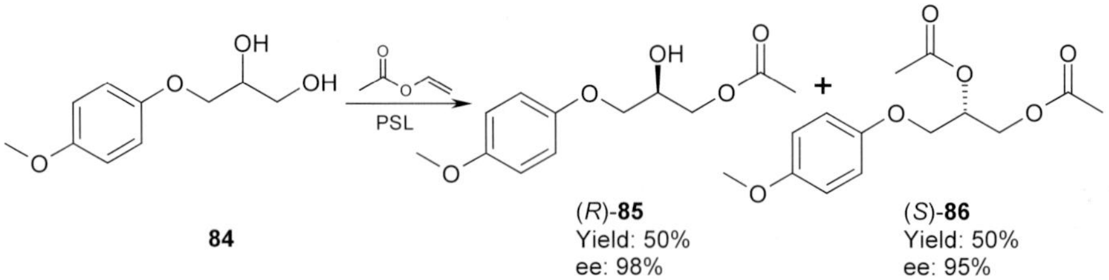

Figure 44: Lipase-catalyzed kinetic resolution of (*R,S*)-3-(4-methoxyphenoxy)propane-1,2-diol (rac-**84**).[86]

A part from vinyl acetate, vinyl benzoate was used as acylating agent in the *Mucor miehei* lipase (MML) and *Candida antarctica* lipase (CAL)-catalysed benzoylation of 1,2-diols in organic solvents **87**.[87] The reaction proceeded with high regioselectivity and moderate enantioselectivity.

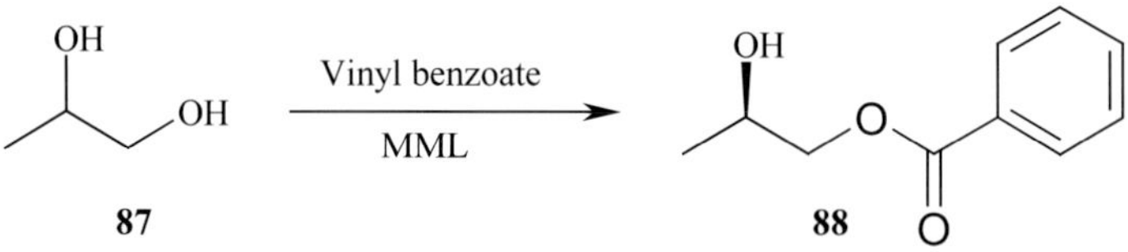

Figure 45: Lipase-catalysed benzoylation of propane-1,2-diol **87**.[87]

An efficient synthesis of (*R*)- and (*S*)-1-amino-2,2-difluorocycloropanecarboxylic acid (DFACC) **91** via lipase-catalyzed desymmetrization of prochiral diols **89** and prochiral diacetates **92** was recently reported.[28] Thus, the lipase-catalyzed transesterification of **89** using vinyl acetate as acyl donor in benzene: di-*i*-propyl ether (20:1) as organic solvent afforded (*R*)-**90** with 91.3 % ee and 96.5 % chemical yield. The reverse enantioselective hydrolysis of **92** in a mixed solvent of acetone and phosphate buffer afforded (*S*)-**90** with 91.7 % ee and 86.2 chemical yield.

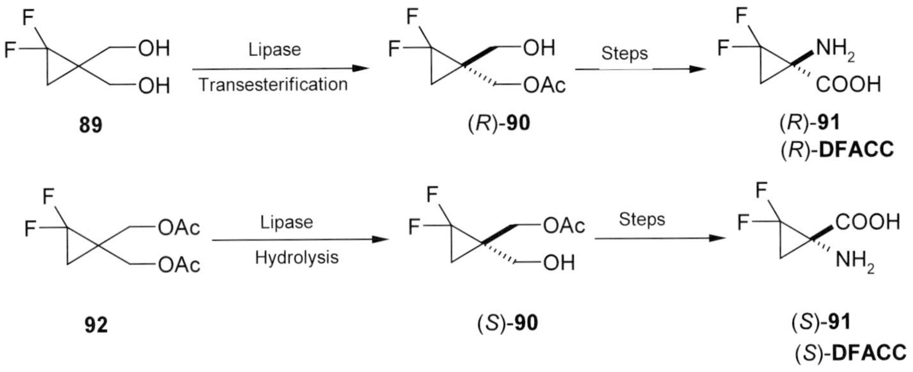

Figure 46: Lipase-catalyzed desymmetrization of prochiral diols **89** and diacetates **92**.[28]

The first enzymatic desymmetrizations of prochiral phosphine oxides was recently reported by Kielbasinski et al.[88] Thus, the prochiral bis(methoxycarbonylmethyl)-phenylphosphine oxide **93** was subjected to the PLE-mediated hydrolysis in buffer affording the chiral monoacetate (*R*)-**94** in 72% ee and 92% chemical yield. In turn, the prochiral bis(hydroxymethyl)phenylphosphine oxide **95** was desymmetrized using either lipase-catalyzed acetylation of **95** with vinyl acetate as acyl donor in organic solvent or hydrolysis of **97** in phosphate buffer and solvent affording the chiral monoacetate **96** with up to 79% ee and 76% chemical yield.

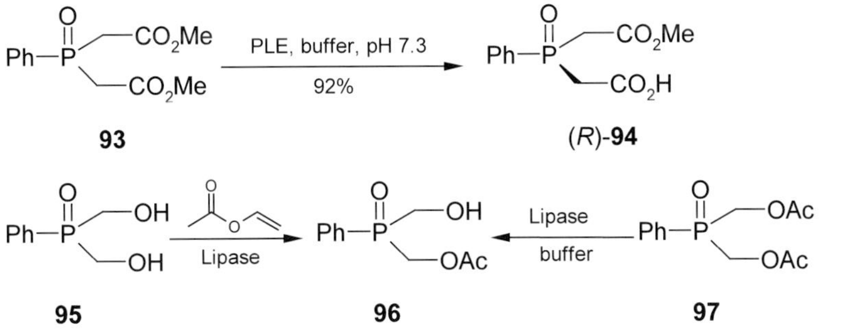

Figure 47: Lipase-catalyzed desymmetrizations of prochiral phosphine oxides.[88]

Neri et al[89] reported the desymmetrization of *N*-Boc-serinol **98** by the selective mono-acetylation using PPL (porcine pancreas lipase) and vinyl acetate as the acylating agent in organic solvent. The mono acetylated product (*R*)-**99** was obtained after 2 hours with 99% ee and isolated in 69% chemical yield. Traces of the diacetylated product **100** were observed. The cyclization of (*R*)-**99** in basic medium afforded the racemic oxazolidinone **101**. The latter was subjected to enzymatic hydrolysis in phosphate buffer affording (*R*)-

102 in up to 93% ee and isolated in chemical yield up to 42%. To avoid basic conditions, (*S*)-**101** was also obtained in one step by cyclization of (*R*)-**99** with thionyl chloride. The reaction proceeded with > 98% ee and 72% yield. The enzymatic hydrolysis of (*S*)-**101** afforded (*S*)-**93** in > 98%ee and 77% yield.

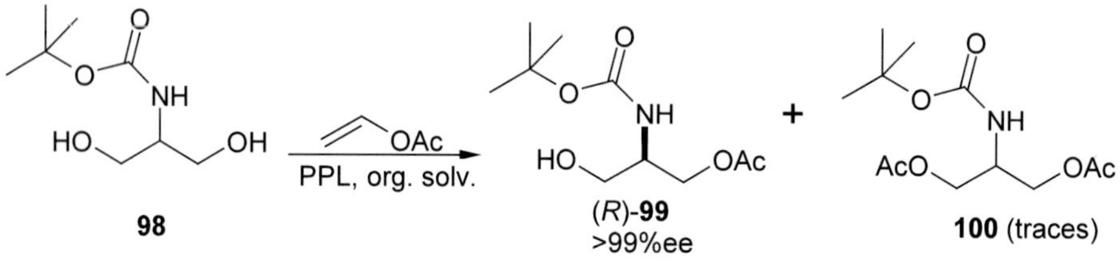

Figure 47: Desymmetrization of N-Boc-serinol **98** by PPL.[89]

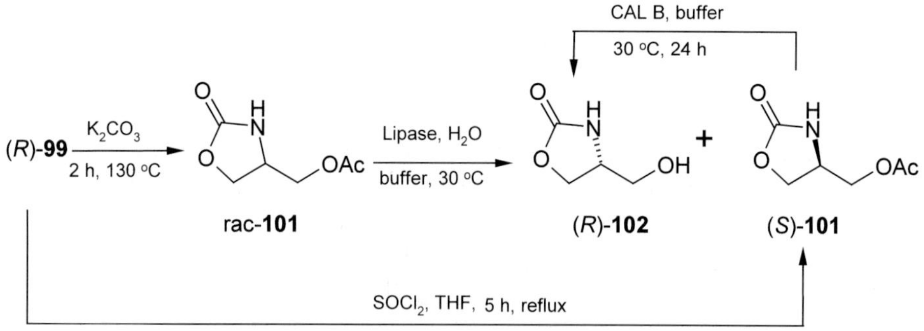

Figure 49: Synthesis of (*S*)-4-acetoxymethyl-2-oxazolidinone **101** and its enzymatic hydrolysis.[89]

The kinetic resolutions of a series of racemic trans-cycloalkane-1,2-diol monoacetates *rac*-**103a-d** were reported using the enantioselctive transesterification mode with vinyl acetate as acyl donor and commercial as well as self-prepared fungal lipases affording the diacetates (*R,R*)-**104a-d** and monoacetates (*S,S*)-**103a-d** in high enantiomeric purity (up to 99% ee). The monoacetates (*R,R*)-**103a-d** were also prepared starting from the racemic diacetates *rac*-**104a-d** by lipase-catalyzed hydrolysis.[90]

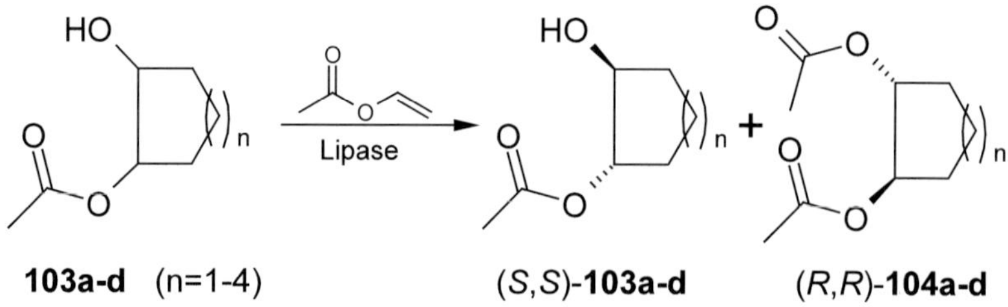

Figure 50: Enzymatic acylation of trans-2-acetoxycycloalkan-1-ols **103**.[90]

5.6. MISCELLANEOUS CASES

Allenes, another class of compounds having interesting properties, have been also resolved by lipases. the kinetic resolution of a variety of racemic 1-ethenyl and ethynyl-substituted 2,3-allenols was reported using a lipase from *Candida antarctica* type B (CAL-B) with vinyl acetate as acyl donors in organic solvent. The biocatalytic resolution afforded (S)-2,3-allenols (S)-**98** and (R)-2,3-allenyl acetates (R)-**99** in chemical yields up to 55% and an enantiomeric excess ee up to 99% for both enantiomers depending on the substituents.[91]

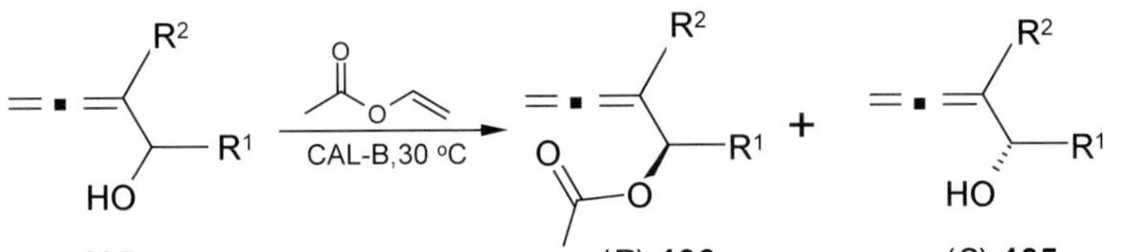

Figure 51: Lipase-catalyzed kinetic resolution of a variety of racemic 1-ethenyl and ethynyl-substituted 2,3-allenols **105**.[91]

Faure et al[92] reported the first enzymatic resolution of phosphane-borane complex. Thus, the borane adduct of (2-hydroxypropyl)diphenylphosphane **107** was resolved using the lipase CAL-B and vinyl acetate as acyl donor in organic solvent.   The remaining unreacted substrate (S)-**107** was recovered with 91 % ee.

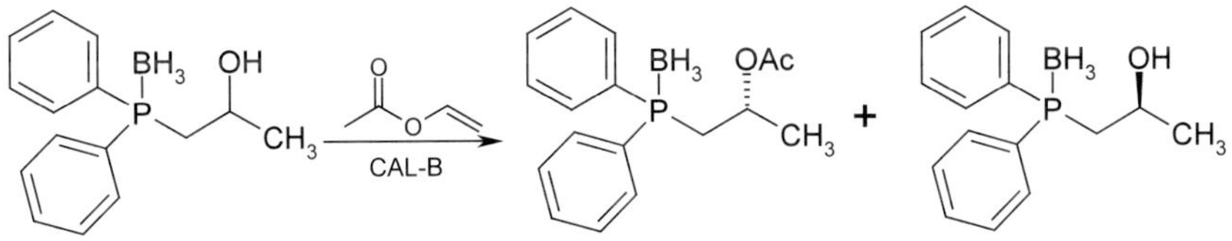

Figure 52: Lipase-catalysed kinetic resolution of a phosphane-borane complex **107**.[92]

The use of *Candida antarctica* lipase B in the kinetic resolution of a series of bicylic 1-heteroaryl primary amines **109** using ethyl acetate as acyl donor in isopropyl ether as organic solvent was studied by Skupinska et al.[93] High yields and enantiomeric excess of either enantiomer could be obtained. The undesired enantiomer could be in some cases be recycled by thermal racemization.

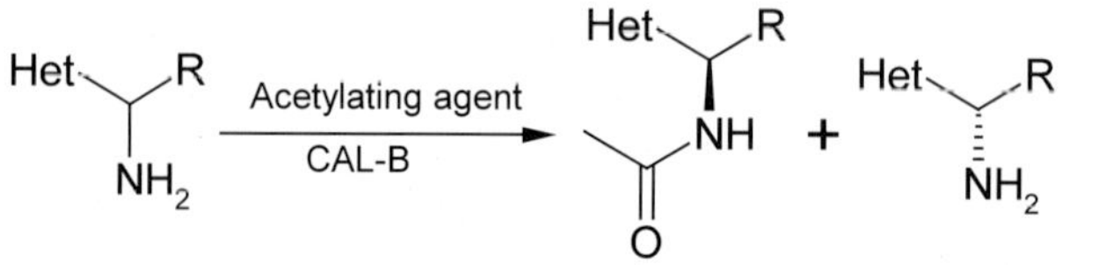

109   (R)-**110**   (S)-**109**

Yield: 47-70%   Yield: 29-48%
ee: 40-98%   ee: 61->99%

Figure 53: Lipase-catalyzed enantioeselctive acetylation of bicyclic 1-heteroaryl primary amines (rac-**109**).[93]

Irurre et al[94] reported the enzymatic resolution of trans-10-Azido-9-acetoxy-9,10-dihydrophenanthrene **111** in gram-scale using *Candida cyclindracea* lipase-catalyzed enantioselective hydrolysis in phosphate buffer. The substrate **111** (the ester) was obtained in 89 % yield and 83 % ee while the product **112** (the alcohol) was obtained in 90 % yield and 98 % ee.

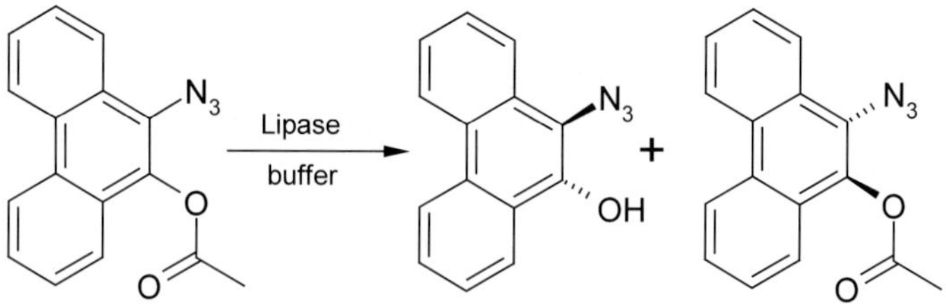

111   (R,R)-**112**   (S,S)-**111**

ee: 98%   ee: 83%

Figure 54: Kinetic resolution of azido acetate **111**.[94]

A practical method for the synthesis of chiral pyridazinone bearing a pyrazolopyridine ring via lipase-catalyzed resolution of 2-(acyloxymethyl)-4,5-dihydro-5-methylpyridazin-3(2H)-one derivatives **113** was reported by Yoshida et al.[95]

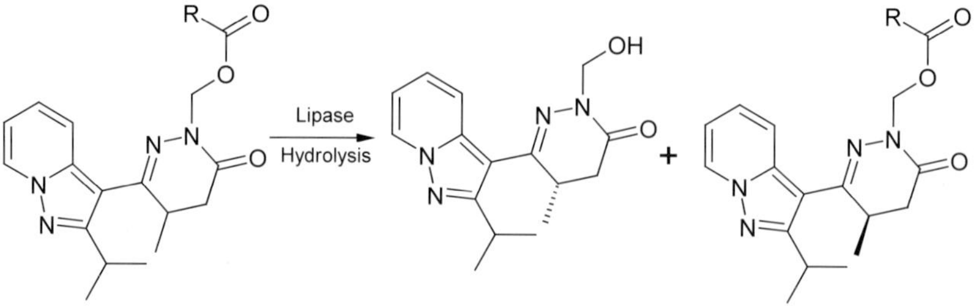

113   (S)-**114**   (R)-**113**

ee: 3-79%   ee: 3-79%

Figure 55: Lipase-catalyzed resolution of 2-(acyloxymethyl)-4,5-dihydro-5-methylpyridazin-3(2H)-one derivatives **113**.[95]

Forro et al[96] reported a very simple method for the synthesis of enantiopure β-amino acids **115a – 118a** (e.g. cispentacin) and β-lactams **115b – 118b** via lipase-catalyzed enantioselective ring opening of unactivated alicyclic β-lactams **119 – 122** in organic media. High enantioselectivity (E> 200) was observed when using CAL-B catalyzed reaction with H_2O (1 equiv) in diisopropyl ether at 60 °C. The products (**115a – 118a**) and the substrates **115b – 118b** were obtained in up to 99% ee with chemical yields ranging from 36 to 47%. Other approaches for the resolution of compounds containing larger alicyclic rings were recently reported.[97]

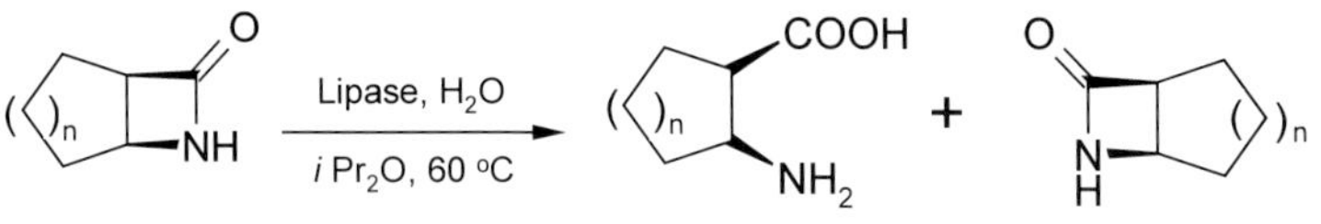

(+/-)-**115 - 118**
n = 1, 2, 3, 4

115a - 118a

115b - 118b

Figure 56: lipase-catalyzed enantioselective ring opening of unactivated alicyclic β-lactams.[96]

6.   Application of the kinetic resolution in industry

As the use of lipase for industrial chemical synthesis becomes easier, several chemical companies have begun to increase significantly their biocatalytic process used in synthetic application. Among these companies, BASF, in which enantiomerically pure alcohols and amines are produced on industrial scale.[98]

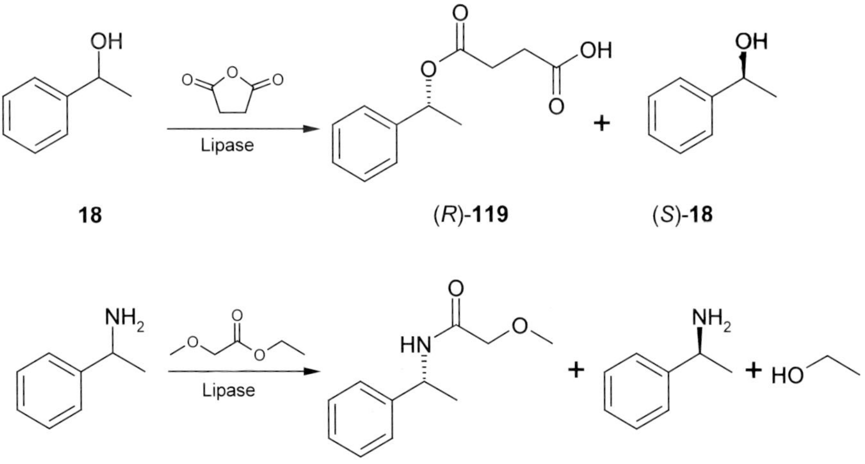

18

(R)-**119**

(S)-**18**

120

(R)-**121**

(S)-**120**

Figure 57: Some of the biocatalytic steps using lipase developed at BASF: Lipase-catalyzed kinetic resolution of a) phenyl ethanol **18** using succinic anhydride, b) Secondary amine **120** using ethyl methoxyacetate as acyl donor.[98]

The enantioselective hydrolysis of the racemic acetamide (R,S)-**122** was developed at

Bayer in the middle of 1990s. The reaction was performed using *Candia Antarctica* lipase B (CAL-B) to afford the free amine (*R*)-**123** in high enantiomeric excess (> 99.5% ee).[99] However, the requirement of high concentration of the catalyst limits the exploitation of the process on industrial scale.[100]

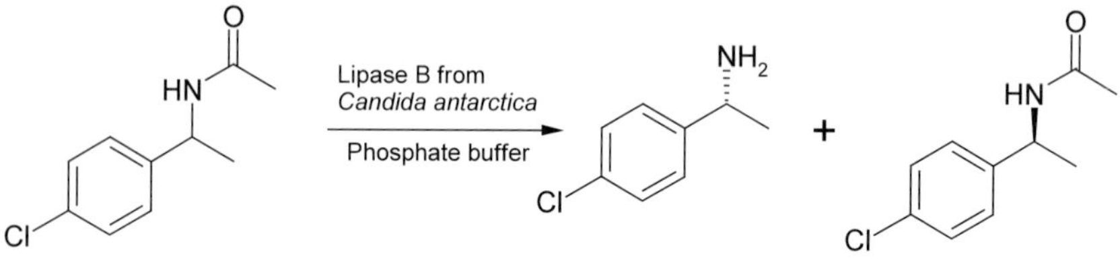

(*R,S*)-**122**   (*R*)-**123**   (*S*)-**122**

Figure 58: Enantioselective hydrolysis of racemic acetamide (*R,S*)-**122** developed at Bayer.[99]

Apart from amines and secondary alcohols, Ladner and Whitesides developed a procedure to resolve racemic glycidylbutyrate (**124**) with porcine pancreatic lipase (PPL) to afford (*R*)-**124** in 89% of the theoretical yield and with 92% ee.[101] This process was further developed and used by Andeno-DSM to produce the epoxy alcohol (*R*)-glycidol [(*R*)-**125**] and (*R*)-**124** on a multitone scale.[102]

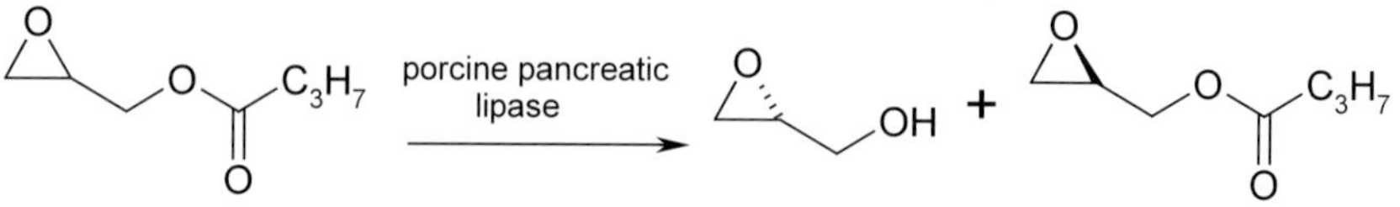

rac-**124**   (*R*)-**125**   (*R*)-**124**

Figure 59: Enantioselective hydrolysis of racemic glycidylbutyrate (**124**) developed at Andeno-DSM.[101]

A broader overview on the industrial methods used for the production of optically active intermediates is recently reviewed by Breuer et al.[100]

7. Conclusions and perspectives

The lipase-catalysed access to enantiomerically pure compounds remains a versatile method for the separation of enantiomers. The selected examples shown in this survey demonstrate the broad applicability of lipases in terms of substrate structures and enantioselectivity. More recently, modern molecular biology methods such as rational protein design and especially directed evolution[103] will further boost the development of tailor-made lipases for future applications in the synthesis of optically pure compounds. It has been already shown that a virtually non-enantioselective lipase (E=1.1 in the resolution of 2-methyldecanoate) could be evolved to become an effective biocatalyst (E>50). Furthermore, variants were identified which showed opposite enantiopreference.

Abbreviations used

Ac	acetyl
ANL	*Aspergillus niger* Lipase
BINOL	1,1'-binaphthyl-2-2'-diol
Boc	*tert*-butyloxycarbonyl
CAL-A	*Candida antarctica* lipase A
CAL-B	*Candida antarctica* lipase B(Novozyme 435)
Cbz	benzyloxycarbonyl
CCL	*Candida cylindracea* lipase
CD	cyclodextrin
Conv.	conversion
CRL	*Candida rugosa* lipase
DFACC	1-amino-2,2-difluorocycloropanecarboxylic acid
DKR	dynamic kinetic resolution
DMF	dimethylformamide
DMSO	dimethylsulfoxide
E	enantiomeric ratio
ee	enantiomeric excess
ee_p	enantiomeric excess of product
ee_s	enantiomeric excess of substrate
Eq.	equation
ES	enzyme-substrate complex
EtOH	ethanol
GC	gas chromatography
HIV	humane immune deficiency virus
HPLC	high performance liquid chromatography
hrs	hours
i-Pr$_2$O	diisopropyl ether
K_{cat}	rate constant
K_M	Michaelis-Menten constant
Min	minute(s)
MML	*Mucor miehei* lipase
MTMS	methyltrimethoxysilane
n.d.	not determined
NMR	nuclear magnetic resonance
Novozyme 525 L	liquid version of novozyme 435
Novozyme IM	immobilized CAL-B
NSAID	non-steroidal anti-inflammatory drugs
P	product
PFL	*Pseudomonas fluorescens* lipase
PLE	Pig liver esterase
PPL	*Porcine pancreas* lipase
PCL	*Pseudomonas cepacia* lipase
PCL-C	*Pseudomonas cepacia* immobilized on ceramics
PCL-D	*Pseudomonas cepacia* immobilized on

	diatomaceous earth
rac.	racemic
rec.	recombinant
S	substrate
t	time
TBDMS	*tert*-butyl dimethyl silyl
TBME	*tert*.- butyl methyl ether
THF	tetrahydrofuran
TMOS	tetramethoxysilane

Acknowledgments

This work was supported by the *Fonds der Chemischen Industrie* and the *Graduate College Analytical Chemistry, University of Tübingen*. The author is indebtly grateful for Prof. Volker Schurig (Institute of Organic Chemistry, University of Tübingen) for his help and support and Prof. Uwe Bornscheuer (*Department of Technical Chemistry & Biotechnology, University of Greifswald*) for his precious advises and for revising as well as writing parts of the manuscript.

References

1. Jones, J. B. Enzymes in organic synthesis. *Tetrahedron* **1986**, 42, 3351-3403.
2. Sih, C. J. and Wu, S.H. Resolution of enantiomers via biocatalysis. *Topics Stereochem.* **1989**, 19, 63-125.
3. Haraldsson, G. The application of lipases in organic synthesis in: The chemistry of acid derivatives edited by S. Patai, John Wiley & Sons Ltd **1992**, vol. 2, 1396-1467.
4. Wong, C.-H. and Whitesides, G. M. *Enzymes in Synthetic Organic Chemistry* (pergamon, Oxford, **1994**).
5. Drauz, K and Waldmann, H. *Enzyme Catalysis in organic Synthesis* (Wiley, Weinheim, **1995**).
6. Schoffers, E., Golebioeski, A. and Johnson, C. R. Enantioselective synthesis through enzymatic asymmetrization. *Tetrahedron* **1996**, 52, 3769-3826.
7. Klibanov, A. M. Why are enzymes less active in organic solvents than in water? *Trends Biotechnol.* **1997**, 15, 87-101.
8. Faber, K. Biotransformation in Organic Chemistry, 3rd ed; Springer Verlag: Heidelberg, Germany **1997**.
9. Schmid, R. D.; Verger, R. Lipases: interfacial enzymes with attractive applications. *Angew. Chem. Int. Ed.* **1998**, 37, 1608-1633.
10. Zaks, A.; Dodds, R. Biotransformations in the discovery and development of pharmaceuticals. *Curr. Opin. Drug. Discov. Dev.* **1998**, 1, 290-303.
11. Roberts, S. M. *Biocatalysis for fine Chemical Synthesis* (Wiley, Weinheim, **1999**).
12. Bornscheuer, U. T.; Kazlauskas, R. J. Hydrolases in organic synthesis; Wiley-VCH: Weinheim, Germany, **1999**.
13. Klibanov, A. M. Improving enzymes by using them in organic solvents. *Nature*, **2001**, vol 409, 241.
14. Koeller, K., M.; Wong, C.-H. Enzymes for chemical synthesis. *Nature* **2001**, vol 409, 232-240.
15. Cotterill, I. C.; Sutherland, A. G.; Robert, S. M.; Grobbauer, R.; Spreitz, J.; Faber, K. Enzymatic resolution of sterically demanding bicyclo[3.2.0]heptanes: evidence for a novel hydrolase in crude porcine pancreatic lipase and the advantages of using organic media for some of the biotransformations. *J. Chem. Soc. Perkin Trans* **1991**, 1, 1365.

16. Muralidhar, RV.; Marchant, R.; Nigam, P. Lipases in racemic resolutions. *J. Chem. Technol. Biotechnol.* **2001**, 76: 3-8.

17. Nakamura, K.; Matsuda, T.; Harada, T. Chiral synthesis of secondary alcohols using *Geotrichum candidum. Chirality*, **2002**, 14:703-708.

18. Miyazawa, T.; Kurita, S.; Shimaoka, M.; Ueji, S.; Yamada, T. Resolution of racemic carboxylic acids via the lipase-catalyzed irreversible transesterification of vinyl esters. *Chirality* **1999**, 11:554-560.

19. Klibanov, A. M. Enzymatic catalysis in anhydrous organic solvents. *Trends Biochem. Sci.* **1989**, 14, 141-144.

20. Persson, B., A.; Larsson, A., L., E.; Le Ray, M., Bäckvall, J.-E. Ruthenium- and enzyme-catalyzed dynamic kinetic resolution of secondary alcohols. *J. Am. Chem. Soc.* **1999**, 121, 1645-1650.

21. Williams, J. M. J.; Parker, R. J.; Neri, C. Enzymatic kinetic resolution in enzyme catalysis in organic synthesis (2nd Edition) **2002**, 1, 287-312.

22. Sheldon, R., A. Chirotechnology: industrial synthesis of optically active compounds. Marcel Dekker, Inc. New York. **1993**.

23. Eliel, E. L.; Wilken, S. H.; Mander, L. N. Stereochemistry of organic compounds; John Wiley, New York, **1994**.

24. Collins, A. N.; Sheldrake, G. N.; Crosby, J. Chirality in industry, John Wiley; Chicheter, U. K, **1997**, vol. 2.

25. Atkinson, R. S. Stereoselective synthesis; John Wiley: Chichester, U. K., **1995**.

26. Carnell, A. J. Desymmetrisation of prochiral ketones using lipases. *J. Mol. Cat. B: Enzymatic* **2002**, 19-20, 83-92.

27. Taschner, M, J.; Black, D. J.; Chen, Q.-Z. The enzymic Baeyer-Villiger oxidation: a study of 4-substituted cyclohexanones. *Tetrahedron: Asymmetry* **1993**, 4, 1387-1390

28. Kirihara, M.; Kawasaki, M.; Takuwa, T.; Kakuda, H.; Wakikawa, T.; Takeuch, Y.; Kirk, K. L. Efficient synthesis of (*R*)- and (*S*)-1-amino-2,2-difluorocyclopropanecarboxylic acid via lipase-catalyzed desymmetrization of prochiral precursors.*Tetrahedron: Asymmetry* **2003**, 14, 1753-1761.

29. Stinson, S. C. Synthetic organic chemistry advances. *Chemical and Engineering News* **2001**, 79, 37.

30. Amiard, G. Resolution by entrainment. The idea of "residual" supersaturation. *Experienta* **1959**, 15, 38-40.

31. Reinhold, D. F.; Firestone, R. A.; Gaines, W. A.; Chemerda, J. M.; Sletzinger, M. Synthesis of L-alpha-methyldopa from asymmetric intermediates. *J. Org. Chem.* **1968**, 33, 1209-1213.

32. Leffingwell, J-C.; Shakelford, R-E. Laevo menthol synthesis and organoleptric properties. *Cosmetic and perfumery* **1974**, 89 (6), 69-89.

33. Scoffers, E.; Golebiowski, A.; Johnson, C. R. Enantioselective synthesis through enzymic asymmetrization. *Tetrahedron* **1996**, 52, 3769-3826.

34. Stecher, H.; Faber, K. Biocatalytic deracemization techniques. Dynamic resolutions and stereoinversions. *Synthesis* **1997**, 1, 1-16.

35. Dijksman, A.; Elzinga, M., J.; Li, Yu-Xin, Arends, I., Sheldon, R., A. Efficient ruthenium-catalyzed racemization of secondary alcohols: application to dynamic kinetic resolution. *Tetrahedron: Asymmetry* **2002**, 13, 879-884.

36. Chen, C.-S.; Wu, S.-H.; Girdaukas, G.; Sih, C. J. Quantitative analyses of biochemical Kinetic resolution of Enantiomers. 2. Enzyme-catalyzed esterification in water-organic solvent biphasic systems, *J. Am. Chem. Soc.* **1987**, 109, 2812-2817.

37. Chen, C.-S.; Sih, C.J., General aspects and optimization of enantioselective biocatalysis in organic solvents: the use of lipases. *Angew. Chem. Int. Ed. Engl.*, **1989**, Vol. 28, 695-707.

38. Prelog, V. Specification of the stereospecificity of some oxidoreductases by diamond lattice sections. *Pure Appl. Chem.* **1964**, 9, 119.

39. Lee, T.; Sakowicz, R.; Martichonoc, V.; Hogan, J. K.; Gold, M.; Jones, J. B. Probing enzyme specificity. *Acta Chem. Scand.* **1996**, 50, 697-706.

40. Fitzpatrick, P. A.; Klibanov, A. M. How can the solvent affect enzyme enantioselectivity? *J. Am. Chem. Soc.* **1991**, 113, 3166.

41. Lemke, K.; Lemke, M.; Theil, F. A Three-dimensional predictive active site model for lipase from *Pseudomonas cepacia*. *J. Org. Chem.* **1997**, 62, 6268.
42. Naemura, K.; Fukuda, R.; Konishi, M.; Hirose, K. Tobe, Y. Lipase YS-catalyzed acylation of alcohols: a predictive active site model for lipase YS to identify which enantiomer of a primary or a secondary alcohol reacts faster in this acylation. *J. Chem. Soc. Perkin Trans 1*, **1994**, 1253.
43. Moreno, Jose M.; Samoza, A.; del Campo, Carmen; Liama, Emilio F.; Sinisterra, Jose V. Organic reactions catalyzed by immobilized lipases. Part I. Hydrolysis of 2-aryl propionic and 2-aryl butyric esters with immobilized *Candida cylindracea* lipase. *J. Mol. Catalysis A: Chemical* **1995**, 95, 179-92.
44. Kazlauskas, R. J.; Weissfloch, A. N. E.; Rappaport, A. T.; Cuccia, L. A. A rule to predict which enantiomer of a secondary alcohol reacts faster in reactions catalyzed by cholesterol esterase, lipase from *Pseudomonas cepacia*, and lipase from *Candida rugosa*. *J. Org. Chem.* **1991**, 56, 2656-2665.
45. Reetz, M. T., Wilensek, S.; Zha Jaeger, K. E. Directed evolution of an enantioselective enzyme through combinatorial multiple-cassette mutagenesis. *Angew, Chem. Int. Ed.* **2001**, 40, 3589-3591.
46. Zha, D., Wilensek, S., Hermes, M.; Jaeger, K. E.; Reetz, M. T. Complete reversal of enantioselectivity of an enzyme-catalyzed reaction by directed evolution. *Chem. Commun.* **2001**, 2664-2665.
47. Koga, Y.; Kato, K.; Nakano, H.; Yamane, T. Inverting enantioselectivity of *Burkholderia cepacia* KWI-56 lipase by combinatorial mutation and high-throughput screening using single-molecule PCR and *in vitro* expression. *J. Mol. Biol.* **2003**, 331, 585-592.
48. Schurig, V. Separation of enantiomers by gas chromatography. *J. Chromatography A* **2001**, 906, 275-299.
49. Ghanem, A.; Ginatta, C. Jian, Z.; Schurig, V. Chirasil-β-dex with a new C11-spacer for enantioselective gas chromatography: application to the kinetic resolution secondary alcohols catalyzed by lipase. *Chromatographia* **2003**, Vol. 57, S-275-281.
50. Aitkem, R. A.; Kilenyi, S. N. Asymmetric synthesis, Chapman and Hall, **1992**.
51. Ghanem, A.; Schurig, V. Lipase-catalysed transesterification of 1-(2-furyl)ethanol using isopropenylacetate. *Chirality* **2001**, 13:118-123.
52. Ghanem, A.; Schurig, V. Per-acetylated β-Cyclodextrin as additive in enzymatic reactions: enhanced reaction rate and enantiomeric ratio in lipase-catalyzed transesterifications in organic solvents. *Tetrahedron: Asymmetry* **2001**, vol 12/19. pp 2761-2766.
53. Lindner, E.; Ghanem, A.; Warad, I.; Eichle, K.; Mayer, E. Schurig, V. Asymmetric hydrogenation of an α,β-unsaturated ketone by diamine(ether-phosphine)ruthenium(II) complexes and lipase-catalyzed kinetic resolution: a consecutive approach. *Tetrahedron: Asymmetry* **2003**, 14, 1045-1053.
54. Ghanem, A. The Utility of cyclodextrins in lipase-catalyzed transesterification in organic solvents: enhanced reaction rate and enantioselectivity. *Org. Biomol. Chem.*, **2003**, 1, 1282 - 1291.
55. Ghanem, A.; Schurig, V. Lipase-catalyzed transesterification of secondary alcohols using isopropenyl acetate as acyl donor in organic solvents. *Monateshefte für chemie*, **2003**, 134, 1151-1157.
56. Ghanem, A.; Schurig, V. Entrapment of *Pseudomonas cepacia* lipase with cyclodextrin in sol-gel: application to the kinetic resolution of secondary alcohols. *Tetrahedron: Asymmetry* **2003**, 14, 2547-2555.
57. Ghanem, A.; Schurig, V. Lipase-catalyzed access to enantiomerically pure (*R*)- and (*S*)-*trans*-4-phenyl-3-butene-2-ol. *Tetrahedron: Asymmetry* **2003**, 14, 57-62.
58. Ghanem, A. The utility of cyclodextrins, sol-gel procedure and gas chromatography in lipase-mediated enantioselective catalysis: kinetic resolution of secondary alcohols. *PhD Thesis, University of Tübingen*, **2002**.
59. Hanefeld, U. Reagents for (ir)reversible enzymatic acylations. *Org. Biomol. Chem.*, **2003**, 1, 2405-2415.
60. Monterde, I, M.; Brieva, R.; Sanchez, M. V., Bayod, M.; Gotor, V. Enzymatic resolution of the

chiral inductor 2-methoxy-2-phenylethanol. *Tetrahedron: Asymmetry* **2002**, 13, 1091-1096.

61. Mezetti, A.; Keith, C.; Kazlauskas, J., R. Highly enantioselective kinetic resolution of primary alcohols of the type Ph-X-CH(CH$_3$)-CH$_2$OH by *Pseudomonas cepacia* lipase: effect of acyl chain lengh and solvent. *Tetrahedron: Asymmetry* **2003**, 14, 3917-3924.

62. Hirose, K.; Naka, H.; Yano, M.; Ohashi, S.; Naemura, K.; Tobe, Y. *Tetrahedron: Asymmetry* **2000**, 11, 1199-1210.

63. Kawasaki, M.; Goto, M.; Kawabata, S.; Kometani, T. The effect of vinyl esters on the enantioselectivity of the lipase-catalyzed transesterification of alcohols. *Tetrahedron Asymmetry* **2001**, 12, 585-596.

64. Sakai, T.; Matsuda, A.; Korenaga, T.; Ema, T. *Bull. Chem. Soc. Jpn.,* **2003**, 76, 1819-1821.

65. Paizs, C.; Tosa, M.; Bodai, V.; Szakacs, G.; Kmecz, I.; Simandi, B.; Majdik, C.; Novak, L.; Irimine, F-D.; Poppe, L. Kinetic resolution of 1-(benzofuran-2-yl)ethanols by lipase-catalyzed enantiomer selective reactions. *Tetrahedron Asymmetry* **2003**, 14, 1943-1949.

66. Pchelka, K. B.;Loupy, A.; Plenkiewicz, J.; Blanco, L. Resolution of racemic 1-azido-3-aryloxy-2-propanols by lipase-catalyzed enantioselective acetylation. *Tetrahedron: Asymmetry* **2000**, 11, 2719-2732.

67. Wielechowska, M.; Plenkiewicz, J. -Alkylthio-3-aryloxypropan-2-ols: synthesis and enantiomer separation by lipase-catalyzed transesterification. *Tetrahedron: Asymmetry* **2003**, 14, 3203-3210.

68. Acherar, S.; Audran, G.; Vanthuyne, N.; Monti, H. Use of lipase-catalyzed kinetic resolution for the enantioselective approach toward sesquiterpenes containing quaternary centers: the cuparane family. *Tetrahedron: Asymmetry* **2003**, 14, 2413-2418.

69. Vorlova, S.; Bornscheuer, U. T.; Gatfield, I.; Hilmer, J-M.; Bertram, H-J.; Schmid, R. D. Enantioselective hydrolysis of *d,l*-menthyl benzoate to *l*-(-)-menthol by recombinant *Candida rugosa* lipase LIP1. *Adv. Synth. Catal.* **2002**, 10, 344.

70. Nascimento, M. G.; Zanotto, S. P.; Melegari, S. P.; Fernandes, L.; Sa, M. M. Resolution of α-methylene-β-hydroxy esters catalyzed by free and immobilized *Pseudomonas sp.* lipase. *Tetrahedron: Asymmetry* **2003**, 14, 3111-3115.

71. Joly, S.; Nair, S. M. Studies on the enzymatic kinetic resolution of β-hydroxy ketones. *J. Mol. Catal. B: Enzymatic* **2003**, 22, 151-160.

72. Baumann, M.; Hauer, H., B.; Bornscheuer, U., T. Rapid screening of hydrolases for the enantioselective conversion of 'difficult-to-resolve' substrates. *Tetrahedron: Asymmetry* **2000**, 11, 4781-4790.

73. Krishna, S., H.; Persson, M.; Bornscheuer, U., T. Enantioselective transesterification of a tertiary alcohol by lipase A from *Candida antarctica*. *Tetrahedron: Asymmetry* **2002**, 13, 2693-2696.

74. Brackenridge, I.; McCague, R.; M.; Roberts, S. M.; Turner, N. J. Enzymic resolution of oxalate of a tertiary alcohol using *porcine pancreatic* lipase. *J. Chem. Soc. Perkin Transactions* 1: **1993**, 10, 1093-4.

75. Chen, S-T-; Fang, J-M. Preparation of optically active tertiary alcohols by enzymatic methods. Application to the synthesis of drugs and natural products. *J. Org. Chem.* **1997**, 62, 4349-4357.

76. Kirke, O.; Christense, M. Lipases from *Candida antarctica*: unique biocatalysts from a unique origin. *Organic Process Research & Development* **2002**, 6, 446-451.

77. Ceynowa, J.; Rauchfleisz, M. High enantioselective resolution of racemic 2-arylpropionic acids in an enzyme membrane reactor. *J. Mol. Catal. B: Enzymatic* **2003**, 23, 43-51.

78. Henke, E.; Schuster, S.; Yang, H.; Bornscheuer, W. T. Lipase-catalyzed resolution of ibuprofen. *Monatshefte für chemie* **2000**, 131, 633-638.

79. Shang, C-S.; Hsu, C-S. Lipase-catalyzed enantioselective esterification of (*S*)-naproxen hydroxyalkyl ester in organic media. *Biotechnology Letters* **2003**, 25: 413-416.

80. Adam, W.; Lazarus, M.; Schmerder, A.; Humpf, H-U, Saha-Möller, C. R.; Schreier, P. synthesis of optically active α-hydroxy acids by kinetic resolution through lipase-catalyzed enantioselective acetylation. *Eur. J. Org. Chem.* **1998**, 2013-2018.

81. Kurokawa, M.; Shindo, T.; Suzuki, M.; Nakajima, N.; Ishihara, K.; Sugai, T. Enzyme-catalyzed enantiomeric resolution of N-Boc-proline as the key-step in an expeditious route towards RAMP. *Tetrahedron: Asymmetry* **2003**, 14, 1323-1333.

82. Liljeblad, A.; Lindborg, J.; Kanerva, A.; Katajisto, J.; Kanerva, L. T. Enantioselective lipase-catalyzed reactions of methyl pipecolinate: transesterification and N-acylation. *Tetrahedron Letters* **2002**, 43, 2471-2474.

83. Fujimoto, Y.; Iwadated, H.; Ikekawa, N. Preparation of optically active 2,2'- dihydroxy-1,1'-binaphthyl via microbial resolution of the corresponding racemic diester. *J. Chem. Soc., Chem. Commun.* **1985**, 1333-1334.

84. J-Hernandez, M.; Johnson, V. D.; Holland, L. H.; McNulty, J.; Capretta, A. Lipase-catalyzed stereoselective resolution and desymmetrization of binaphthols. *Tetrahedron: Asymmetry* **2003**, 14, 289-291.

85. Sanfilippo, C.; Nicolosi, G.; Delogu, G.; Fabbri, D. ; Dettori, M. A. Access to optically active 2,2'-dihydroxy-6,6'-dimethoxy-1,1'-biphenyl by a simple biocatalytic procedure. *Tetrahedron: Asymmetry* **2003**, 14, 3267-3270.

86. Theil, F. Enantioselective lipase-catalyzed transesterifications in organic solvents. *Methods in Biotechnology*, **2001**, 15, 277-289.

87. Ciuffreda, P.; Alessandrini, L.; Terranea, G.; Santaniello, E. Lipase-catalyzed selective benzoylation of 1,2-diols with vinyl benzoate in organic solvents. *Tetrahedron: Asymmetry* **2003**, 14, 3197-3201.

88. Kielbasinski, P.; Zurawinski, R.; Albrycht, M.; Mikolajczyk, M. The first enzymatic desymmetrizations of prochiral phosphine oxides. Tetrahedron: Asymmetry **2003**, 14, 3379-3384.

89. Neri, C.; Williams, J. M. J. New routes to chiral Evans auxiliaries by enzymatic desymmetrization and resolution strategies. *Adv. Synth. Catal.* **2003**, 345, 835-848.

90. Bodai, V.; Orovecz, O.; Szakacs, G.; Novak, L.; Poppe, L. Kinetic resolution of trans-2-acetoxycycloalkan-1-ols by lipase-catalyzed enantiomerically selective acylation. *Tetrahedron: Asymmetry* **2003**, 14, 2605-2612.

91. Xu, D.; Li, Z.; Ma, S. Novozym-435-catalyzed efficient preparation of (1S)-ethenyl and ethynyl 2,3-allenols and (1R)-ethenyl and ethynyl 2,3-allenyl acetates with high enantiomeric excess. *Tetrahedron: Asymmetry* **2003**, 14, 3657-3666.

92. Faure, B.; Iacazio, G.; Maffei, M. First enzymatic resolution of a phosphane–borane complex *J. Mol. Cat. B* : Enzymatic **2003**, 26, 29-32.

93. Skupinska, K., A.; McEachern, E., J.; Baird, I., R.; Skerlj, R., T.; Bridger, G., J. Enzymatic resolution of bicyclic 1-heteroarylamines using *Candida antarctica* lipase B. *J. Org. Chem.* **2003**, 68, 3546-3551.

94. Irurre, J.; Riera, M.; Guixa, M. Gram-scale synthesis of (+)- and (-)-trans-10-amino-9-hydroxy-9,10-dihydrophenanthrene by enzymic hydrolysis. *Chirality* **2002**, 14: 490-494.

95. Yoshida, N.; Aono, M.; Tsubuki, T.; Awano, K.; Kobayashi, T. Enantioselective synthesis of a chiral pyridazinone derivative by lipase-catalyzed hydrolysis. *Tetrahedron: Asymmetry* **2003**, 14, 529-535.

96. Forro, E.; Fülöp, F. Lipase-catalyzed enantioselective ring opening of unactivated alicyclic-fused β-lactams in an organic solvent. *Organic Letters* **2003**, 5(8), 1209-1212.

97. Gyarmati, Z. Cs.; Liljeblad, A.; Rintola, M.; Bernath, G.; Kanerva, T. L. Lipase- catalyzed kinetic resolution of 7-, 8- and 12-membered alicyclic β-amino esters and N-hydroxymethyl-β-lactam enantiomers. *Tetrahedron: Asymmetry* **2003**, 14, 3805-3814.

98. Schmid, A.; Dordick, S. J.; Hauer, B.; Kiener, A.; Wubbolts, M.; Witholt, B. Industrial biocatalysis today and tomorrow. *Nature*, January **2001**, Vol. 409, 258-268.

99. Smidt, H.; Fischer, A.; Fischer, P.; Schmid, R. D.; Stelzer, U. Production of optically active amines. *Ger. Offen*, **1996**, 11.

100. Breuer, M.; Ditrich, K.; Habicher, T.; Hauer, B.; Kesseler, M.; Stuemer, R.; Zelinski, T. Industrial methods for the production of optically active intermediates. *Angew. Chem. Int. Ed.* **2004**, 43, 788-824.

101. Ladner, W. E.; Whitesides, G. M. Lipase-catalyzed hydrolysis as a route to esters of chiral epoxy alcohols. *J. Am. Chem. Soc.* **1984**, 106, 7250-7251.

102. Sheldon, R. A. in *Specialty Chemicals*, Elsevier, London, **1991**.

103. Reetz, M. T.; Jaeger, K. E. Superior biocatalysts by directed evolution. *Topics in Current Chemistry* **1999**, 200 (Biocatalysis: From Discovery to Application), 31-57.

ENZYMATIC KINETIC RESOLUTION

KAORU NAKAMURA[a] and TOMOKO MATSUDA[b]
[a]*Institute for Chemical Research, Kyoto University, Uji, Kyoto 611-0011, Japan*
[b]*Department of Materials Chemistry, Faculty of Science and Technology, Ryukoku University, Otsu, Shiga 520 - 2194, Japan*

1. Introduction

Biocatalysts, mainly hydrolytic enzymes and oxidoreductases, have been used for organic reactions due to their excellent enantioselectivities and environmentally friendliness.[1] Typical enzymatic reactions used for the organic synthesis are shown in Figure 1. Especially, hydrolytic enzymes for kinetic resolutions of racemates have been utilized widely because of their high stabilities, wide substrate specificities, lack of cofactor requirements and high availabilities.

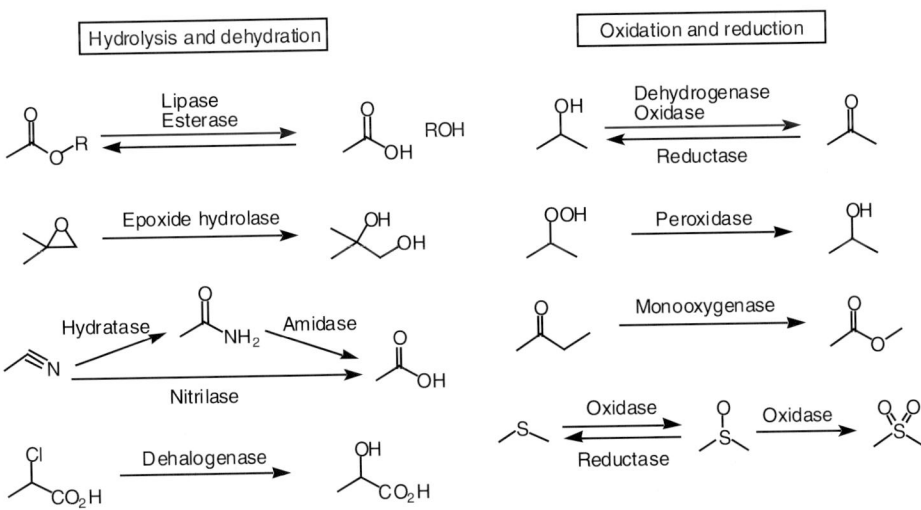

Figure 1. Biocatalytic reactions useful for organic synthesis

One of the most studied hydrolytic enzymes is chymotrypsin which represents a group of serine proteases. It catalyzes the hydrolysis of peptides to amino acids. The reaction mechanism of the reaction is shown in Figure 2. Two amino acid residues,

231

F. Toda (ed.), Enantiomer Separation, 231-266.

Asp and His, of the enzyme locate together to help a nucleophilic attack of Ser on the carbonyl carbon of the substrate. The reaction proceeds through a tetrahedral transition state, cleavage of the peptide bond, rapid diffusion of amine moiety leaving acyl-enzyme intermediate followed by the hydrolysis, giving an acid product.

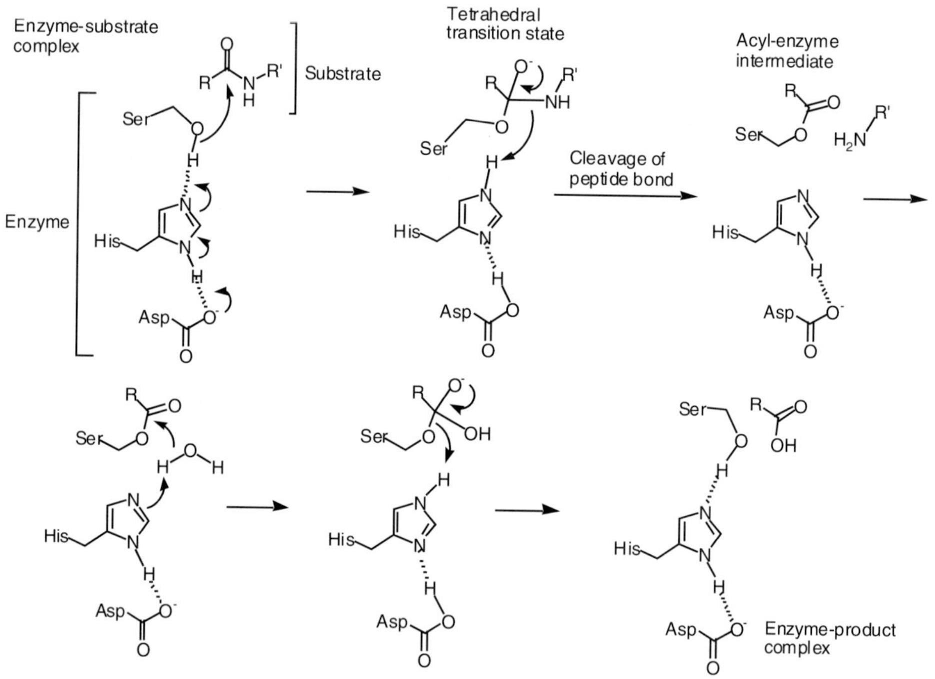

Figure 2.   Reaction mechanism of hydrolysis of a peptide by chymotrypsin

1.1 ENANTIOSELECTIVITY

Most of the enzymes show extremely strict chiral recognitions, and only one of the enantiomers can be the substrate of the enzyme. For example, chymotrypsin incorporates L-peptides only to the enzyme-substrate binding site to form enzyme-substrate complex, so it shows very high enantioselectivity (Figure 3 (a)). Oxidoreductases also form the enzyme-substrate complex of only one enantiomer, so enantioselectivities are high when isolated enzymes are used for reactions instead of whole cells containing both (R)- and (S)-specific enzymes, which leads to overall low enantioselectivities.

In the case of lipases and esterases, chiral recognitions are not so strict. Both enantiomers were incorporated to the enzyme to form the substrate-enzyme complex. However, the slow reacting enantiomer lacked the necessary hydrogen-bonding interaction, for example in the hydrolysis of menthol acetate, between the substrate menthol and the enzyme histidine group for the reaction to proceed further (Figure 3(b)).[2,3] The explanation was also supported by the observation in the esterification reaction of 1-phenylethanol by lipases.[4] Km values of the slow and fast reacting enantiomer were almost same (Km(slow) : Km(fast) = 1 : 1 - 0.3).   However, Vmax values were significantly different (Vmax(slow) : Vmax(fast) = 1 : 150 - 500).

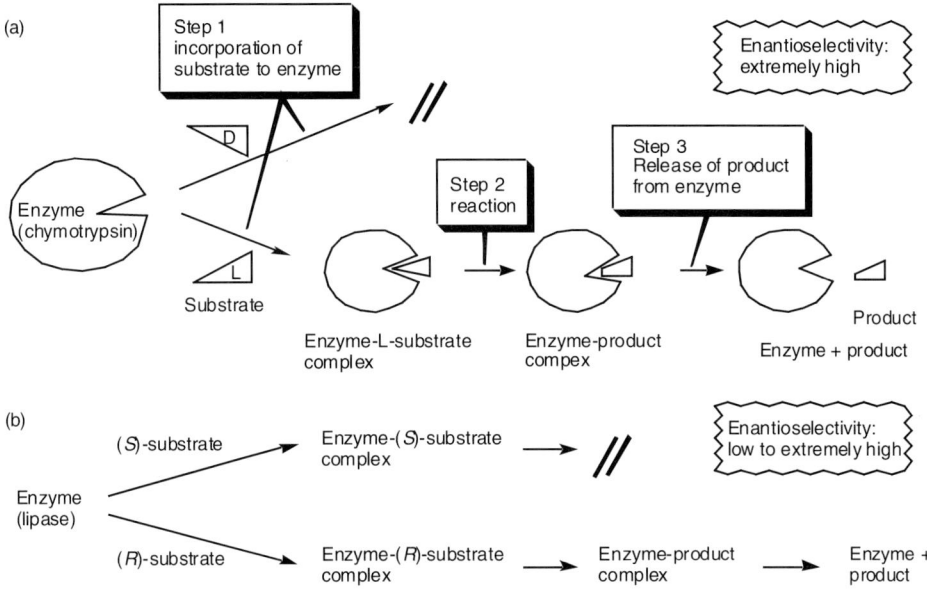

Figure 3.   Chiral recognitions by the enzymes

1.2 RECOGNITION OF CHIRALITY REMOTE FROM REACTION CENTER

One of the most distinguishable features of the biocatalytic reactions compared with the chemical ones is that the biocatalysts can recognize a remote chiral center apart from the reaction center of substrates.   As shown in Figure 4, biocatalysts can recognize a chirality separated by six bonds as well as at the reaction center.

1.3 EVALUATION OF ENATIOSELECTIVITY: E-VALUE

To evaluate the enantioselectivity of the reaction, usually, enantiomeric excess, ee, of products is used.   However, when a racemic substrate is reacted, ee of the product as well as that for the substrate changes depending on the conversion as shown in Figure 5(a).   To evaluate the enantioselectivity of this type of reaction, the ratio of the specificity constants of the enantiomers, E-value, was introduced (Figure 5(b)).[5]   The methods of the calculation of E for the reaction of racemates is shown in Figure 5(c).

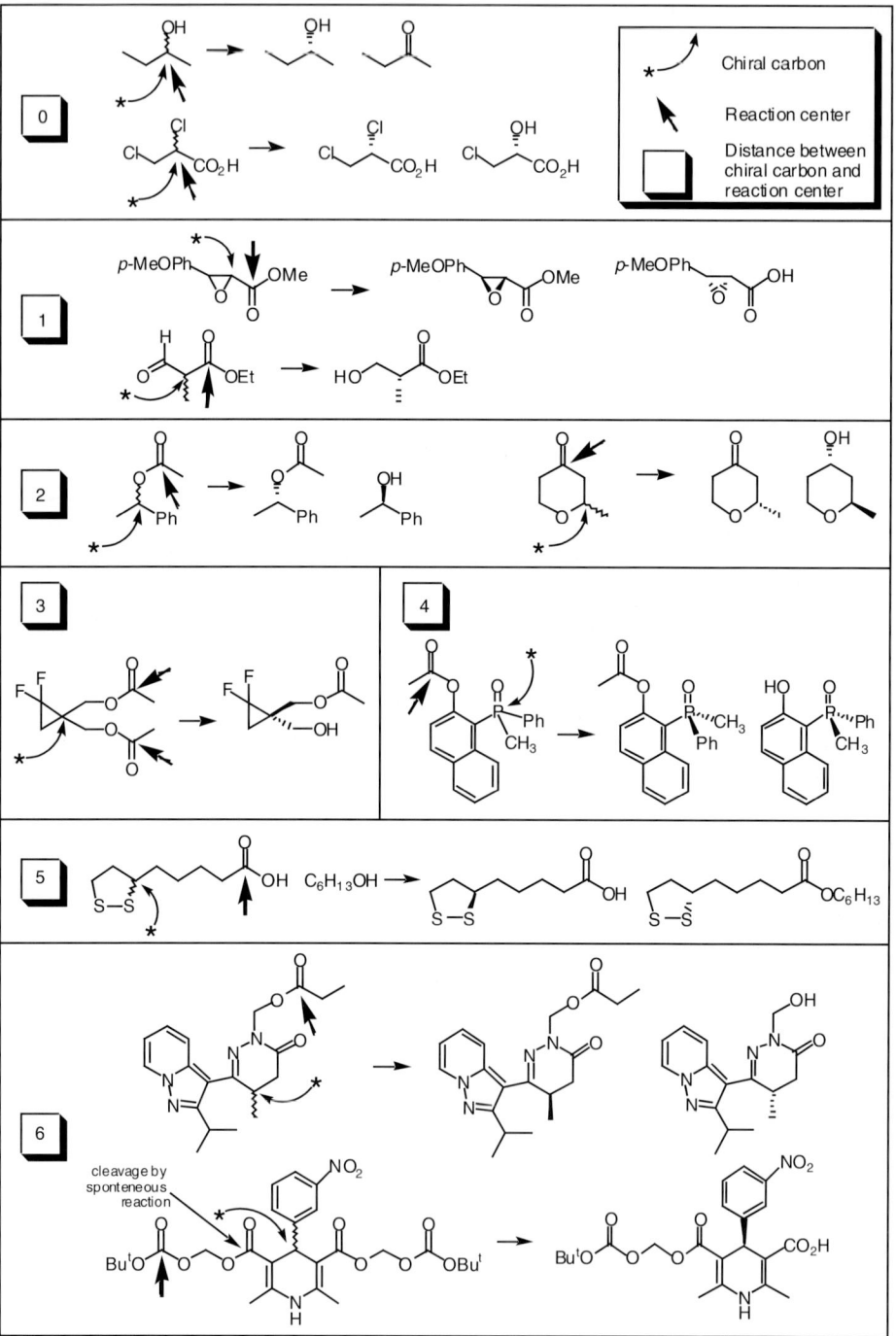

Figure 4. Recognition of chiralities at various distances from reaction centers

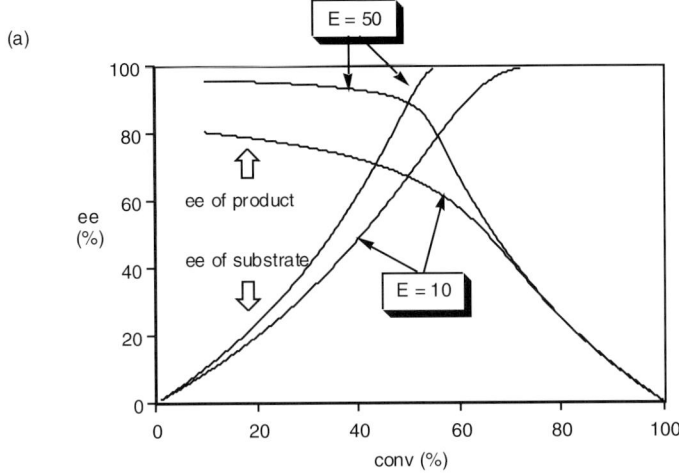

(a)

(b) When S-substrate reacts faster than R-substrate:

$$E = \frac{V_S / K_S}{V_R / K_R}$$

V: Maximal velocity
K: Michaelis constant

For the reaction of racemates:

$$E = \frac{\text{initial reaction rate of } S}{\text{initial reaction rate of } R}$$

(c) Calculation of E from conversion and ee of substrate:

$$E = \frac{\ln[\,(1 - conv) \times (1 - ee(substrate)\,)\,]}{\ln[\,(1 - conv) \times (1 + ee(substrate)\,)\,]}$$

Calculation of E from conversion and ee of product:

$$E = \frac{\ln[\,1 - conv \times (1 + ee(product)\,)\,]}{\ln[\,1 - conv \times (1 - ee(product)\,)\,]}$$

Figure 5.   Evaluation of enantioselectivity: E-values[5]

2. Esterification of alcohols and hydrolysis of esters

Various lipases and esterases have been used for the enantioselective esterification of alcohols and hydrolysis of esters.   For example, *Burkholderia cepacia* lipases (PS, Amano Enzyme Inc.) and *Candida antarctica* lipase (CAL, Novozymes) have been widely used for its wide substrate specificities, high activities and chemo, regio and enantioselectivities.   Fundamentals and some selected applications are shown in this section.   The origins and abbreviations of lipases introduced here are as follows.

PS: *Burkholderia cepacia*
AK: *Pseudomonas sp.*
AH: *Pseudomonas sp.*
LIP: *Pseudomonas aeruginosa*
CRL: *Candida rugosa*
CAL: *Candida antarctica*
QL: *Alcaligenes sp.*

2.1 REVERSIBILITY OF REACTION: ESTERIFICATION vs. HYDROLYSIS

Hydrolytic enzymes such as lipases catalyze hydrolysis of esters in aqueous media, but when used in non-aqueous media such as organic solvents, ionic liquids and supercritical fluids, they catalyze reverse reactions; the synthesis of esters.   For example, lipases in natural environment catalyze the hydrolysis of fatty acid esters as shown in Figure 6(a).   However, when they are used in organic solvents, they catalyze the esterification reaction (Figure 6(b)).

Various methods have been developed to shift the equilibrium between esterification and hydrolysis toward the ester formation.[6]   For example, as an acetyl donor, vinyl acetate and isopropenyl acetate have been used because the resulting unnecessary alcohol, vinyl or isopropenyl alcohol, spontaneously changes to the corresponding aldehyde or ketone as shown in Figure 6(c). [6a]

Acid anhydrides have been also used due to the irreversibities of the reaction and the advantage in downprocessing.[6b]   For example, the esterification of a racemic alcohol with succinic anhydride gives (S)-alcohol and (R)-ester (Figure 6(d)).   The latter has an acid moiety so can be separated easily from the former due to the solubility difference in weakly basic water.

7

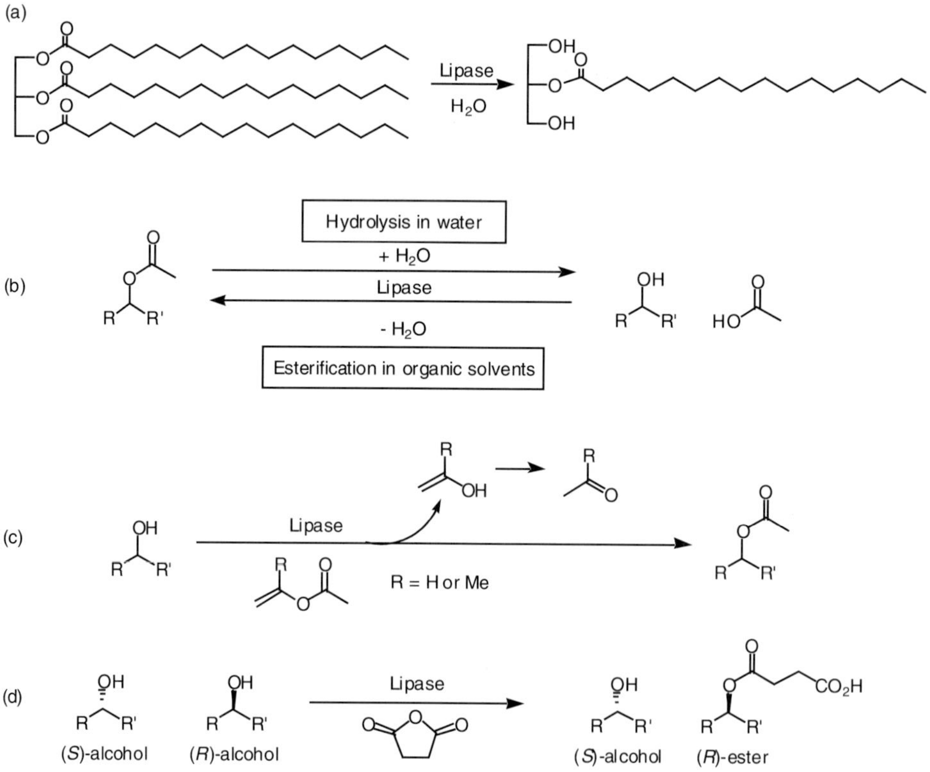

Figure 6.   Forward (natural) and reverse reactions catalyzed by hydrolytic enzymes, lipases

2.2 SOLVENTS FOR ESTERIFICATION AND HYDROLYSIS REACTIONS

Various kinds of solvents have been used for reactions catalyzed by hydrolytic enzymes.[7] For hydrolysis, aqueous buffer solutions as well as mixtures of water and organic solvents have been used. For esterifications, organic solvents[7a-d] (Figure 7(a), (b)), supercritical fluids[7e-g] (Figure 7(c), (d)), and ionic liquids[7h] (Figure 7(e)) have been used. The enantioselectivities largely depend on the solvent used. Examples are as follows.

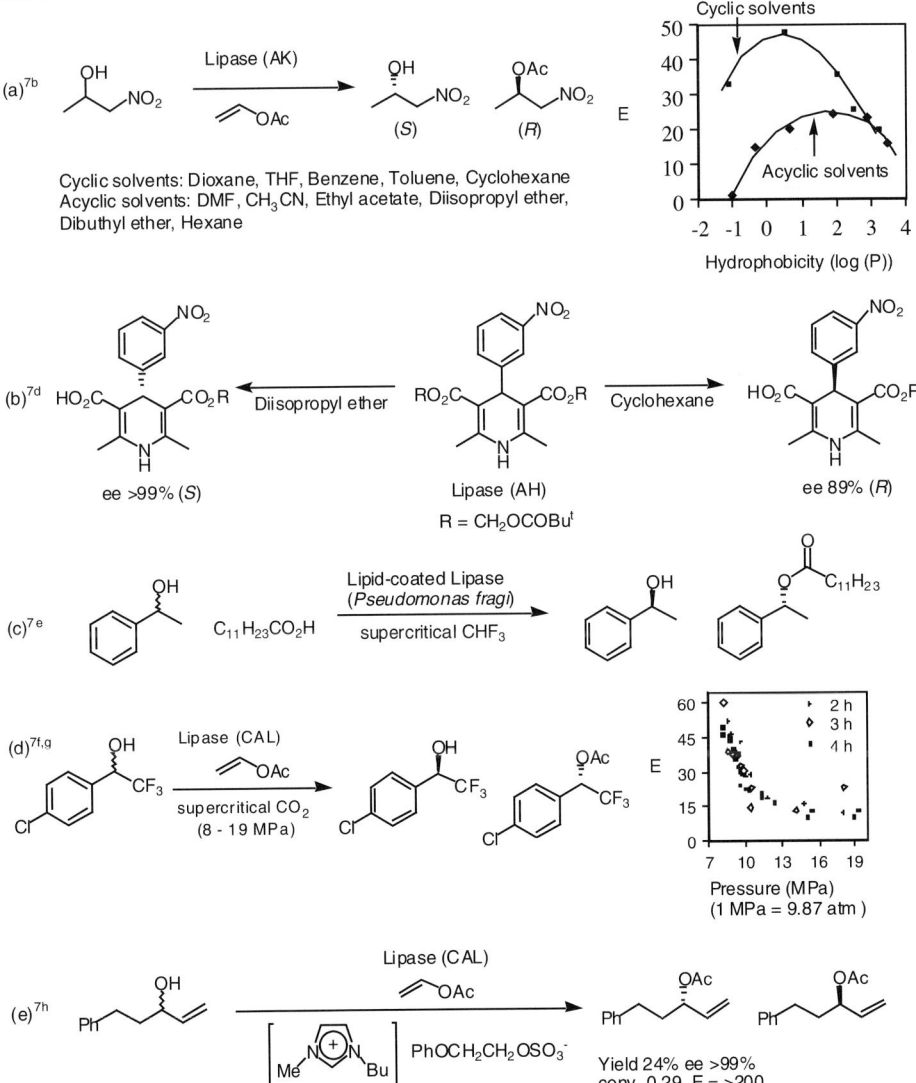

Figure 7. Solvents for hydrolytic enzymes

Example 1: Enantioselectivity of the acetylation of a nitro alcohol by lipase (AK) was affected by the hydrophobicity parameter (log (P)). However, cyclic and acyclic solvents showed different absolute values (Figure 7(a)).[7b]
Example 2: Enantioselectivities of the hydrolysis of 1,4-dihydropyridine, cardiovascular drug, by lipase (AH) in diisopropyl ether (ee >99% (S)) was completely opposite to that in cyclohexane (ee 89% (R)) (Figure 7(b)).[7d]
Example 3: Enantioselectivity of the esterification of 1-(p-chlorophenyl)-2,2,2-trifluoroethanol in supercritical carbon dioxide was largely affected by pressure which changes the density of the solvent (Figure 7(d)).[7f,g]

2.3 GENETIC ENGINEERING OF ENZYMES FOR HYDROLYSIS

Genetic engineering of hydrolytic enzymes has been widely conducted to improve the activities, stabilities, and enantioselectivities. Here, examples for the point mutation and directed evolution are shown in Figure 8.[8] A mutant of lipase (PS) replaced three amino acids (F221L, V226L, L287I) by site-specific mutagenesis showed the inversion of enantioselectivity in the hydrolysis of 1,4-dihydropyridine (Figure 8(a)).[8a] The positions of two of the mutations, V226L, L287I, were at the vicinity of reaction center (catalytic triads) of the lipase, but F221L located at the end of β-sheet remote from the reaction center. The double mutant (V226L, L287I) showed no change in enantioselectivity.

The improvement in enantioselectivity by the directed evolution of *Pseudomonas aeruginosa* lipase is shown in Figure 8(b).[8b] The combination of different mutagenesis methods (error-prone PCR and site-specific saturation mutagenesis) improved the enantioselectivity from E=1.1 in wild type to E=25.8.

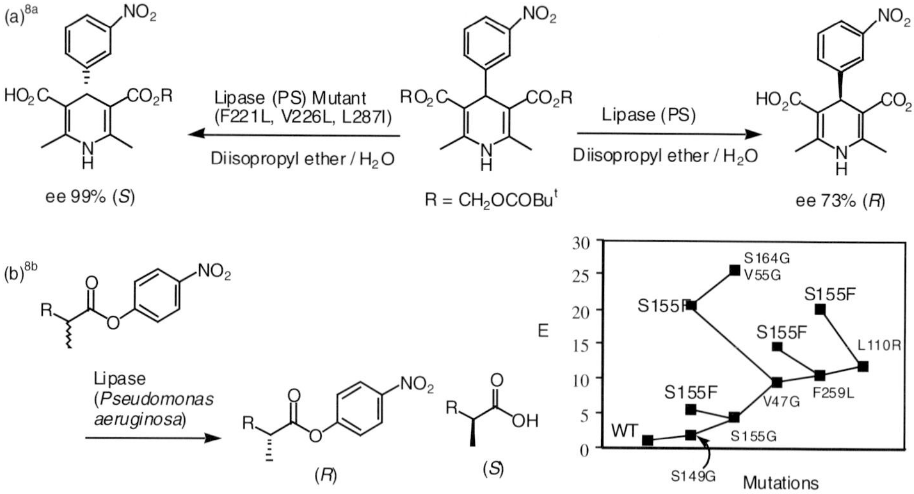

Figure 8.   Genetic engineering of enzymes for hydrolysis of esters

2.4 HOLLOW-FIBER MEMBRANE REACTOR FOR THE LIPASE CATALYZED HYDROLYSIS: SYNTHESIS OF DILTIAZEM

A key intermediate in the synthesis of a useful coronary vasodilator, Diltiazem, (-)-3-phenylglycidic acid ester, (-)-MPGM, was synthesized through lipase-catalyzed optical resolutions (Figure 9).[9] Hollow-Fiber membrane reactor was developed for this purpose.[9c] The reactor had a toluene layer, sponge layer containing enzyme, and aqueous layer (Figure 9 (b)). (+/-)-MPGM was hydrolyzed to the corresponding (2S, 3R)-acid remaining (-)-MPGM. The unnecessary acid was removed efficiently by the spontaneous decarboxylation followed by the formation of water-soluble complex with $NaHSO_3$. The pressure of the toluene layer was kept slightly higher than that of the aqueous layer, so that the substrate moved to the sponge layer but the water did not moved to the toluene layer, preventing the decomposition of MPGM.

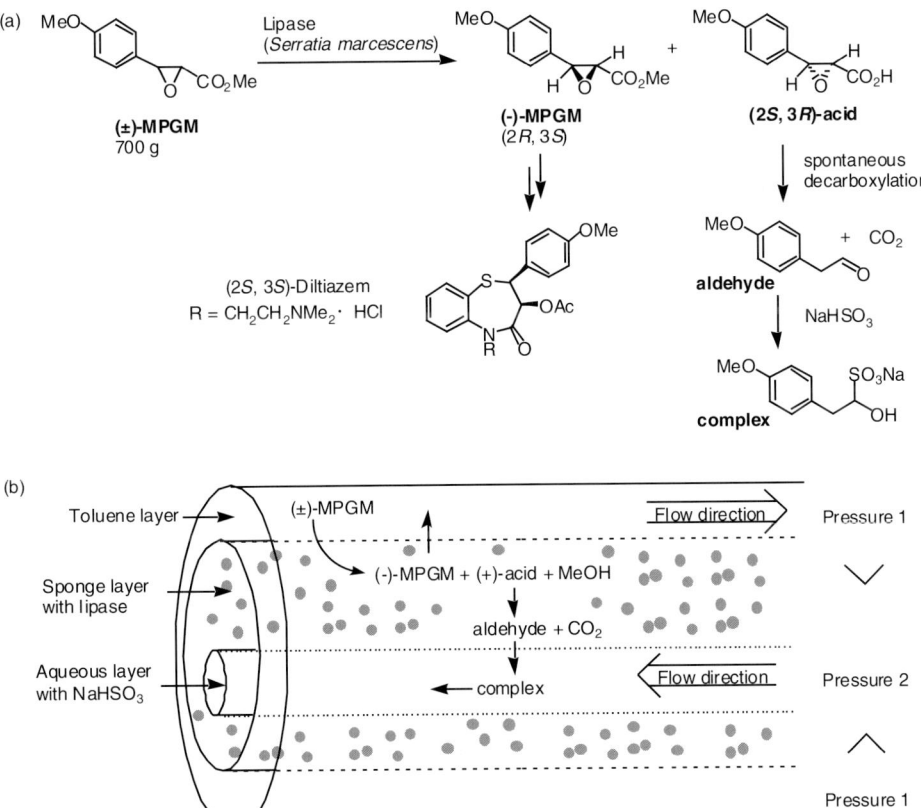

Figure 9. Optical resolution of 3-phenylglycidic acid methyl ester (MPGM) by lipase and separation of the optically active product, (-)-MPGM, from unnecessary acid using Hollow-Fiber membrane reactor[9]

2.5 LIPASE CATALYZED OPTICAL RESOLUTION COUPLED WITH IN SITU INVERSION: SYNTHESIS OF PRALLETHRIN (PYRETHROID) etc.

Lipase catalyzed optical resolution of the racemate gives 50% of the products at 100% ee. However, when the resolutions are coupled with reactions which proceed with retention with the substrate but inversion with product, or vise versa, then in total, 100% yield and 100% ee can be achieved theoretically.[10]

This strategy was taken for the synthesis of pyrethroids using microbial lipases (Figure 10 (a)).[10a,b] Hydrolysis of 4-acetoxy-3-methyl-2-(2-propynyl)-cyclopent-2-enone was catalyzed by *Arthrobacter* lipase in a two-liquid phase reaction system of water and the insoluble substrate. This hydrolysis proceeded even at a substrate concentration of 80 w/v%. After the enzymatic hydrolysis, the resulting mixture was mesylated and hydrolyzed with aqueous $CaCO_3$ with the inversion of (R)-mesylate and retention of (S)-acetate, affording the corresponding (S)-alcohol in 83% yield and 91.4% ee from the racemic acetate. An efficient process was developed for the total conversion of the racemic substrate to the (S)-product which was an important intermediate for the preparation of Prallethrin.

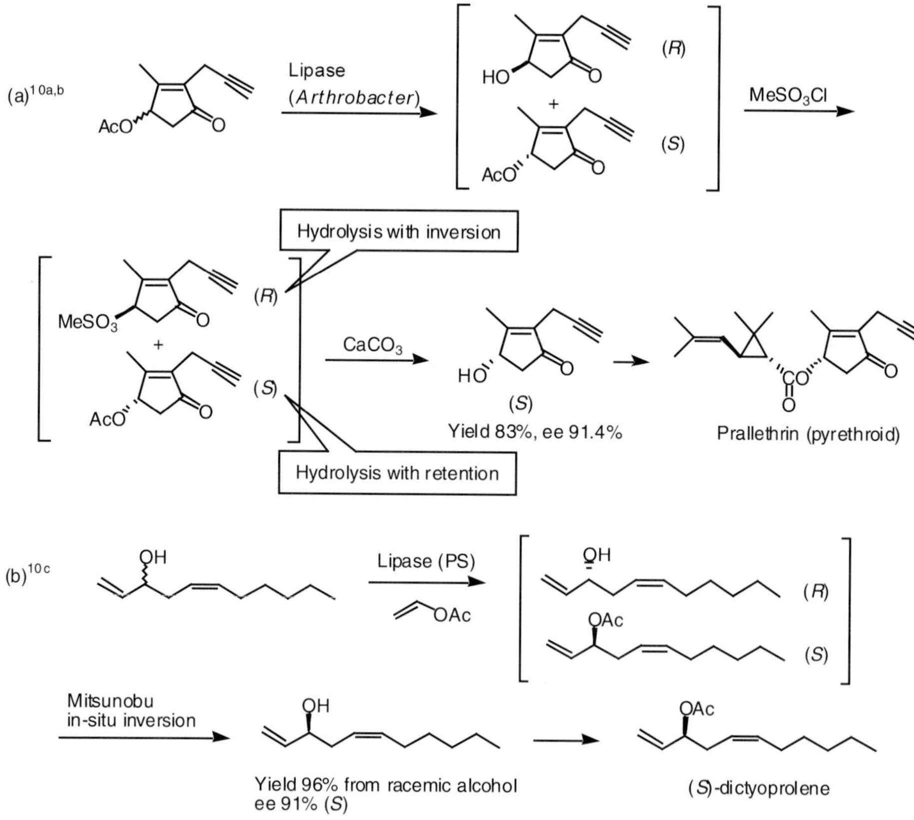

Figure 10. Lipase catalyzed resolution coupled with a chemical inversion

Another example employed Mitsunobu reaction for the inversion reaction (Figure 10(b)).[10c] A single enantiomer of a (stereo)chemically labile allylic-homoallylic alcohol was obtained in 96% yield and 91% ee from the racemate through a lipase-catalyzed kinetic resolution coupled with in situ inversion under carefully controlled (Mitsunobu) conditions. Using this reaction, the algal fragrance component, (S)-dictyoprolene, was synthesized.

2.6 DYNAMIC KINETIC RESOLUTION USING HYDROLYTIC ENZYMES

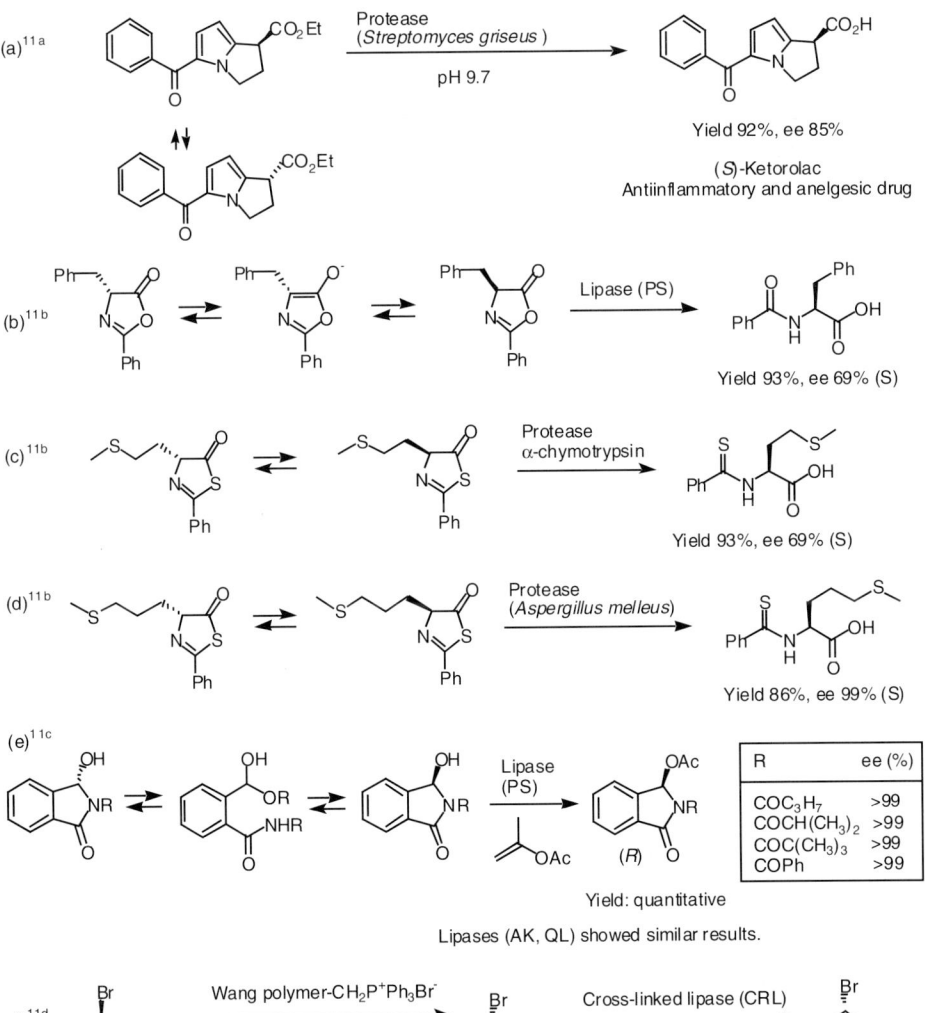

Figure 11. Dynamic kinetic resolutions of esters and alcohols

Dynamic kinetic resolution of racemates to obtain 100% yield of products with 100% ee theoretically are possible using enantioselective hydrolytic enzymes when the substrate racemizes but the product does not under the reaction conditions. There are various hydrolysis and esterification examples for this type of reactions (Figure 11).[11] For example, this method was applied for the synthesis of (S)-Ketorolac, antiinflammatory and anelgesic drug (Figure 11(a). The hydrogen at the α-position of Ketorolac ethyl ester substrate is more acidic than that of the product acid.[11a] Therefore, at pH 9.7, the substrate racemized but the product did not. By taking the advantage of this acidity difference, the protease catalyzed enantioselective hydrolysis of the ester was conducted giving (S)-Ketorolac in 92% yield and 85% ee.

2.7 DYNAMIC KINETIC RESOLUTIONS BY BIOCATALYSTS COUPLED WITH METAL CATALYSTS

Dynamic kinetic resolutions of secondary alcohols and amines have been achieved by the combination of biocatalysts with metal catalysts.[12] For example, a metal catalyst was used to racemize the substrate, phenylethanol, and a lipase was used for the enantioselective esterification as shown in Figure 12. The yield was improved from 50% in kinetic resolution without racemization of the substrate to 100% with metal catalyzed racemization.

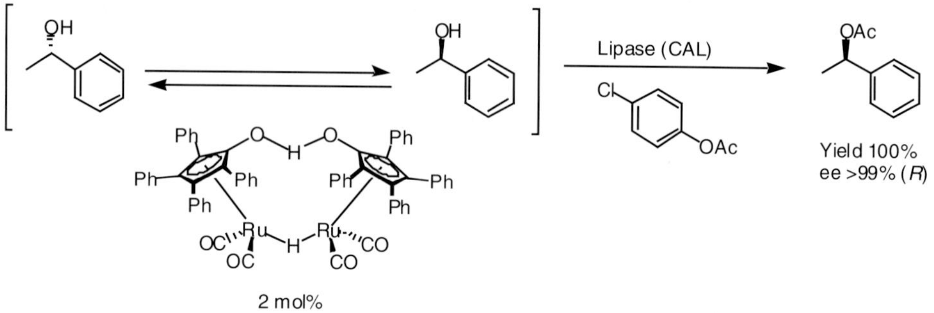

Figure 12. Dynamic kinetic resolution by biocatalyst coupled with ruthenium catalyst[12b]

2.8 EXAMPLES FOR ESTERIFICATIONS AND HYDROLYSIS OF ESTERS

There are a numerous amount of examples for optical resolutions through hydrolytic-enzyme-catalyzed esterifications[13] and hydrolysis.[14] Only selected reactions are shown in Figure 13-16. For example, in a large scale optical resolution of 1-phenylethanol by lipase (PS) with vinyl acetate, 67.5 Kg of (R)-1-phenylethanol was successfully synthesized with the flow system as shown in Figure 13 (a).

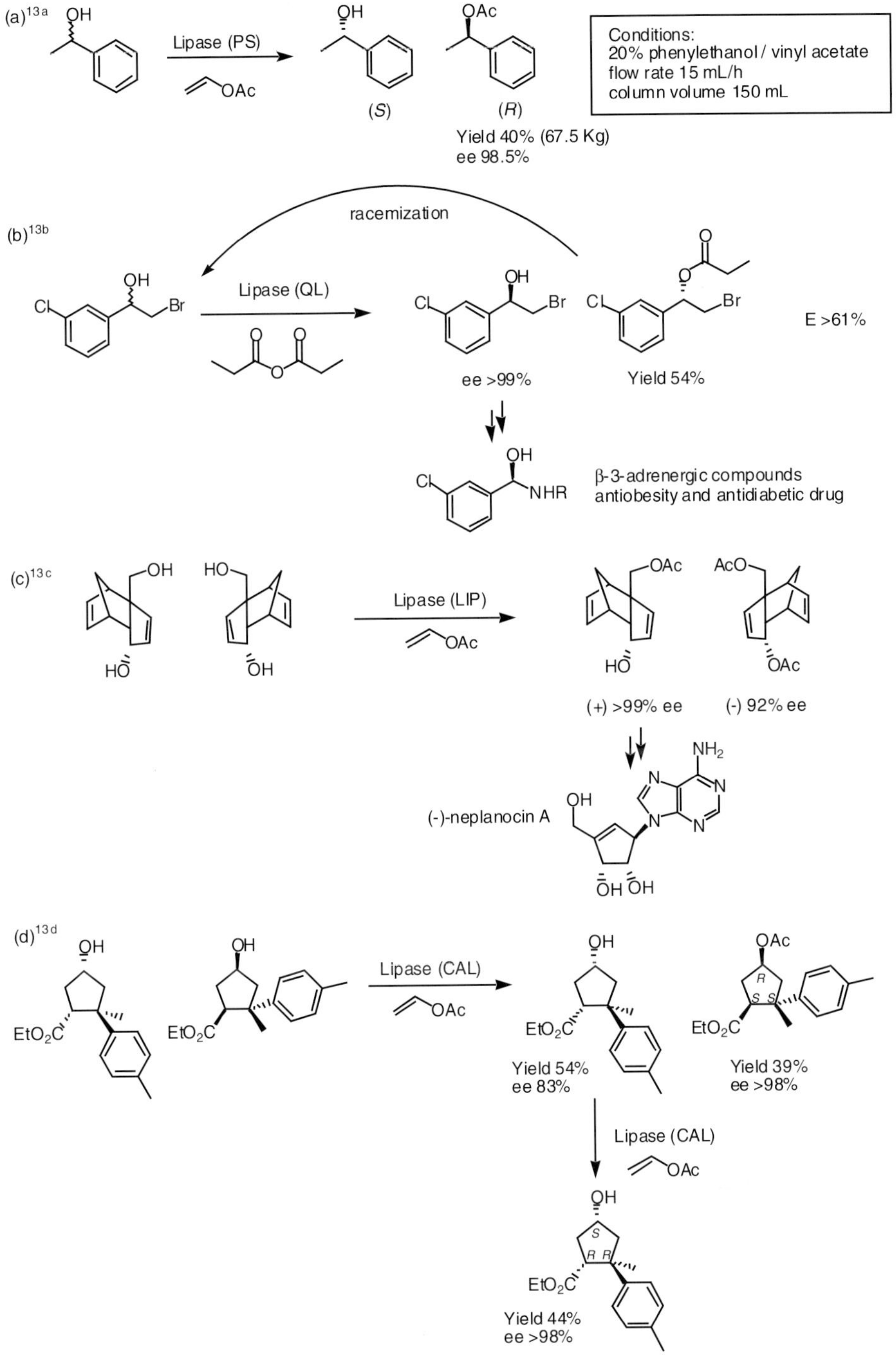

Figure 13. Examples for optical resolutions through esterifications

For the hydrolysis of cyano acetate with lipase, a method to improve the enantioselectivity was developed. The addition of a thiacrown ether largely improved the enantioselectivity to E = >700 from E = 53 without any additives (Figure 14(a)).

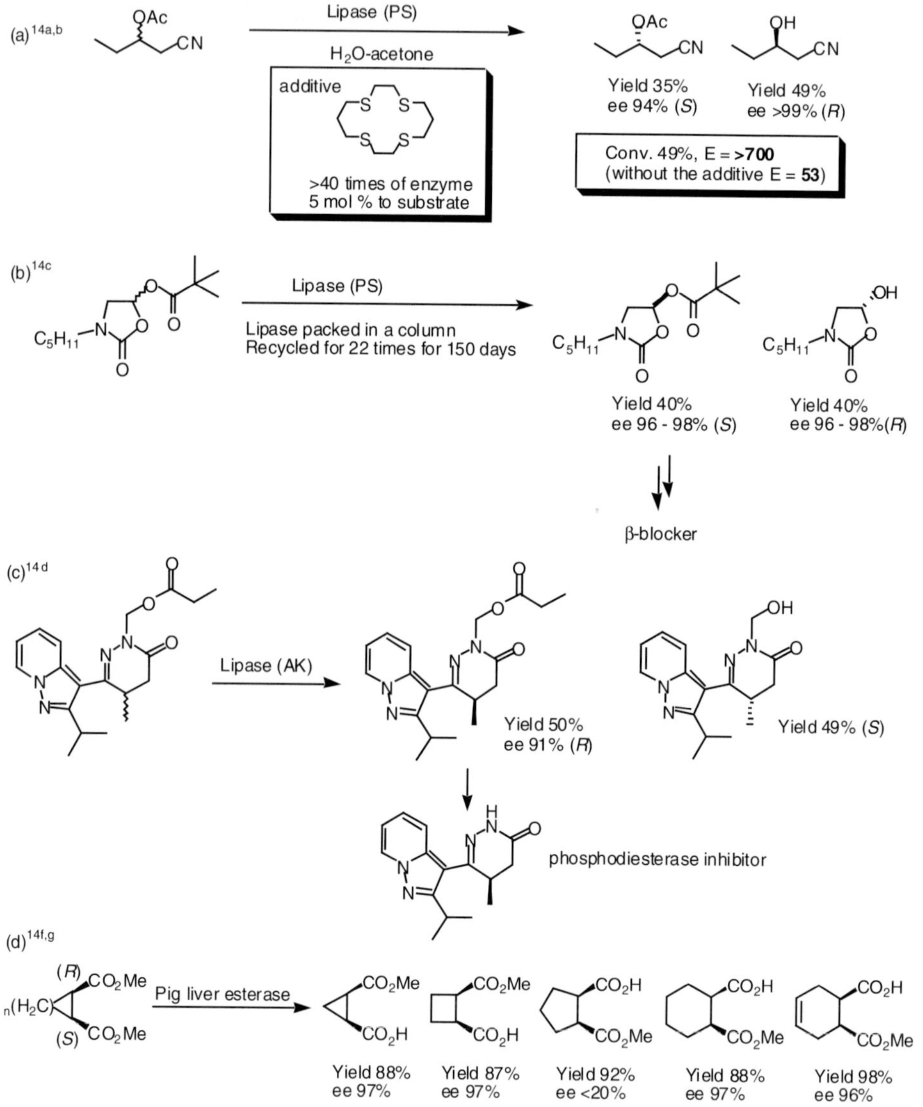
(a)14a,b
OAc
CN
Lipase (PS)
H2O-acetone
additive
>40 times of enzyme
5 mol % to substrate
OAc
CN
OH
CN
Yield 35%
ee 94% (S)
Yield 49%
ee >99% (R)
Conv. 49%, E = >700
(without the additive E = 53)
(b)14c
C5H11
Lipase (PS)
Lipase packed in a column
Recycled for 22 times for 150 days
Yield 40%
ee 96 - 98% (S)
Yield 40%
ee 96 - 98%(R)
β-blocker
(c)14d
Lipase (AK)
Yield 50%
ee 91% (R)
Yield 49% (S)
phosphodiesterase inhibitor
(d)14f,g
(R) CO2Me
n(H2C)
(S) CO2Me
Pig liver esterase
Yield 88%
ee 97%
Yield 87%
ee 97%
Yield 92%
ee <20%
Yield 88%
ee 97%
Yield 98%
ee 96%

Figure 14. Examples for optical resolutions through hydrolysis of esters

Chiral fluorinated compounds have been also synthesized due to the increasing attentions on the unique physical and biological features induced by fluorine atoms. Therefore, various racemic fluorinated alcohols were optically resolved (Figure 15).[15]

For the resolution of cyanopentafluorophenylethanol with lipase, the reaction temperature was decreased to improve the enantioselectivities (Figure 15(b)). The reactions of fluorinated meso compounds were conducted to obtain fluorinated amino acid (Figure 15(d, e)).   Esterification of meso alcohol gave the corresponding (R)-amino acid, whereas the hydrolysis gave the corresponding (S)-product.

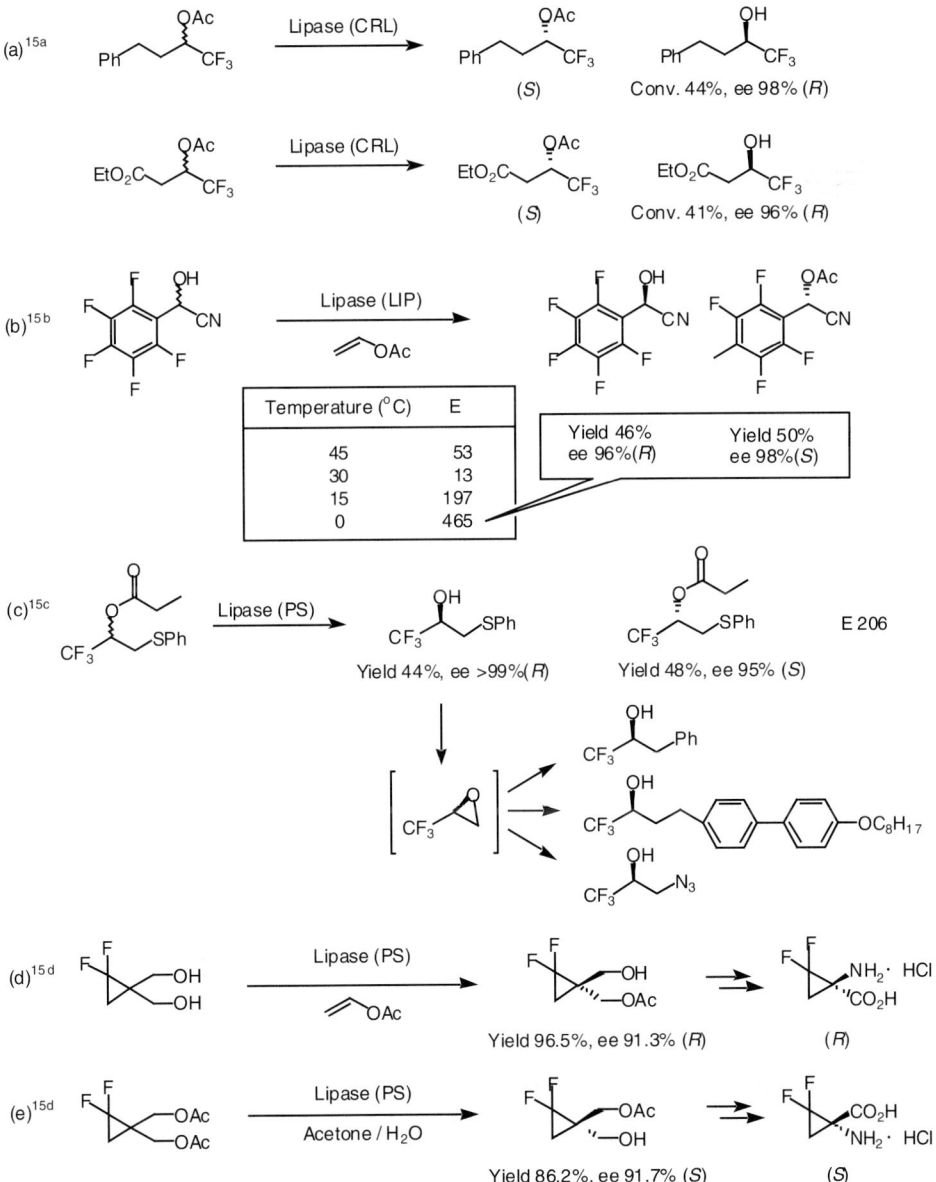

Figure 15.   Examples for optical resolutions through esterifications and hydrolysis of fluoro compounds

Hydrolytic enzymes can also catalyzed the esterification of alcohols or acids with hetero atoms.[16,17]   Some examples for the reactions of phosphorous and sulfur compounds by lipases are shown in Figure 16.   By the repeated enantioselective acylation and hydrolysis of a hydroxyl phosphonate and its acetate with lipase AH, phosphonic acid analogue of carnitine (essential cofactor of fatty acid metabolism), (R)-phosphocarnitine, and its enantiomer were synthesized as shown in Figure 16 (b).

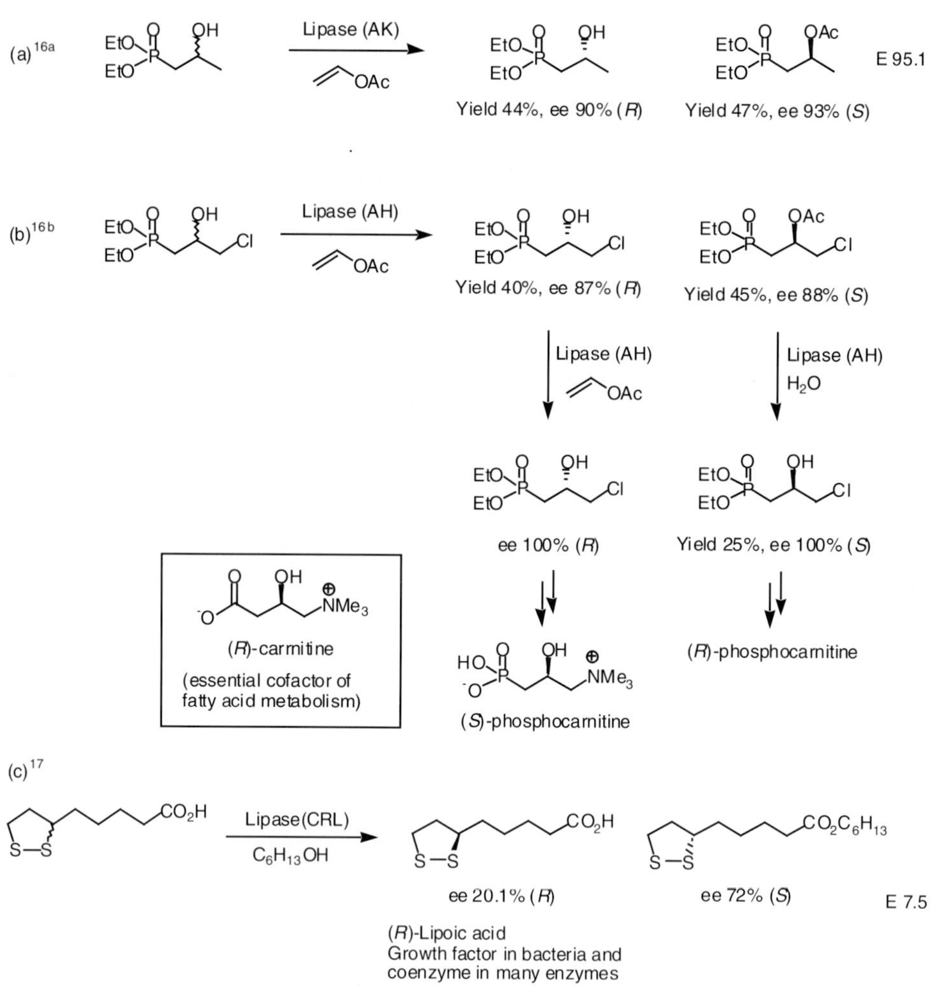

Figure 16.   Examples for optical resolutions through esterifications and hydrolysis of phosphorous and sulfur compounds

2.9 P, S AND HELICAL-CHIRAL COMPOUNDS

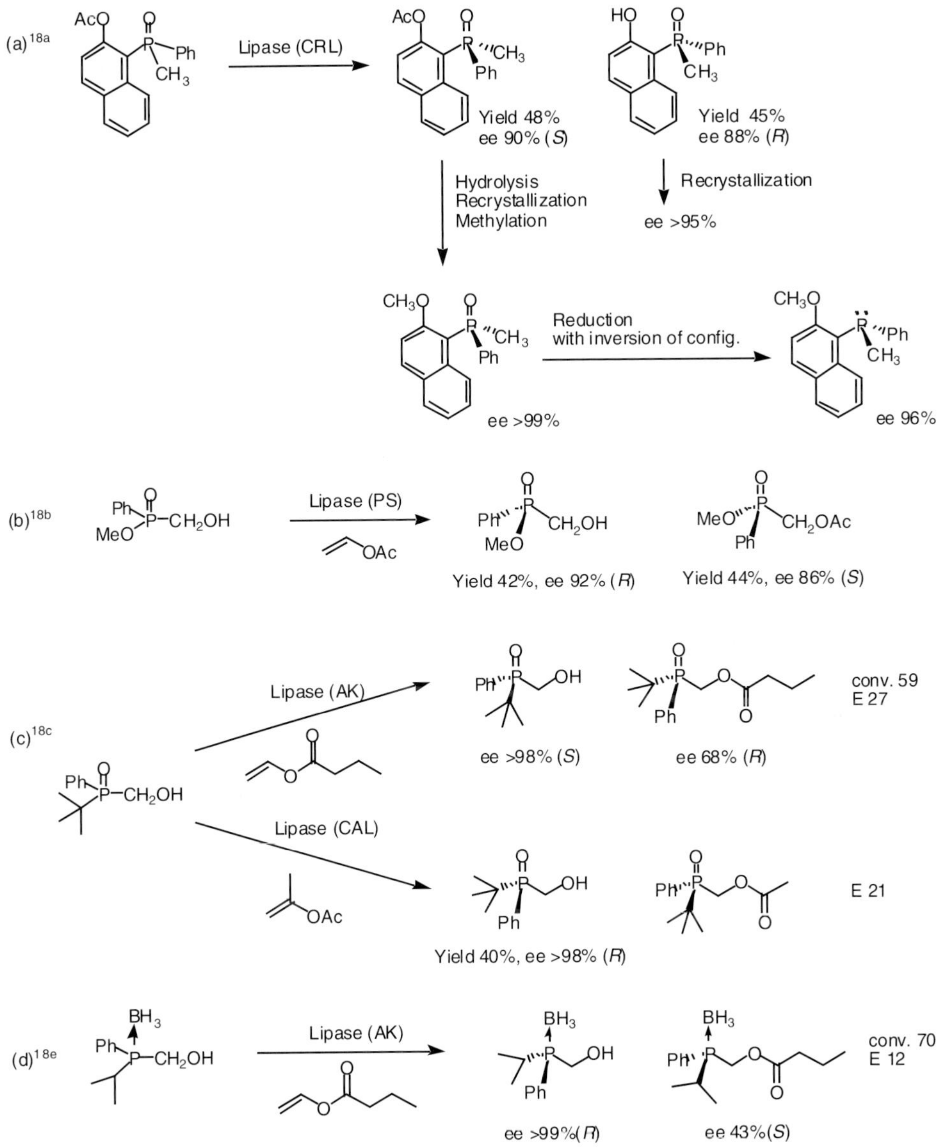

Figure 17. Optical resolution of P-chiral compounds with hydrolytic enzymes

P-Chiral phosphine and phosphine oxide compounds were synthesized through lipase-catalyzed optical resolution of the corresponding racemic substrate (Figure 17).[18] For example, lipase from *Candida rugosa* (CRL) was used for the enantioselective hydrolysis of acetoxynaphtyl phosphine oxide compounds (Figure 17 (a)). (R)-Enantiomer was hydrolyzed selectively, leaving (S)-acetoxy compounds, which was hydrolyzed and recrystallized to improve the ee to >95%. The methylation of these chiral phosphine oxides followed by the reduction with triethyl amine/trichlorosilane with inversion of the configuration yielded chiral phosphines, potential ligands for transition metal catalyzed reactions.

P-Chiral phosphine-borane compounds were synthesized through lipase-catalyzed optical resolution (Figure 17 (d)).[18e] Alkyl(1-hydroxymethyl)phosphine-boranes (up to ee 99%) were obtained using lipase AK or CAL.

S-chiral compounds were synthesized through lipase-catalyzed reactions. For example, chiral sulfoxide was synthesized through lipase-catalyzed hydrolysis of the ester to give (R)-ester and (S)-acid (Figure 18).[19]

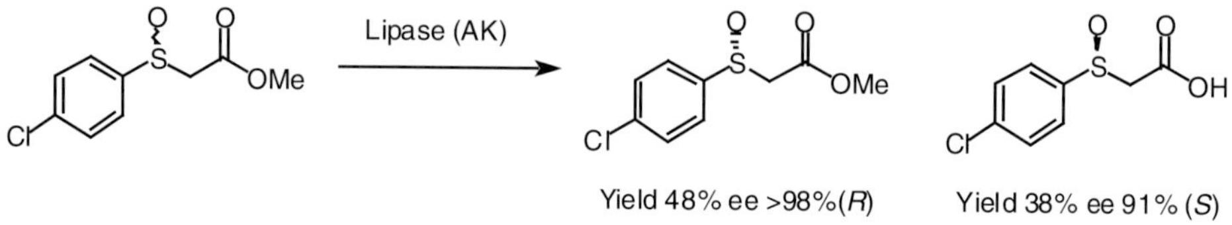

Yield 48% ee >98%(R)   Yield 38% ee 91% (S)

Figure 18.   Optical resolution of S-chiral compounds with hydrolytic enzymes[19]

The substrate specificities of hydrolytic enzymes, especially lipase is very wide and large molecules such as helicene can be also resolved using lipase PS and CAL.[20] PS recognized (M)-isomer giving (M)-acetate and leaving (P)-diol, whereas CAL recognized and catalyzed the esterification of the opposite isomer, (P)-isomer, leaving (M)-diol (Figure 19).

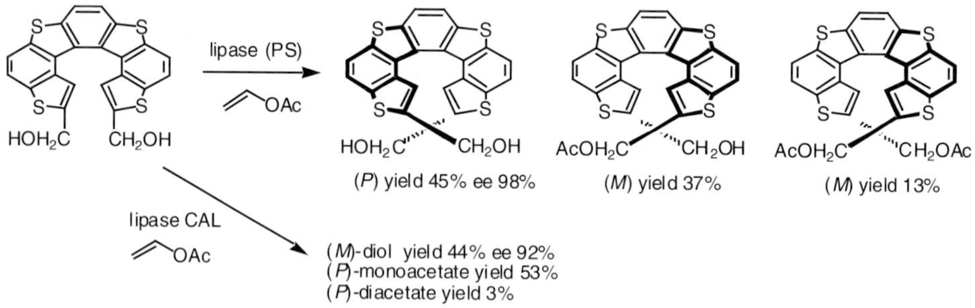

(P) yield 45% ee 98%   (M) yield 37%   (M) yield 13%

lipase CAL

OAc → (M)-diol yield 44% ee 92%
         (P)-monoacetate yield 53%
         (P)-diacetate yield 3%

Figure 19.   Optical resolution of S-chiral compounds with hydrolytic enzymes[20]

3. Hydrolysis of epoxide

Epoxide is an important intermediate for various bioactive compounds, so the demand for the chiral epoxide is increasing. Epoxide hydrolase can hydrolyze epoxide enantioselectively (Figure 20).[21] For example, *Aspergillus niger* was used for the hydrolysis of carvone epoxide (Figure 20(a)).[21a] In the reaction of styrene oxide, the nucleophilic reaction with *B. sulfurescents* occurred on the chiral carbon through inversion and that with *A. niger* occured on the adjacent carbon through retention of the configuration at the chiral carbon.   Therefore, the reaction of racemic styrene oxide with the combination of these two microorganisms gave the corresponding (*R*)-diol in 92% yield and 89% ee (Figure 20(b)).[21b]

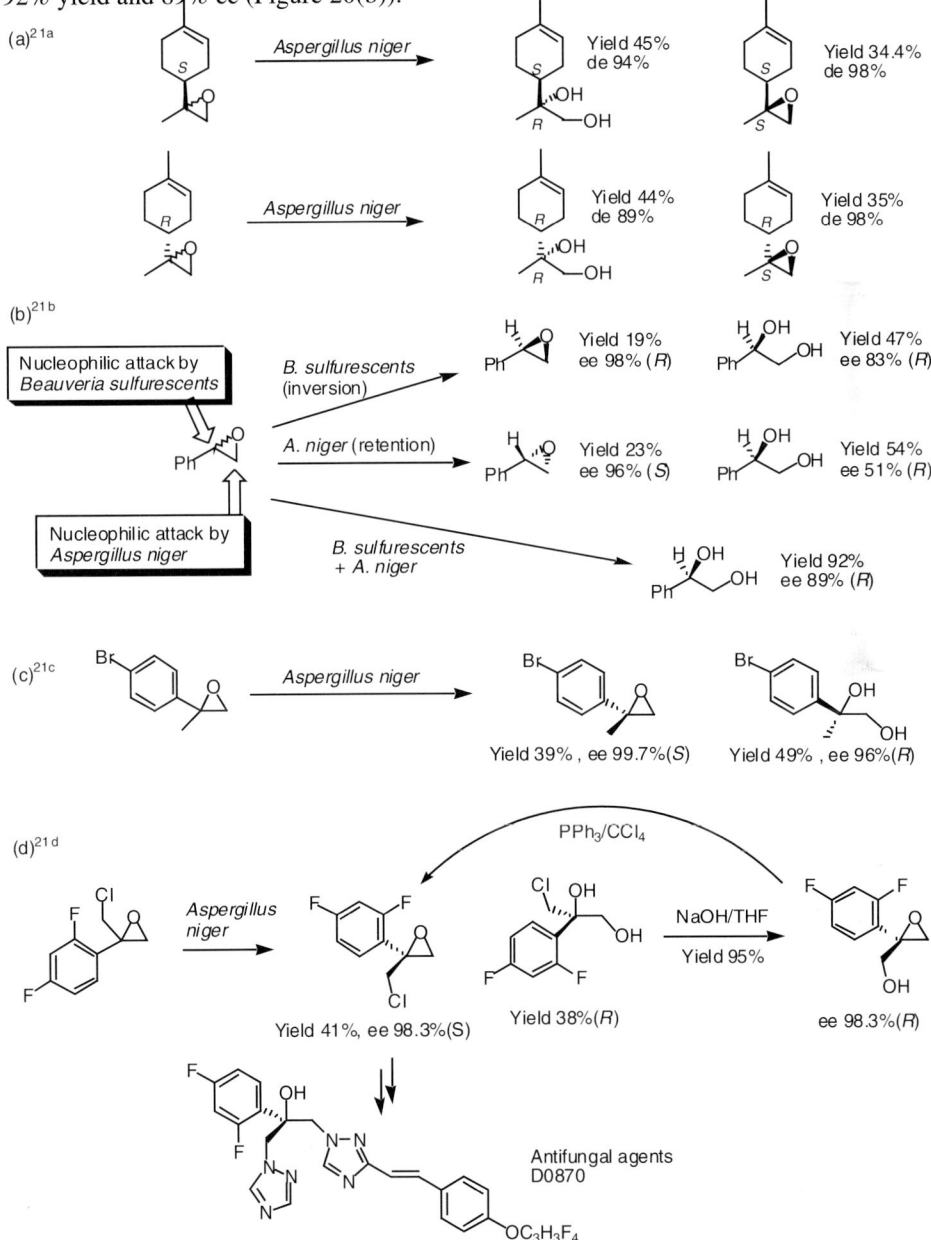

Figure 20.   Epoxide hydrolase-catalyzed resolutions

4. Acylation of amine and hydrolysis of nitrile and amide

Enantioselective acylation of amine and hydrolysis of amide are widely studied. These reactions are catalyzed by acylases, amidases and lipases. Some examples are shown in Figure 21.[22] Aspartame, artificial sweetener, is synthesized by a protease, thermolysin (Figure 21(a)).[22a] In this reaction, the L-enantiomer of racemic phenylalanine methyl ester reacted specifically with the α-carboxyl group of N-protected L-aspartate. Both the separation of the enantiomers of the phenylalanine and the protection of the γ-carboxyl group of the L-aspartate were unnecessary, which simplified the synthesis.

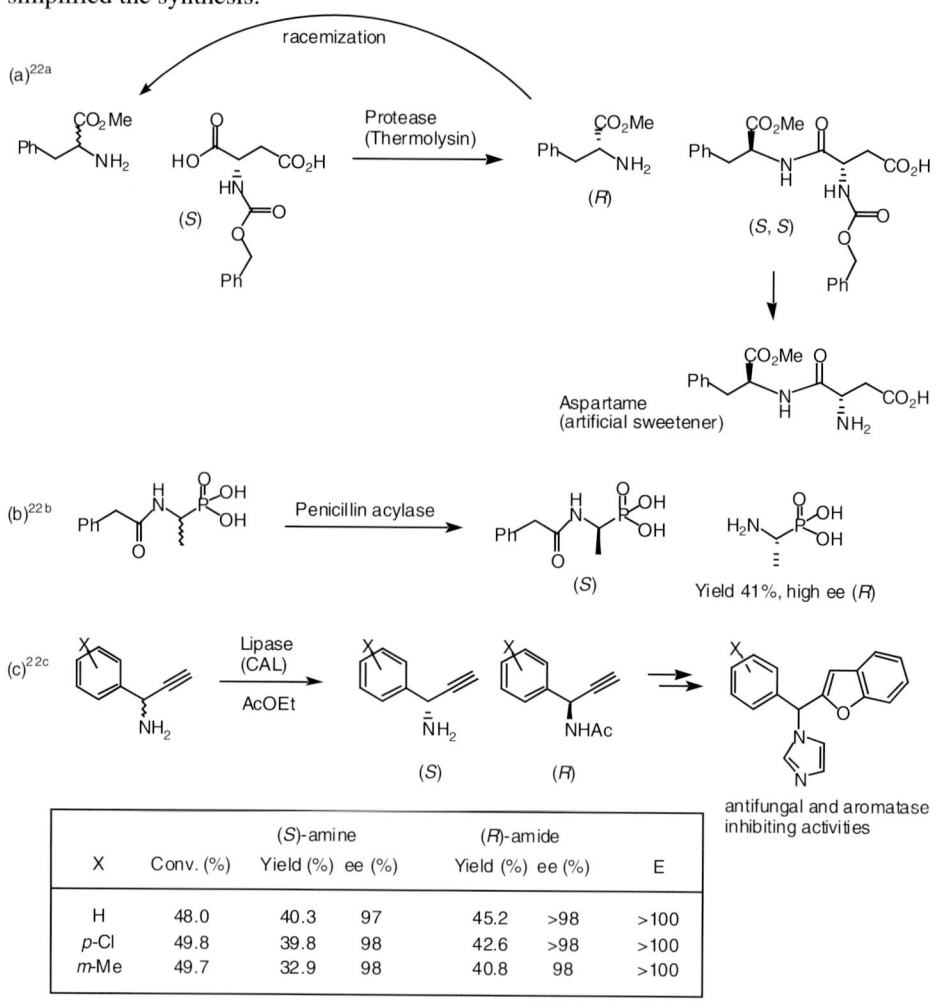

Figure 21.   Enantioselective acylation of amines and hydrolysis of an amide

Enantioselective hydrolysis of nitriles has been catalyzed by various *Rhodococus* and *Pseudomonas species* etc. to give amides or acids (Figure 22).[23] Meso compounds were also hydrolyzed to give the corresponding acid in high yield and ee (Figure 22 (d, e, g)).[23b,f,g,h] In the reaction of substituted malononitrile with *Rhodococcus rhodochrous*, the hydrolysis of the nitrile proceeded without enantio discrimination but the hydrolysis of the diamide proceeded with high enantioselectivities, giving (*R*)-acid in 92% and 96% ee (Figure 22 (g)).[23i]

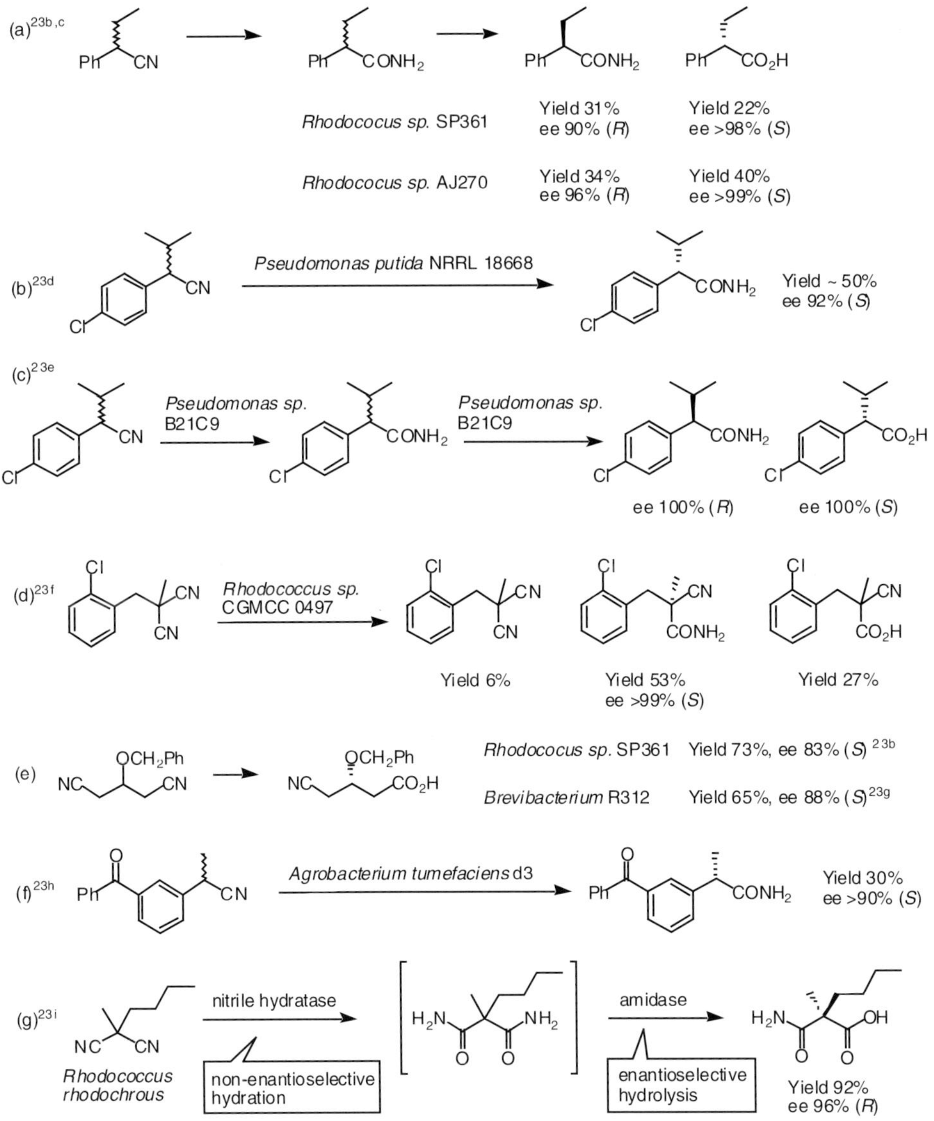

Figure 22. Enantioselective hydrolysis of nitriles

5. Dehalogenation

Novel approach for optically pure alcohol from racemic compounds is the use of dehalogenases.[24] For example, L-2-halo acid dehalogenase *Pseudomonas putida* was used for the synthesis of D-3-chlorolactic acid from racemic 2,3-dichloropropionic acid (Figure 23(a)).[24a-d] The enzyme catalyzed hydrolytic release of halogen from 2-halocarboxylic acids and produces 2-hydroxy acids with inversion of the configuration. L-2-Halo acid dehalogenase acted on the L-isomer of 2-halo acids and produces D-2-hydroxy acid with an excellent enantioselectivity.

Alcaligenes sp. and *Pseudomonas* sp. have been used for the production of (*R*) or (*S*)-2,3-dichloro-1-propanol by the enantioselective degradation of (*S*) or (*R*) enantiomers, respectively (Figure 23(b)).[24e,f] The enantioselectivities were excellent but the yield did not exceed 50% in this case because half of the starting materials were degradated.

In the reaction using halohydrin dehalogenase from *Agrobacterium radiobacter* (Figure 23(c)), the (*R*)-enantiomer was converted to the corresponding epoxide which was further converted to (*S*)-diol (ee 91%) by epoxide hydrolase from the same organism to prevent the attack of the chloride at the β-position.[24g] The remaining (*S*)-dichloropropanol was also obtained in optically pure form (ee >99%).

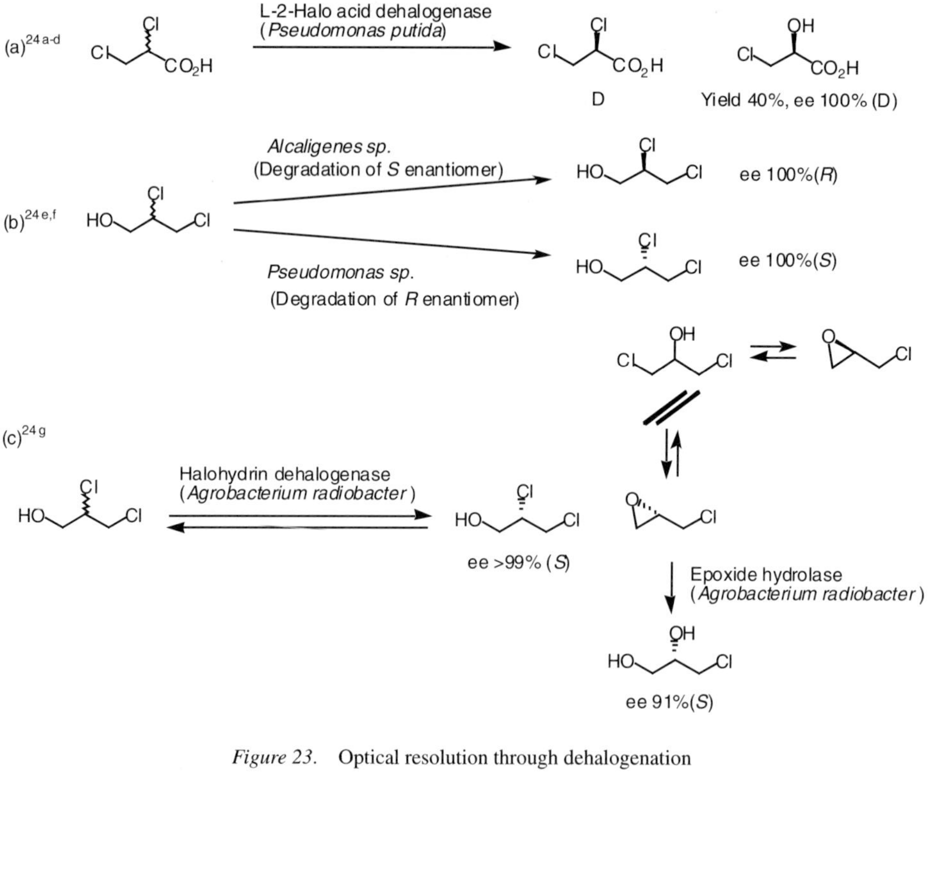

Figure 23. Optical resolution through dehalogenation

6. Oxidation of alcohols and reduction of ketones

Enantioselective oxidation of racemic alcohols as well as reduction of racemic ketones and aldehydes have been widely applied to obtain optically active alcohols.[25-27] The enzymes catalyzing these reactions are alcohol dehydrogenase, oxidases, and reductases etc. Coenzymes (NADH, NADPH, flavine etc) are usually necessary for theses enzymes. For example, for the oxidation of alcohols, NAD(P)$^+$ are used. The hydride removed from the substrate is transferred to the coenzyme bound in the enzyme, as shown in Figure 24. There are four stereochemical patterns, but only three types of the enzymes are known.

E1:   *Pseudomonas sp.* alcohol dehydrogenase[25a]
      Lactobacillus kefir alcohol dehydrogenase[25b]
E2:   *Geotrichum candidum* glycerol dehydrogenase[25c-e]
      Mucor javanicus dihydroxyacetone reductase[25f]
E3:   Yeast alcohol dehydrogenase[25g]
      Horse liver alcohol dehydrogenase[25h-j]
      Moraxella sp. alcohol dehydrogenase[27c]
E4:   Unknown

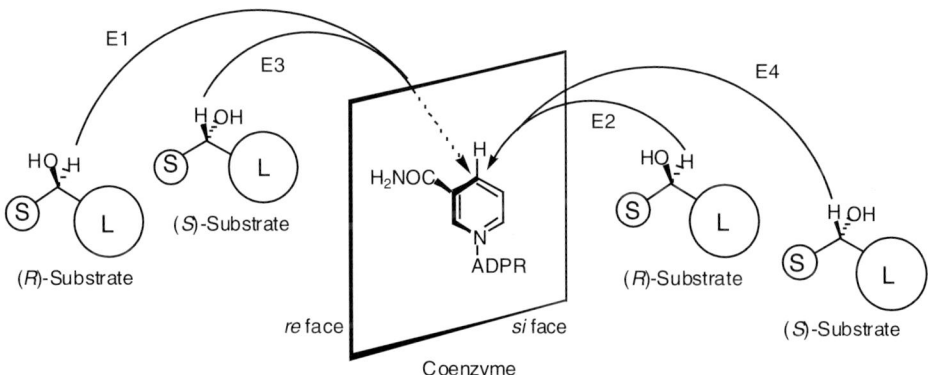

Figure 24.   Hydride transfer from the substrate alcohols to NAD(P)$^+$
(S is a small group, and L is a large group)

6.1 ENANTIOSELECTIVE OXIDATION OF RACEMIC ALCOHOLS

Enantioselective oxidation of racemic alcohols is a useful method to obtain chiral alcohols. Various microorganisms as well as isolated enzymes have been used. For example, *Geotrichum candidum* cell was used to oxidize aromatic alcohols (Figure 25).[26] Although the cell catalyzed the highly enantioselective oxidation of (S)-alcohol in water, the reactivities was not high enough. The efficiency was improved by spreading the cell on the surface of the water-absorbing polymer. The use of the organic-aqueous two-layer system improved the efficiency further more because the substrate alcohol was more soluble in the aqueous layer in the cell where the enzyme located than the product ketone, so the equilibrium between the oxidation of alcohol and the re-reduction of the product ketone was pushed to be favorable for the oxidation. As a results, in the oxidation of phenylethanol derivatives, chiral alcohols were obtained in up to >99% ee.   Ortho substituted derivatives were inert to the oxidation.

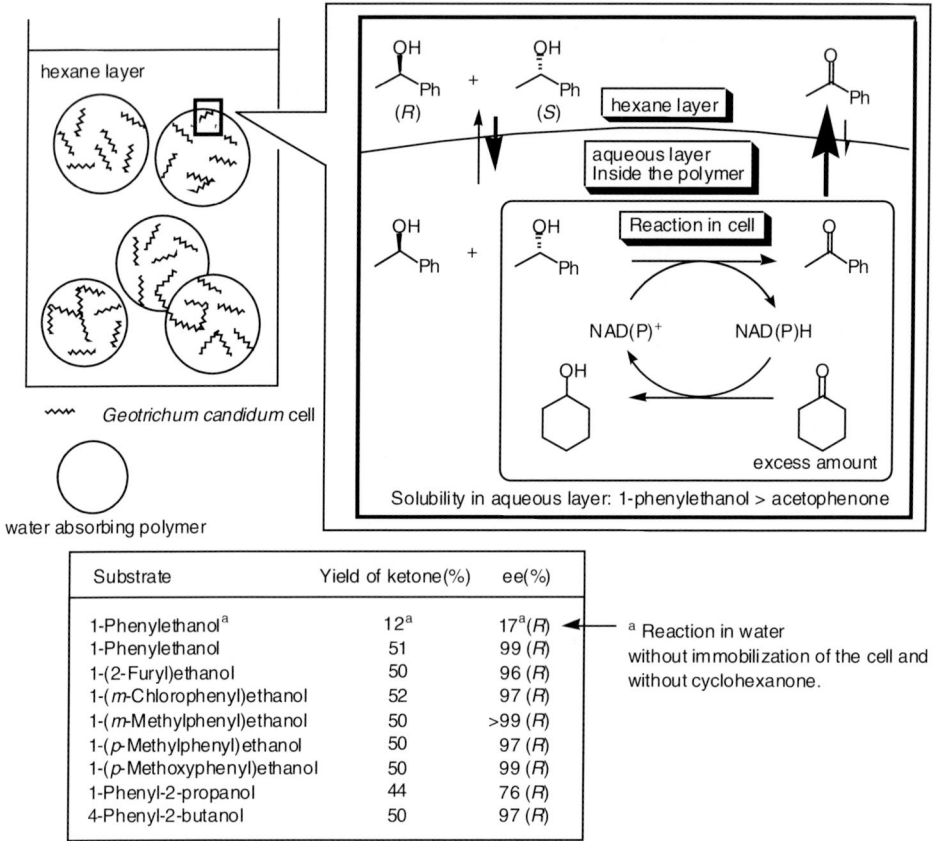

Substrate	Yield of ketone(%)	ee(%)
1-Phenylethanol[a]	12[a]	17[a](R)
1-Phenylethanol	51	99 (R)
1-(2-Furyl)ethanol	50	96 (R)
1-(m-Chlorophenyl)ethanol	52	97 (R)
1-(m-Methylphenyl)ethanol	50	>99 (R)
1-(p-Methylphenyl)ethanol	50	97 (R)
1-(p-Methoxyphenyl)ethanol	50	99 (R)
1-Phenyl-2-propanol	44	76 (R)
4-Phenyl-2-butanol	50	97 (R)

[a] Reaction in water without immobilization of the cell and without cyclohexanone.

Figure 25.   Oxidation of phenylethanol derivatives by the immobilized cell of *Geotrichum candidum* in hexane[26]

Other microorganisms and isolated enzymes have been also used for the oxidation of alcohols (Figure 26).[27]   Horse liver alcohol dehydrogenase was used for the oxidation of meso diols (Figure 26 (d, e)).[27e,f]   When one of the hydroxyl groups was oxidized, cyclization proceeded spontaneously, followed by the enzyme-catalyzed oxidation, giving chiral lactones.

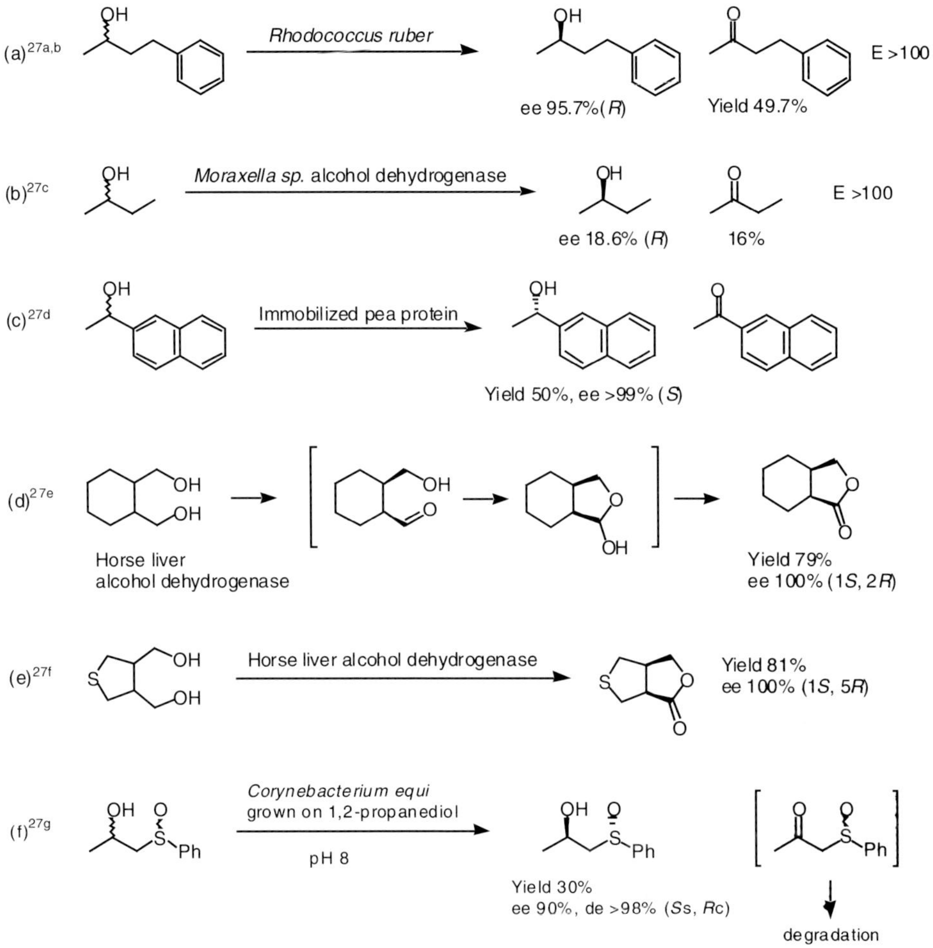

Figure 26. Enantioselective oxidations of alcohols

6.2 OPTICAL RESOLUTION THROUGH REDUCTION OF CARBONYL COMPOUNDS

Optical resolution of racemic carbonyl compounds is catalyzed by several microorganisms as well as isolated enzymes (Figure 27).[28] Various compounds including organometallic compounds (Figure 27(d))[28d] were resolved successfully.

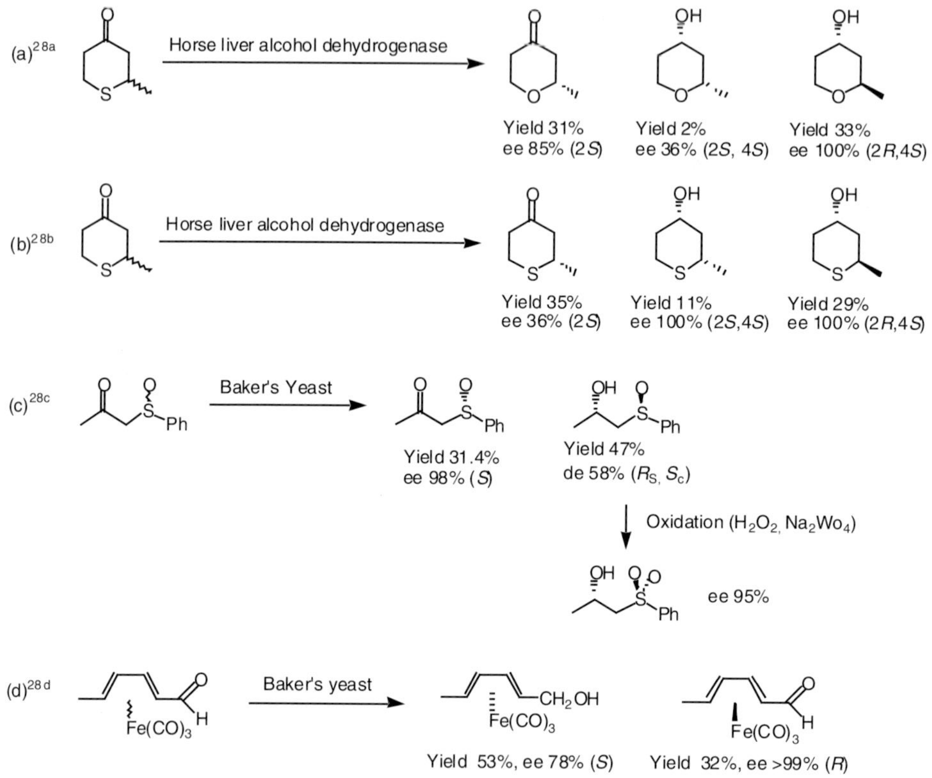

$(a)^{28a}$ Horse liver alcohol dehydrogenase

Yield 31%
ee 85% (2S)

Yield 2%
ee 36% (2S, 4S)

Yield 33%
ee 100% (2R,4S)

$(b)^{28b}$ Horse liver alcohol dehydrogenase

Yield 35%
ee 36% (2S)

Yield 11%
ee 100% (2S,4S)

Yield 29%
ee 100% (2R,4S)

$(c)^{28c}$ Baker's Yeast

Yield 31.4%
ee 98% (S)

Yield 47%
de 58% (R_S, S_C)

Oxidation (H_2O_2, Na_2Wo_4)

ee 95%

$(d)^{28d}$ Baker's yeast

Yield 53%, ee 78% (S)   Yield 32%, ee >99% (R)

Figure 27. Kinetic resolutions through reduction of carbonyl compounds

6.3 DYNAMIC KINETIC RESOLUTION OF RACEMIC KETONES THROUGH ASYMMETRIC REDUCTION

Dynamic kinetic resolution of racemic ketones proceeds through asymmetric reduction when the substrate does racemize and the product does not under the applied experimental conditions.[29]   For example, baker's yeast reduction of (R/S)-2-(4-methoxyphenyl)-1,5-benzothiazepin-3,4(2H,5H)-dione gave only (2S, 3S)-alcohol as a product out of four possible isomers as shown in Figure 28 (a).[29a] Only (S)-ketone was recognized by the enzyme as a substrate and reduction of the ketone proceeded enantioselectively.   The resulting product was used for the synthesis of (2S, 3S)-Diltiazem, a coronary vasodilator.

Dynamic kinetic resolution of α-alkyl-β-keto ester was conducted successfully using biocatalysts (Figure 28(b)).[29c]   For the reduction of ethyl 2-methyl-3-oxobutanoate, extensive screening methodology was used to find the suitable microorganism.   As a result, Klebsiella pneumoniae IFO 3319 out of 450 bacterial strains was found to give the corresponding (2R, 3S)-hydroxy ester with 99% de and >99% ee in Kg scale quantitatively.

The dynamic resolution of an aldehyde is shown in Figure 28(c).[29c-g]   The racemization of starting aldehyde and enantioselective reduction of carbonyl group by bakers' yeast resulted in the formation of chiral carbon centers.   The ee of the product was improved from 19% to 90% by changing the ester moiety from the isopropyl group to the neopentyl group.[29e]

Other biocatalysts were also used to perform the dynamic kinetic resolution through reduction (Figure 28(c)).   For example, *Thermoanaerobium brockii* reduced the aldehyde with a moderate enantioselectivity,[29f] and *Candida humicola* was found, as a result of screening from 107 microorganisms, to give the (R)-alcohol with 98.2% ee when ester group was methyl.[29g]

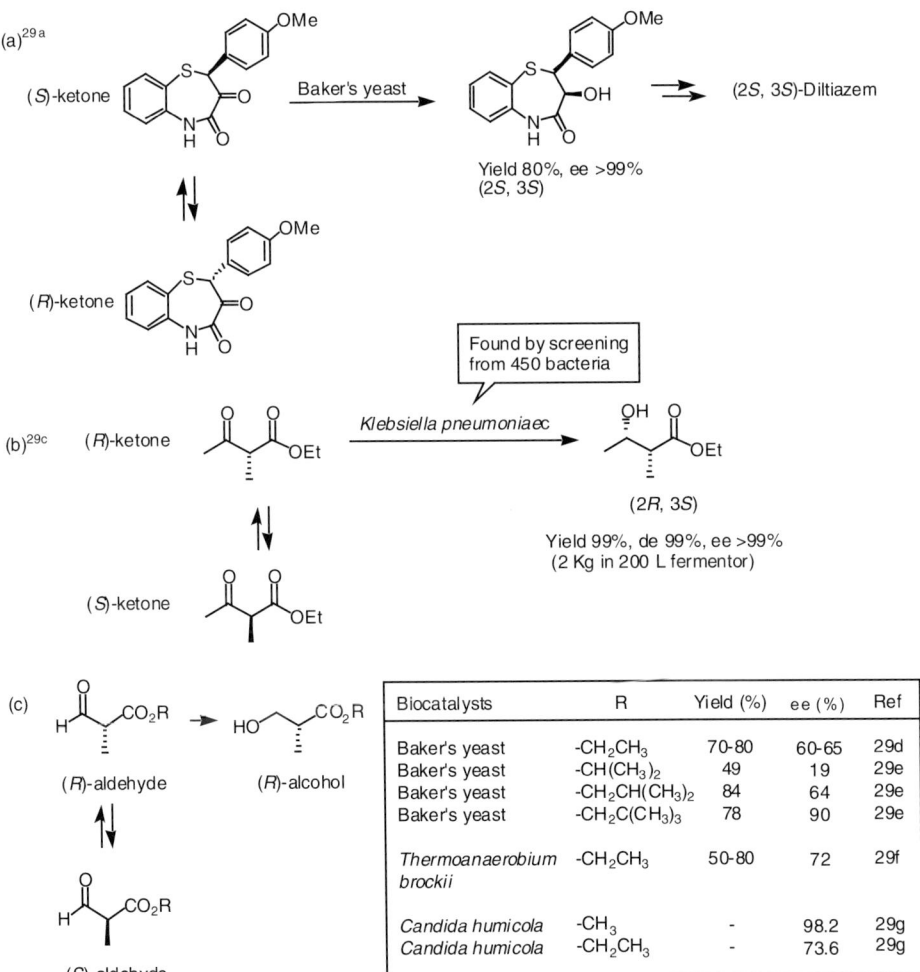

Biocatalysts	R	Yield (%)	ee (%)	Ref
Baker's yeast	$-CH_2CH_3$	70-80	60-65	29d
Baker's yeast	$-CH(CH_3)_2$	49	19	29e
Baker's yeast	$-CH_2CH(CH_3)_2$	84	64	29e
Baker's yeast	$-CH_2C(CH_3)_3$	78	90	29e
Thermoanaerobium brockii	$-CH_2CH_3$	50-80	72	29f
Candida humicola	$-CH_3$	-	98.2	29g
Candida humicola	$-CH_2CH_3$	-	73.6	29g

Figure 28.   Dynamic kinetic resolutions by reduction of ketones and aldehydes

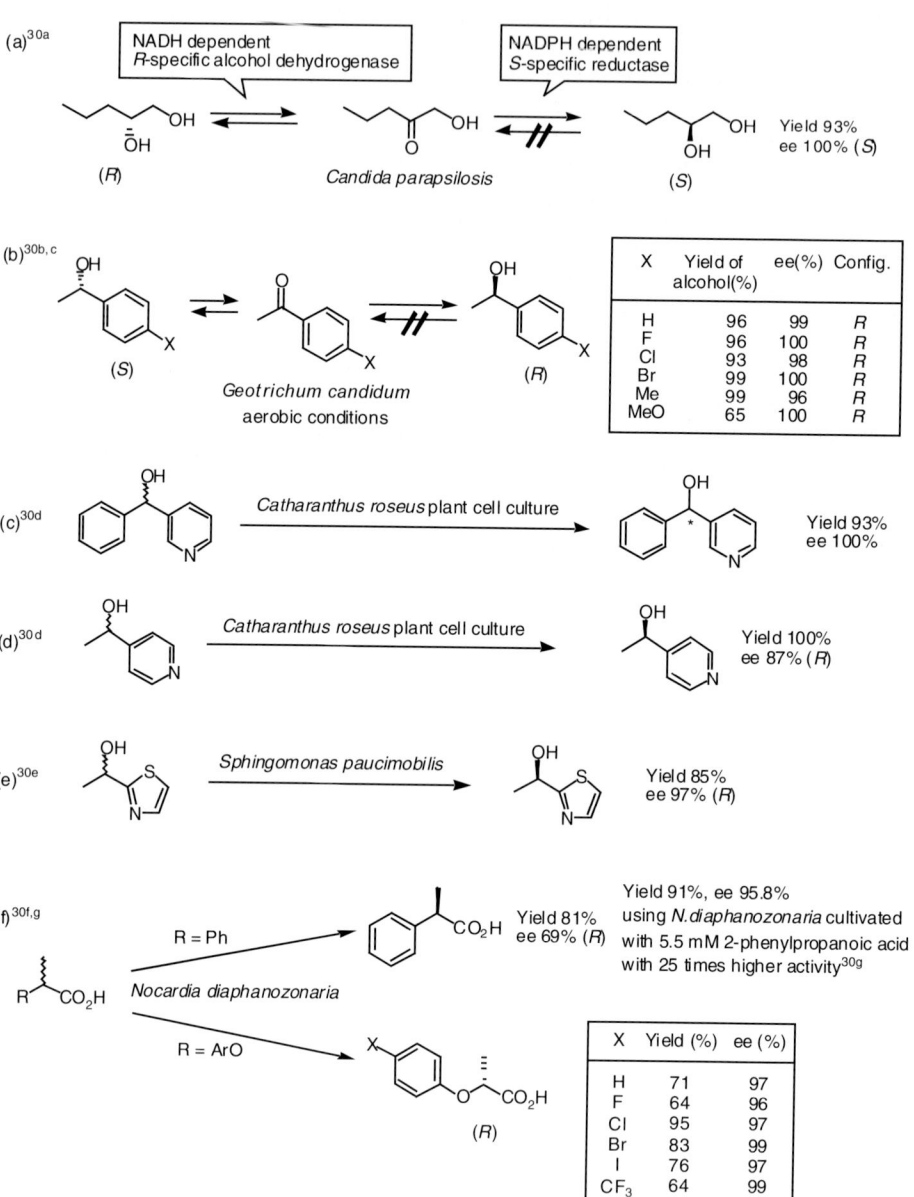

Figure 29.   Deracemization reactions

6.4 DERACEMIZATION THROUGH OXIDATION AND REDUCTION

Deracemization reaction, which converts the racemic compounds into chiral form in one step in one pot without changing their chemical structures, can be performed using microorganism containing several different stereochemical enzymes (Figure 29).[30] For example, for deracemization of 1, 2-pentandiol (Figure 29(a)), the (*R*)-specific NADH-enzyme in *Candida parapsilosis* was reversible and (*S*)-specific NADPH-enzyme in the same microorganisms was irreversible.[30a] Therefore, whole cell reaction of racemic 1, 2-pentandiol gave (*S*)-diol in 93% yield and 100% ee.

For the deracemization of phenylethanol derivatives using *Geotrichum candidum* under aerobic conditions (Figure 29(b)), the (*S*)-specific enzyme was reversible and (*R*)-enzyme was irreversible, so (*R*)-alcohol accumulated when the cell and racemic alcohols were mixed.[30b,c] Para-substituted phenylethanol derivatives gave better results than metha-substituted derivatives.

Racemic acid was deracemized using *Nocardia diaphanozonaria* (Figure 29(f)).[30f,g] In this reaction, subtle difference in the structure of the substrate determined the enantioselectivity.[30f]

7. Kinetic resolution of hydroperoxides

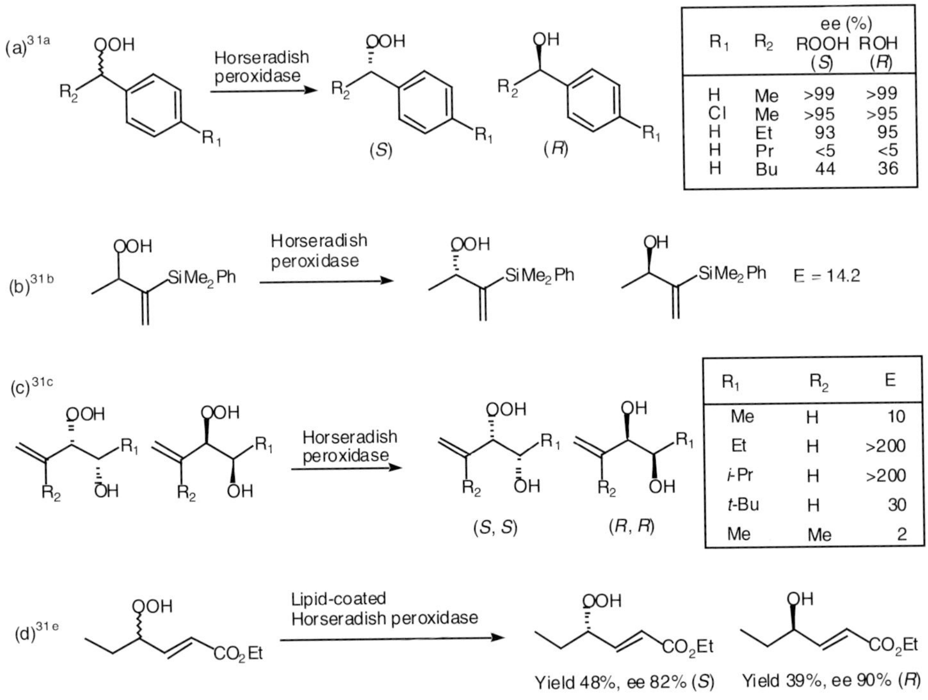

R_1	R_2	ee (%) ROOH (*S*)	ee (%) ROH (*R*)
H	Me	>99	>99
Cl	Me	>95	>95
H	Et	93	95
H	Pr	<5	<5
H	Bu	44	36

R_1	R_2	E
Me	H	10
Et	H	>200
i-Pr	H	>200
t-Bu	H	30
Me	Me	2

Figure 30. Enantioselective reductions of hydroperoxides to alcohols for the synthesis of chiral hydroperoxides

Peroxidases have been used for the kinetic resolution of racemic peroxides to give chiral peroxides.[31]   For example, a horseradish peroxidase selectively recognized sterically uncumbered (R)-alkyl aryl hydrogenperoxides, which allowed kinetic resolution to provide (R)-alcohol and (S)-peroxide (Figure 30(a)). [31a]   However, poor enzyme recognition was observed with hydroperoxides possessing larger R2 groups such as propyl or butyl moiety.   This reaction was performed on a preparative scale conveniently to provide optically pure hydroperoxides.

Aliphatic peroxides were also resolved using the horseradish peroxidase (Figure 30 (b)-(d)).[31b,c,e]   (R)-Enantiomers were recognized enantioselectively to give the corresponding (R)-alcohols remaining (S)-peroxides.

Another strategy to obtain optically active hydroperoxides is the enantioselective acetylation of peroxides by lipases.[32]   The examples are shown in Figure 31.   By using this methods, (S)-methyl 13-hydroperoxy-9Z, 11E-octadecadienoate, an important substance for medical studies concerning toxicity and other physiological actions, was synthesized successfully (Figure 31 (b)).[32b]

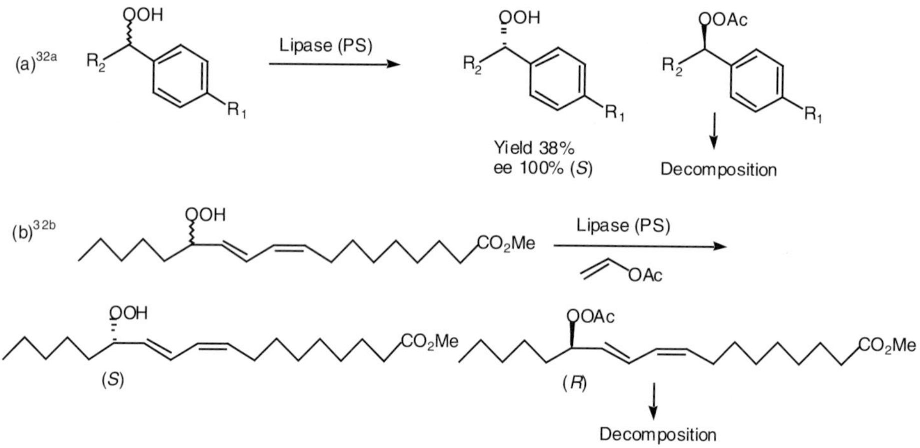

Figure 31.   Enantioselective acetylation of hydroperoxide for the synthesis of chiral peroxides

8. Baeyer-Villiger oxidations

Monooxygenases have been used for the Baeyer-Villiger oxidations to obtain optically active lactones (Figure 32).[33] A cyclohexanone monooxygenase from *Acinetobacter calcoaceticus*   has   been   widely   used.   In   the   oxidation   of 2-oxabicyclo[3,2,0]-heptan-7-one by *Acinetobacter calcoaceticus* (Figure 32 (a), n = 0, m = 2), (S, S)-substrate leaded to the normal lactone, whereas the (R, R)-substrate leaded to the abnormal lactone with 90% ee and >98% ee, respectively.   Other sustrates proceeded with a similar stereochemical course.

Recently, the enzyme was expressed in baker's yeast and used for the asymmetric oxidation of alkyl cyclohexanones.[33c]   The use of designer baker's yeast combined the advantages of using purified enzymes (single catalytic species, no overmetabolism) with the benefits of whole-cell reactions (experimentally simple, no cofactor regeneration necessary). The enantioselectivity of the recombinant enzyme was same as that in purified-original enzyme.[33g]

Dynamic kinetic resolution process was also applied to biocatalytic Baeyer-Villiger oxidations. The recombinant *E. coli* expressing the cyclohexanone monooxygenase from *A. calcoaceticus* was used for the oxidation of racemic 2-benzyloxymethylcyclopentanone giving the corresponding (*R*)-lactone in 85% yield and 96% ee (Figure 32(d)).[33h]

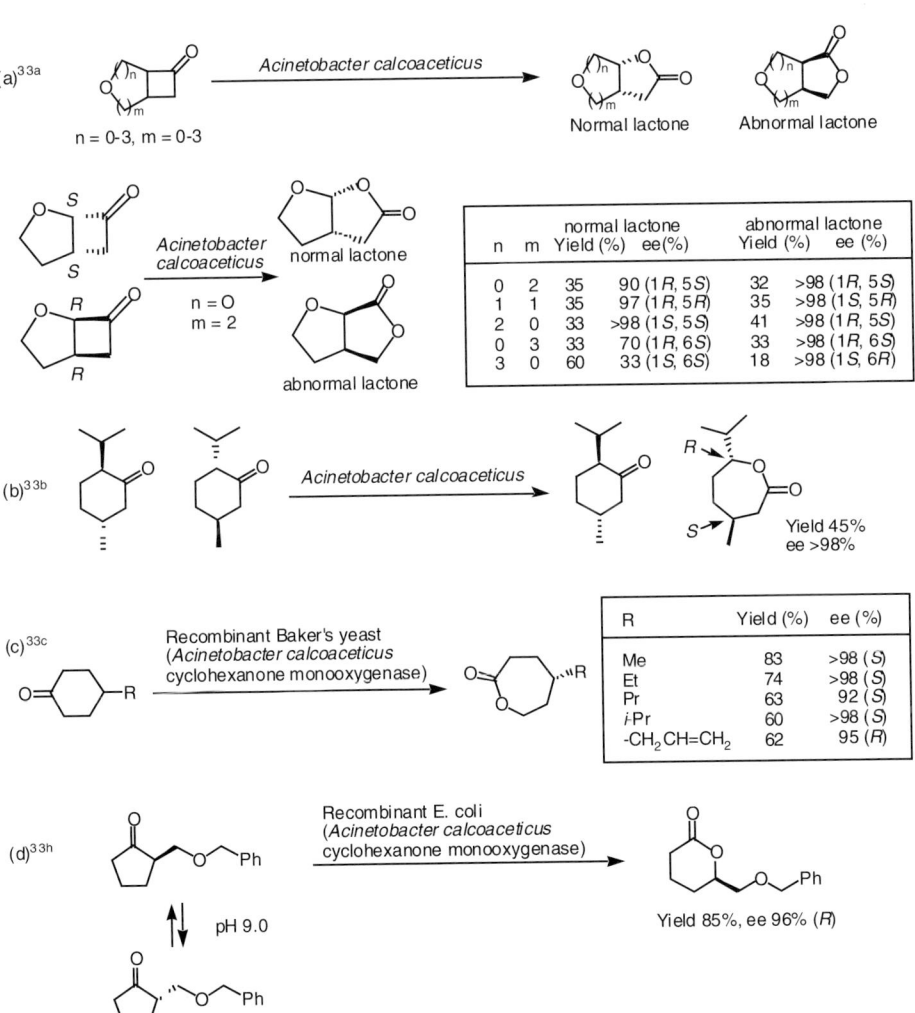

		normal lactone		abnormal lactone	
n	m	Yield (%)	ee(%)	Yield (%)	ee (%)
0	2	35	90 (1*R*, 5*S*)	32	>98 (1*R*, 5*S*)
1	1	35	97 (1*R*, 5*R*)	35	>98 (1*S*, 5*R*)
2	0	33	>98 (1*S*, 5*S*)	41	>98 (1*R*, 5*S*)
0	3	33	70 (1*R*, 6*S*)	33	>98 (1*R*, 6*S*)
3	0	60	33 (1*S*, 6*S*)	18	>98 (1*S*, 6*R*)

R	Yield (%)	ee (%)
Me	83	>98 (*S*)
Et	74	>98 (*S*)
Pr	63	92 (*S*)
i-Pr	60	>98 (*S*)
-CH$_2$CH=CH$_2$	62	95 (*R*)

Figure 32.   Kinetic resolutions of ketones through Baeyer-Villiger oxidation

9. Oxidation and reduction of sulfur compounds

Asymmetric synthesis of sulfoxides can be achieved by biocatalytic oxidation of sulfides and reduction of sulfoxides (Figure 33).[34,27g] One example is the reduction of alkyl aryl sulfoxides by intact cells of *Rhodobacter sphaeroides f.sp. denitrificans* (Figure 33 (a)).[34a] In the reduction of methyl *p*-substituted phenyl sulfoxides, (*S*)-enantiomers were exclusively deoxygenated while enantiomerically pure (*R*)-isomers were recovered in good yield. For poor substrates such as ethyl phenyl sulfoxide, the repetition of the incubation after removing the toxic product was effective in enhancing the ee of recovered (*R*)-enantiomers to 100%.

Rhodococcus equi has been used for oxidation of sulfinic acid (Figure 33(b)).[34b] Incubation of racemic arenesulfinic acid esters with the bacterial strain gave the corresponding sulfonates leaving optically pure (*R*)-substrate (>99% ee).

For the reaction of phenylsulfinyl propanone with *Rhodococcus equi*, both oxidation of sulfoxide and reduction of carbonyl group proceeded.[27g] The carbonyl group of (*R*)-sulfoxide was reduced to the corresponding (Rs, Sc)-alcohols without the reaction of sulfoxide group, while (*S*)-sulfoxide was oxidized first to the corresponding sulfone and then the carbonyl group was reduced to the corresponding (*S*)-alcohol.

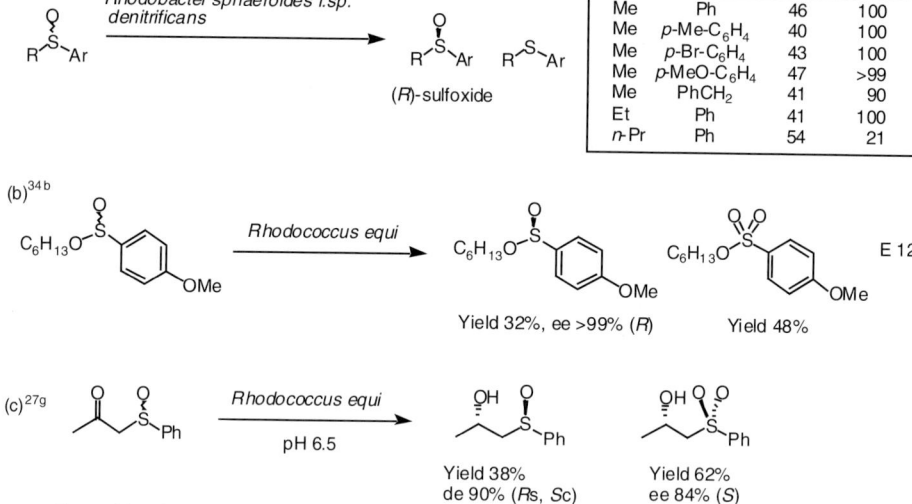

R	Ar	Yield (%) (*R*)-sulfoxide	ee (%)
Me	Ph	46	100
Me	*p*-Me-C$_6$H$_4$	40	100
Me	*p*-Br-C$_6$H$_4$	43	100
Me	*p*-MeO-C$_6$H$_4$	47	>99
Me	PhCH$_2$	41	90
Et	Ph	41	100
n-Pr	Ph	54	21

Figure 33. Oxidation and reduction of sulfur compounds for the synthesis of S-chiral compounds

10. Conclusions

Optical resolution of racemic compounds by biocatalysts has been a useful method as shown in this review. For this purpose, two types of biocatalysts are mainly used; hydrolytic enzymes and oxidoreductases.

Hydrolytic enzymes such as proteases, esterases and lipases are widely used for kinetic resolution of alcohols, amines, nitriles etc. Among them, lipases are most frequently utilized due to its high commercial availabilities and wide substrate specificities with high enantioselectivities. Other hydrolytic enzymes such as amino acylases and nitrilase are also used often because these exhibit high enantioselectivity. However, availability of these enzymes is relatively poor comparing with lipases and substrate specificity of these enzymes is not as wide as that of lipases.

Oxidoreductases such as dehydrogenases are also used for kinetic resolution. Selectivity of oxidoreductases is relatively high compared with that of hydrolytic enzymes. However, these enzymes have a drawback that the number of available and useful enzymes or microorganisms is still limited, and finding of new types of oxidoreductases that exhibit wide substrate specificities with high selectivities is awaiting.

To evolve useful enzymes, genetic engineering technology has been applied increasingly to improve stability of enzymes, enantioselectivity, extension of substrate specificity for kinetic resolution of racemic compounds. Novel enzymes created by this technique will be available in large quantities and varieties within a next few years. In the near future, a lot of useful enzymes will be on the market and expanding number of chemists can use enzymes more freely than present due to the improvement in the simplification of experimental procedures.

11. References

1   (a) A. Schmid, J. S. Dordick, B. Hauer, A. Kiener, M. Wubbolts, B. Witholt, *Nature*, **2001**, 409, 258 - 268.
    (b) K. M. Koeller, C.-H. Wong, *Nature*, **2001**, 409, 232 - 240.
    (c) W. -D. Fessner (Ed.) "Biocatalysis from discovery to application" Springer, Berlin, 2000.
    (d) K. Drauz, H. Waldmann (Eds.) "Enzyme catalysis in organic synthesis a comprehensive handbook" Wiley-VCH Verlag GmbH, Weinheim, 2002.
2   R. J. Kazlauskas, A. N. E. Weissfloch, A. T. Rappaport, L. A. Cuccia, *J. Org. Chem.* **1991**, 56, 2656 - 2665.
3   M. Cygler, P. Grochulski, R. J. Kazlauskas, J. D. Schrag, F. Bouthillier, B. Rubin, A. N. Serreqi, A. K. Gupta, *J. Am. Chem. Soc.* **1994**, 116, 3180 - 3186.
4   (a) K. Nakamura, M. Kawasaki, A. Ohno, *Bull. Chem. Soc. Jpn.* **1996**, 69, 1079 - 1085.
    (b) T. Ema, J. Kobayashi, S. Maeno, T. Sakai, M. Utaka, *Bull. Chem. Soc. Jpn.* **1998**, 71, 443 - 453.
5   C.-S. Chen, Y. Fujimoto, G. Girdaukas, C. J. Sih, *J. Am. Chem. Soc.* **1982**, 104, 7294-7299.
6   (a) Y.-F, Wang, J. J. Lalonde, M. Momongan, D. E. Bergbreiter, C.-H. Wong, *J. Am. Chem. Soc.* **1988**, 110, 7200 - 7205.
    (b) Y. Terao, K. Tsuji, M. Murata, K. Achiwa, T. Nishio, N. Watanabe, K. Seto, *Chem. Pharm. Bull.* **1989**, 37, 1653 - 1655.
7   (a) T. Sakurai, A. L. Margolin, A. J. Russell, A. M. Klibanov, *J. Am. Chem. Soc.* **1988**, 110, 7236 - 7237.
    (b) K. Nakamura, Y. Takebe, T. Kitayama, A. Ohno, *Tetrahedron Lett.* **1991**, 32, 4941 - 4944.
    (c) T. Kitayama, T. Rokutanzono, R. Nagao, Y. Kubo, M. Takatani, K. Nakamura, T. Okamoto, *J. Mol. Catal, B: Enzymatic,* **1999**, 7, 291 - 297.
    (d) Y. Hirose, K. Kariya, I. Sasaki, Y. Kurono, H. Ebiike, K. Achiwa, *Tetrahedron Lett.* **1992**, 33, 7157 - 7160.
    (e) T. Mori, M. Funasaki, A. Kobayashi, Y. Okahata, *Chem. Commun,* **2001**, 1832 - 1833.
    (f) T. Matsuda, R. Kanamaru, K. Watanabe, T. Kamitanaka, T. Harada, K. Nakamura, *Tetrahedron: Asymm.* **2003**, 14, 2087 - 2091.
    (g) T. Matsuda, R. Kanamaru, K. Watanabe, T. Harada, K. Nakamura, *Tetrahedron Lett.* **2001**, 42, 8319 - 8321.
    (h) T. Itoh, N. Ouchi, S. Hayase, Y. Nishimura, *Chem. Lett.* **2003**, 32, 654 - 655.

8   (a) Y. Hirose, K. Kariya, Y. Nakanishi, Y. Kurono, K. Achiwa, *Tetrahedron Lett.* **1995**, 36, 1063 - 1066.
      (b) K. Liebeton, A. Zonta, K. Schimossek, M. Nardini, D. Lang, B. W. Dijkstra, M. T. Reetz, K.-E. Jaeger, *Chemistry and Biology*, **2000**. 7, 709 - 718.
9   (a) T. Shibatani, K. Omori, H. Akatsuka, E. Kawai, H. Matsumae, *J. Mol. Catal, B: Enzymatic*, **2000**, 10, 141 - 149.
      (b) T. Furutani, M. Furui, T. Mori, T. Shibatani, *Appl. Biochem. Biotechnol*, **1996**, 59, 319 - 328.
      (c) M. Furui, T. Furutani, T. Shibatani, Y. Nakamoto, T. Mori, *J. Ferment. Bioeng.* **1996**, 81, 21 - 25.
      (d) H. Matsumae, T. Shibatani, *J. Ferment. Bioeng.* **1994**, 77, 152 - 158.
      (e) H. Matsumae, M. Furui, T. Shibatani, T. Tosa, *J. Ferment. Bioeng.* **1994**, 78, 59 - 63.
      (f) H. Matsumae, M. Furui, T. Shibatani, *J. Ferment. Bioeng.* **1993**, 75, 93 - 98.
      (g) S. B. Desai, N. P. Argade, K. N. Ganesh, *J. Org. Chem.* **1996**, 61, 6730 - 6732.
10   (a) S. Mitsuda, T. Umemura, H. Hirohara, *Appl. Microbiol. Biotechnol.* **1988**, 29, 310 - 315.
      (b) S. Mitsuda, S. Nabeshima, *Recl. Trav. Chim. Pays-Bas*. **1991**, 110, 151 - 154.
      (c) A. Wallner, H. Mang, S. M. Glueck, A. Steinreiber, S. F. Mayer, K. Faber, *Tetrahedron: Asymm.* **2003**, 14, 2427 - 2432.
11   (a) G. Fülling, C. J. Sih, *J. Am. Chem. Soc.* **1987**, 109, 2845 - 2846.
      (b) J. Z. Crich, R. Brieva, P. Marquart, R.-L. Gu, S. Flemming, C. J. Sih, *J. Org. Chem.* **1993**, 58, 3252 - 3258.
      (c) M. Sharfuddin, A. Narumi, Y. Iwai, K. Miyazawa, S. Yamada, T. Kakuchi, H. Kaga, *Tetrahedron: Asymm.* **2003**, 14, 1581 - 1585.
      (d) M. M. Jones, J. M. J. Williams, *Chem. Commun.* **1998**, 2519 - 2520.
12   (a) M.-J. Kim, Y. Ahn, J. Park, *Curr. Opin. Biotechnol.* **2002**, 13, 578-587
      (b) B. A. Persson, A. L. E. Larsson, M. L. Ray, J.-E. Bäckvall, *J. Am. Chem. Soc.* **1999**, 121, 1645 - 1650.
      (c) B. A. Persson, F. F. Huerta, J.-E. Bäckvall, *J. Org. Chem.* **1999**, 64, 5237 - 5240.
13   (a) Y. Hirose, 4th Kouso Ouyou Symposium, Nagoya, Japan, 2003.
      (b) K. Tanaka, M. Yasuda, *Tetrahedron: Asymm.* **1998**, 9, 3275 - 3282.
      (c) N. Yoshida, T. Kamikubo, K. Ogasawara, *Tetrahedron Lett.* **1998**, 39, 4677 – 4678.
      (d) S. Acherar, G. Audran, N. Vanthuyne, H. Monti, *Tetrahedron: Asymm.* **2003**, 14, 2413 - 2418.
14   (a) T. Itoh, Y. Takagi, T. Murakami, Y. Hiyama, H. Tsukube, *J. Org. Chem.* **1996**, 61, 2158-2163.
      (b) T. Itoh, K. Mitsukura, W. Kanphai, Y. Takagi, H. Kihara, H. Tsukube, *J. Org. Chem.* **1997**, 62, 9165-9172.
      (c) S. Hamaguchi, M. Asada, J. Hasegawa, K. Watanabe, *Agric. Biol. Chem.* **1985**, 49, 1661 - 1667.
      (d) N. Yoshida, M. Aono, T. Tsubuki, K. Awano, T. Kobayashi, *Tetrahedron: Asymm.* **2003**, 14, 529 - 535.
      (e) S. Iriuchijima, N. Kojima, *Agric. Biol. Chem.* **1982**, 46, 1153 - 1157.
      (f) G. Sabbioni, J. B. Jones, *J. Org. Chem.* 1987, 52, 4565 – 4570.
      (g) S. Kobayashi, K. Kamiyama, T. Iimori, M. Ohno, *Tetrahedron Lett.* 1984, 25, 2557 – 2560.
15   (a) J. T. Lin, T. Yamazaki, T. Kitazume, *J. Org. Chem.* **1987**, 52, 3211-3217.
      (b) T. Sakai, Y. Miki, M. Tsuboi, H. Takeuchi, T. Ema, K. Uneyama, M. Utaka, *J. Org. Chem.* **2000**, 65, 2740 - 2747.
      (c) M. Shimizu, K. Sugiyama, T. Fujisawa, *Bull. Chem. Soc. Jpn.* **1996**, 69, 2655 - 2659.
      (d) M. Kirihara, M. Kawasaki, T. Takuwa, H. Kakuda, T. Wakikawa, Y. Takeuchic K. L. Kirk, *Tetrahedron: Asymm.* **2003**, 14 1753 – 1761.
16   (a) R. Zurawinski, K. Nakamura, J. Drabowicz, P. Kielbasinski, M. Mikolajczyk, *Tetrahedron: Asymm.* **2001**, 12, 3139 - 3145.
      (b) M. Mikolajczyk, J. Luczak, P. Kielbasinski, *J. Org. Chem.* **2002**, 67, 7872 – 7875.
17   N. W. Fadnavis, K. Koteshwar, *Tetrahedron: Asymm.* **1997**, 8, 337 - 339.
18   (a) A. N. Serreqi, R. J. Kazlauskas, *J. Org. Chem.* **1994**, 59, 7609 - 7615.
      (b) P. Kielbasinski, J. Omelanczuk, M. Mikolajczyk, *Tetrahedron: Asymm.* **1998**, 9, 3283 - 3287.
      (c) K. Shioji, Y. Ueno, Y. Kurauchi, K. Okuma, *Tetrahedron Lett.* **2001**, 42, 6569 - 6571.
      (d) K. Shioji, A Tashiro, S. Shibata, K. Okuma, *Tetrahedron Lett.* **2003**, 44, 1103 - 1105.
      (e) K. Shioji, Y. Kurauchi, K. Okuma, *Bull. Chem. Soc. Jpn.* **2003**, 76, 833 - 834.
19   K. Burgess, I. Henderson, *Tetrahedron Lett.* **1989**, 30, 3633 - 3636.
20   K. Tanaka, Y. Shogase, H. Osuga, H. Suzuki, K. Nakamura, *Tetrahedron Lett.* **1995**, 36, 1675 - 1678.
21   (a) X.-J. Chen, A. Archelas, R. Furstoss, *J. Org. Chem.* **1993**, 58, 5528 - 5532.
      (b) S. Pedragosa-Moreau, A. Archelas, R. Furstoss, *J. Org. Chem.* **1993**, 53, 5533 - 5536.
      (c) M. Cleij, A. Archelas, R. Furstoss, *Tetrahedron:Asymm.* **1998**, 9, 1839 - 1842.

(d) N. Monfort, A. Archelas, R. Furstoss, *Tetrahedron: Asymm.* **2002**, 13, 2399 - 2401.
22 (a) S. M. Roberts, N. J. Turner, A. J. Willetts, M. K. Turner, "Introduction to biocatalysis using enzymes and micro-organisms" Cambridge University Press, Cambridge, 1995, pp 175 - 178.
(b) V. A. Solodenko, T. N. Kasheva, V. P. Kukhar, E. V. Kozlova, D. A. Mironeko, V. K. Svedas, *Tetrahedron*, **1991**, 47, 3989 - 3998.
(c) F. Messina, M. Botta, F. Corelli, M. P. Schneider, F. Fazio, *J. Org. Chem.* **1999**, 64, 3767 - 3769.
23 (a) H. Yamada, S. Shimizu, M. Kobayashi, *The Chemical Record*, **2001**, 1, 152 - 161.
(b) T. Beard, M. A. Cohen, J. S. Parratt, N. J. Turner, *Tetrahedron: Asymm.* **1993**, 4, 1085 - 1104.
(c) M.-X. Wang, G. Lu, G.-J. Ji, Z.-T. Huang, O. Meth-Cohn, J. Colby, *Tetrahedron: Asymm.* **2000**, 11, 1123 - 1135.
(d) R. D. Fallon, B. Stieglitz, I. Turner Jr. Appl. *Microbiol. Biotechnol.* **1997**, 47, 156 - 161.
(e) S. Masutomo, A. Inoue, K. Kumagai, R. Murai, S. Mitsuda, *Biosci. Biotech. Biochem.* **1995**, 59, 720 - 722.
(f) Z.-L. Wu, Z.-Y. Li, *Tetrahedron: Asymm.* **2003**, 14, 2133 - 2142.
(g) S. J. Maddrell, N. J. Turner, A. Kerridge, A. J. Willetts, J. Crosby, *Tetrahedron Lett.* **1996**, 37, 6001 - 6004.
(h) R. Bauer, H.-J. Knackmuss, A. Stolz, *Appl. Microbiol. Biotechnol.* **1998**, 49, 89 - 95.
(i) M. Yokoyama, T. Sugai, H. Ohta, *Tetrahedron: Asymm.* **1993**, 4, 1081 - 1084.
24 (a) M. Onda, K. Motosugi, H. Nakajima, *Agric. Biol. Chem.* **1990**, 54, 3031 - 3033.
(b) K. Motosugi, N. Esaki, K. Soda, *Biotechnol. Bioeng.* **1984**, 26, 805 - 806.
(c) K. Motosugi, N. Esaki, K. Soda, *Agric. Biol. Chem.* **1982**, 46, 837 - 838.
(d) K. Motosugi, N, Esaki, K. Soda, *J. Bacteriol.* **1982**, 150, 522 - 527.
(e) N. Kasai, T. Suzuki, Y. Furukawa, *Chirality*, **1998**, 10, 682 - 692.
(f) N. Kasai, T. Suzuki, *Adv. Synth. Catal.* **2003**, 345, 437 - 455.
(g) J. H. L. Spelberg, J. E. T. v. H. Vlieg, T. Bosma, R. M. Kellogg, D. B. Janssen, *Tetrahedron: Asymm.* **1999**, 10, 2863 - 2870.
25 (a) C. W. Bradshaw, H. Fu, G.-J. Shen, C.-H. Wong, *J. Org. Chem.* **1992**, 57, 1526 - 1532.
(b) C. W. Bradshaw, W. Hummel, C.-H. Wong, *J. Org. Chem.* **1992**, 57, 1532 - 1536.
(c) K. Nakamura, T. Shiraga, T. Miyai, A. Ohno, *Bull. Chem. Soc. Jpn.* **1990**, 63, 1735 - 1737.
(d) K. Nakamura, S. Takano, K. Terada, A. Ohno, *Chemistry Lett.* **1992**, 951 - 954.
(e) K. Nakamura, T. Yoneda, T. Miyai, K. Ushio, S. Oka, A. Ohno, *Tetrahedron Lett.* **1988**, 29, 2453 - 2454.
(f) H. Dutler, J. L. Van Der Baan, E. Hochuli, Z. Kis, K. E. Taylor, V. Prelog, *Eur. J. Biochem.* **1977**, 75, 423 - 432.
(g) V. Prelog, *Pure Appl. Chem.* **1964**, 9, 119 - 130.
(h) J. B. Jones, *Tetrahedron*, **1986**, 42, 3351 - 3403.
(i) J. B. Jones, J. F. Beck in Applications of Biochemical Systems in Organic Chemistry (Ed.: J. B. Jones, C. J. Sih, D. Perlman), John Wiley and Sons, New York, 1976, p. 107 - 401.
(j) L. K. P. Lam, I. A. Gair, J. B. Jones, *J. Org. Chem.* **1988**, 53, 1611 - 1615.
26 (a) K. Nakamura, Y. Inoue, T. Matsuda, I. Misawa, *J. Chem. Soc. Perkin Trans. 1*, **1999**, 2397 -2402.
(b) K. Nakamura, Y. Inoue, A. Ohno, *Tetrahedron Lett.* **1994**, 35, 4375 - 4376.
(c) K. Nakamura, T. Matsuda, T. Harada, *Chirality*, **2002**, 14, 703 - 708.
27 (a) W. Stampfer, B. Kosjek, C. Moitzi, W. Kroutil, K. Faber, *Angew. Chem. Int. Ed.* **2002**, 41, 1014 - 1017.
(b) W. Stampfer, B. Kosjek, K. Faber, W. Kroutil, *J. Org. Chem.* **2003**, 68, 402 - 406.
(c) K. Velonia, I. Tsigos, V. Bouriotis, I. Smonou, *Bioorg. Med. Chem. Lett.* **1999**, 9, 65 - 68.
(d) H. Nagaoka, *Biotechnol. Prog.* **2003**, 19, 1149 - 1155.
(e) I. J. Jakovac, H. B. Goodbrand, K. P. Lok, J. B. Jones, *J. Am. Chem. Soc.* **1982**, 104, 4659 - 4665.
(f) G. S. Y. Ng, L-C. Yuan, I. J. Jakovac, J. B. Jones, *Tetrahedron*, **1984**, 40, 1235 - 1243.
(g) H. Ohta, Y. Kato, G. Tsuchihashi, *J. Org. Chem.* **1987**, 52, 2735 - 2739.
28 (a) J. A. Haslegrave, J. B. Jones, *J. Am. Chem. Soc.* **1982**, 104, 4666 - 4671.
(b) J. Davies, J. B. Jones, *J. Am. Chem. Soc.* **1979**, 101, 5405 - 5410.
(c) S. Iriuchijima, N. Kojima, *Agric. Biol. Chem.* **1978**, 42, 451 - 455.
(d) J. A. S. Howell, M. G. Palin, G. Jaouen, S. Top, H. E. Hafa, J. M. Cense, *Tetrahedron : Asymm.* **1993**, 4, 1241 – 1252.
29 (a) T. Kometani, Y. Sakai, H. Matsumae, T. Shibatani, R. Matsuno, *J. Ferment. Bioeng.* **1997**, 84, 195 - 199.
(b) H. Matsumae, H. Douno, S. Yamada, T. Nishida, Y. Ozaki, T. Shibatani, T. Tosa, *J. Ferment. Bioeng.*

1995, 79, 28 - 32.

(c) H. Miya, M. Kawada, Y. Sugiyama, *Biosci. Biotech. Biochem.* **1996**, 60, 95 - 98.

(d) M. F. Züger, F. Giovannini, D. Seebach, *Angew. Chem. Int. Ed. Engl.* **1983**, 22, 1012 - 1012.

(e) K. Nakamura, T. Miyai, K. Ushio, S. Oka, A. Ohno, *Bull. Chem. Soc. Jpn.* **1988**, 61, 2089 - 2093.

(f) D. Seebach, M. F. Züger, F. Giovannini, B. Sonnleitner, A. Fiechter, *Angew. Chem. Int. Ed. Engl.* **1984**, 23, 151 - 151.

(g) P. K. Matzinger, H. G. W. Leuenberger, *Appl. Microbiol. Biotechnol.* **1985**, 22, 208 - 210.

30   (a) J. Hasegawa, M. Ogura, S. Tsuda, S. Maemoto, H. Kutsuki, T. Ohashi, *Agric. Biol. Chem.* **1990**, 54, 1819 – 1827.

(b) K. Nakamura, Y. Inoue, T. Matsuda, A. Ohno, *Tetrahedron Lett.* **1995**, 36, 6263 - 6266.

(c) K. Nakamura, M. Fujii, Y. Ida, *Tetrahedron: Asymm.* **2001**, 12, 3147 - 3153.

(d) M. Takemoto, K. Achiwa, *Phytochemistry*, **1998**, 49, 1627 - 1629.

(e) G. R. Allan, A. J. Carnell, *J. Org. Chem.* **2001**, 66, 6495 - 6497.

(f) D. Kato, S. Mitsuda, H. Ohta, *Org. Lett.* **2002**, 4, 371 - 373.

(g) K. Mitsukura, T. Yoshida, T. Nagasawa, *Biotechnol. Lett.* **2002**, 24, 1615 - 1621.

31   (a) W. Adam, U. Hoch, M. Lazarus, C. R. Saha-Möller, P. Schreier, *J. Am. Chem. Soc.* **1995**, 117, 11898 - 11901.

(b) W. Adam, C. Mock-Knoblauch, C. R. Saha-Möller, *Tetrahedron: Asymm.* **1997**, 8, 1947 - 1950.

(c) W. Adam, M. Lazarus, U. Hoch, M. N. Korb, C. R. Saha-Möller, P. Schreier, *J. Org. Chem.* **1998**, 63, 6123 - 6127.

(d) W. Adam, B. Boss, D. Harmsen, Z. Lukacs, C. R. Saha-Möller, P. Schreier, *J. Org. Chem.* **1998**, 63, 7598 - 7599.

(e) H. Nagatomo, Y. Matsushita, K. Sugamoto, T. Matsui, *Tetrahedron: Asymm.* **2003**, 14, 2339 - 2350.

32   (a) N. Baba, M. Mimura, J. Hiratake, K. Uchida, J. Oda, *Agric. Biol. Chem.* **1988**, 52, 2685 - 2687.

(b) N. Baba, K. Tateno, J. Iwasa, J. Oda, *Agric. Biol. Chem.* **1990**, 54, 3349 - 3350.

33   (a) F. Petit. R. Furstoss, *Tetrahedron: Asymm.* **1993**, 4, 1341 - 1352.

(b) V. Alphand, R. Furstoss, *Tetrahedron: Asymm.* **1992**, 3, 379 - 382.

(c) J. D. Stewart, K. W. Reed, C. A. Martinez, J. Zhu, G. Chen, M. M. Kayser, *J. Am. Chem. Soc.* **1998**, 120, 3541 - 3548.

(d) J. D. Stewart, K. W. Reed, J. Zhu, G. Chen, M. M. Kayser, *J. Org. Chem.* **1996**, 61, 7652 - 7653.

(e) M. M. Kayser, G. Chen, J. D. Stewart, *J. Org. Chem.* **1998**, 63, 7103 - 7106.

(f) M. D. Mihovilovic, G. Chen, S. Wang, B. Kyte, F. Rochon, M. M. Kayser, J. D. Stewart, *J. Org. Chem.* **2001**, 66, 733 - 738.

(g) M. J. Taschner, D. J. Black, Q.-Z. Chen, *Tetrahedron: Asymm.* **1993**, 4, 1387 - 1390.

(h) N. Berezina, V. Alphand, R. Furstoss, *Tetrahedron: Asymm.* **2002**, 13, 1953 - 1955.

34   (a) M. Abo, A. Okubo, S. Yamazaki, *Tetrahedron: Asymm.* **1997**, 8, 345 - 348.

(b) T. Kawasaki, N. Watanabe, T. Sugai, H. Ohta, *Chem. Lett.* **1992**, 1611 - 1614.

PREPARATIVE-SCALE SEPARATION OF ENANTIOMERS ON CHIRAL STATIONARY PHASES BY GAS CHROMATOGRAPHY

Volker Schurig

Institute of Organic Chemistry, University of Tübingen
Auf der Morgenstelle 18, 72076 Tübingen, Germany

volker.schurig@uni-tuebingen.de

Abstract: The state-of-the-art of enantioselective (semi)preparative gas chromatography (GC) is reviewed. The treatise illuminates the historical background, early trials and recent advances. A comprehensive compendium of references is provided. Enantiomers are separated by three types of molecular interactions: hydrogen-bonding, coordination and (*inter alia*) inclusion. Consequently, chiral stationary phases are comprised of dipeptides, metal chelates and (modified) cyclodextrins. Single enantiomers commanding interest in different areas of chemical sciences have been isolated. Absolute configurations of the separated enantiomers are determined by chiroptics and/or by anomalous X-ray diffraction. The amounts of obtained enantiomers suffice for biological trials.

Key Words: Preparative-scale GC, enantiomeric separation, chiral stationary phases, SMB-GC, nitrogen invertomer, inhalation anesthetics, terpenoids, flavours.

1.   INTRODUCTION

Preparative-scale gas chromatography (GC) is inherently restricted to thermally stable and volatile compounds. Contrary to liquid chromatography, the recovery of the isolated enantiomers from the gaseous mobile phase (carrier gas) is straightforward when arerosol and mist formation are prevented by specially designed collection vessels (Hupe, 1971). The early 1970s witnessed some remarkable approaches to (achiral) preparative-scale GC utilizing 2 m x 1 - 10 cm (i.d.) packed columns producing plate heights of 1 - 3 mm and furnishing daily production rates of 50 mg (in case of small

F. Toda (ed.), Enantiomer Separation, 267-300.
© 2004 *Kluwer Academic Publishers, Printed in the Netherlands.*

separation factors) and up to 500 g in a single run (due to an exceptional separation factor of $\alpha = 16$, benzene/cyclohexane) (Bayer et al., 1963). The early state-of-the-art was reviewed by Pescar, 1971. The operating principles and basic characteristics of large diameter gas-chromatographic columns (125, 300, and 400 mm i.d.) used for isolation of pure compounds from essential oils were described and daily production rates from 100 g up to 100 kg were realized and cost breakdowns were given along with optimum operating conditions (Bonmati et al., 1984).

The preparative-scale separation of enantiomers on chiral stationary phases (CSPs) by GC cannot match the overwhelming success achieved in the realm of liquid chromatography (LC) (Francotte, 1994, 1996 and 2001). Modern commercial instrumentation for preparative-scale GC is not readily available. In contrast to LC, separation factors α in enantioselective GC are usually small ($\alpha = 1.01$ - 1.20). This is beneficial for fast analytical separations but detrimental to preparative-scale separations. Only in rare instances are large chiral separation factors ($\alpha > 1.5$) observed in enantioselective GC. Only in one instance, a separation factor as high as $\alpha = 10$ was detected in enantioselective GC for a chiral fluorinated diether and a modified γ-cyclodextrin (Schurig and Schmidt, 2003) (vide supra).

Preparative-scale chromatography relies on a compromise between three variables (cf. Figure 1): (i) component resolution (determined by selectivity, efficiency and retention factor), (ii) speed of analysis and (iii) column sample capacity (Pescar, 1971). Any two of the desired goals may be realized only at the expense of the third. If a large amount of sample is required in a short time, resolution must be high. If resolution is insufficient, either the column load is limited or the time required for separation is long.

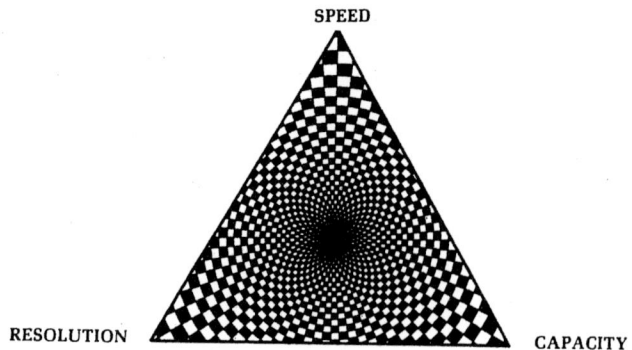

Figure 1. Magic triangle: speed vs. resolution vs. capacity in preparative-scale chromatography (Pescar, 1971).

Throughput has been defined as the amount of purified material per unit of time and per unit of mass of stationary phase, depending on different factors such as loading capacity, column efficiency, selectivity, temperature, column dimensions and flow-rate (Francotte, 2001). The present state-of-the-art of preparative-scale enantioselective GC separations follows rather empirical trial-and-error approaches. Except for one advance (Staerk et al., 1995), no systematic optimization strategies have been developed for throughput in enantioselective GC. Well established ancillary strategies in enantioselective preparative-scale LC such as closed-loop recycling techniques and peak-cutting procedures ('peak-shaving') (Dingenen and Kinkel, 1994) have not been optimized in GC.

No reports are yet available on the separation of enantiomerically enriched fractions arising from incomplete asymmetric reactions. It should be noted that enantiomerically enriched mixtures may be fractionated in racemate and excess enantiomer also on achiral stationary phases (Gil-Av and Schurig, 1994; Nicoud et al., 1996) by adopting the principle of recrystallization of mixtures enriched in one enantiomer leading to the pure enantiomer.

Preparative-scale enantioselective GC is still in a state of infancy and the sample throughput is limited at present. In contrast to LC, the separation of 1 000 kg of a single enantiomer is precluded by enantioselective GC. No reports are even available on the separation of 1 kg amounts of single enantiomers by GC. Yet for the study of chirality-activity relationships in pheromone and flavour research, the isolation of mg amounts of pure enantiomers by semipreparative-scale GC is usually sufficient for physiological and biological trials. One may even consider the separation of enantiomers in the sub-nanogram regime as a micropreparative trial when the separated enantiomers elicit a response in a biological detection array. Such systems have indeed been known from the outset of achiral GC. Thus, three sex attractants present in extracts of the abdomens of female silk worms were analyzed by GC at 132°C and male silk worms were used as a biological detector with a detection limit of 10^{-11} g (Bayer, 1958). Micropreparative-scale separations of enantiomers with CSP-coated widebore capillary columns (Bicchi et al., 1997 (& ref. cited therein); Bicchi et al., 1998) are outside the scope of this review. The present survey is restricted to the use of packed columns coated with chiral stationary phases for the (semi)preparative-scale separation of enantiomers.

2.   Classification of Chiral Stationary Phases (CSPs) for Gas Chromatography

Three principal CSPs, distinguishable by the mode of enantioselective interaction, *i.e.*, hydrogen bonding, coordination and inclusion, have thoroughly been investigated using gas chromatography (Schreier et al., 1995; Schurig, 2001):

- enantiomeric separation on chiral amino acid derivatives *via* hydrogen-bonding (Gil-Av et al., 1966; König, 1987)

- enantiomeric separation on chiral metal coordination compounds *via* complexation (Schurig, 2002)

- enantiomeric separation on cyclodextrin derivatives *via (inter alia)* inclusion (Schurig and Nowotny, 1990; König, 1992; Snopek et al., 1996)

The early development of enantioselective GC witnessed the use of packed columns containing CSPs and (in principle) apt for semipreparative-scale separations, i.e. in hydrogen-bonding-type CSPs (Gil-Av and Feibush, 1967), coordination-type CSPs (Golding et al., 1977) and inclusion-type CSPs (Kościelski et al., 1983).

3.   EXAMPLES

3.1.   A DIPEPTIDE AS CSP

The year 1966 constitutes the advent of enantioselective GC. Following the first direct capillary-gas-chromatographic separation of derivatized α-amino acid enantiomers on an optically active stationary phase (Gil-Av et al., 1966), the use of a preparative-scale packed column was subsequently reported in one of the follow-up publications (Gil-Av and Feibush, 1967). Thus, racemic N-TFA-alanine *tert.*-butyl ester was resolved at 100°C on a 2 m x 1 mm (i.d.) column containing 5 % of the dipeptide stationary phase N-TFA-(L)-valine-(L)-valine cyclohexyl ester coated on Chromosorb W (cf. Figure 2, left). The two fractions of the enantiomers were subjected to an optical rotatory dispersion (ORD) measurement (cf. Figure 2, right). The two traces with opposite rotation angles clearly established that enantiomers had indeed been separated.

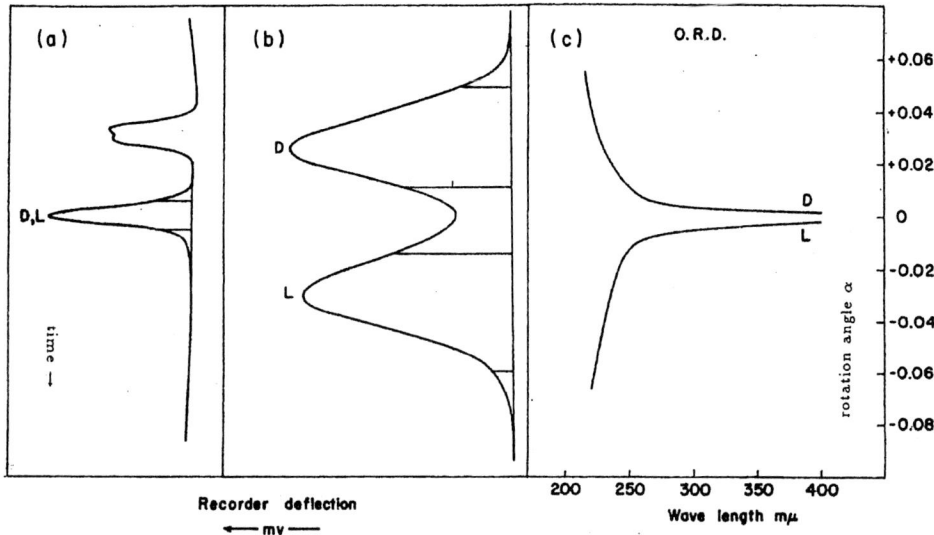

Figure 2. Left: Preparative-scale enantiomeric separation of *N*-TFA-alanine *tert.*-butyl ester on *N*-TFA-(*L*)-valine-(*L*)-valine cyclohexyl ester (experimental conditions cf. text). Right: ORD spectra of the two isolated fractions corresponding to the marked areas (Gil-Av and Feibush, 1967).

3.2.   METAL CHELATES AS CSPs

3.2.1.   *Methyloxirane, 2,2-dimethyl-3-phenyloxirane*

Methyloxirane (epoxypropane) represents one of the smallest chiral cyclic molecules. Its resolution therefore represents a particular challenge. Whereas the enantiomeric separation of various alkyl-substituted oxiranes, oxetanes, oxolanes and oxanes have been performed by capillary complexation GC on 11 different nickel(II) *bis*[α-(heptafluorobutanoyl)-terpeneketonates] (Schurig et al., 1989a), methyl- and ethyloxirane were originally resolved on a 2 m x 2.2 mm (i.d.) packed stainless steel column containing 15 % (w/w) of a 0.133 M solution of anhydrous europium(III) *tris*[3-(trifluoroacetyl)-(*1R*)-camphorate] (Eu-CAM$_3$) in squalane coated on Chromosorb W (HP, 100-120 mesh) in less than 10 min at 40°C and N$_2$ as carrier gas (Golding et al., 1977) (cf. Figure 3). Eu-CAM$_3$ is well known as a versatile chiral NMR shift reagent causing chemical shift nonequivalence for externally enantiotopic nuclei in racemic compounds. The successful enantiomeric separation by GC was verified by spiking the racemic sample with single enantiomers.

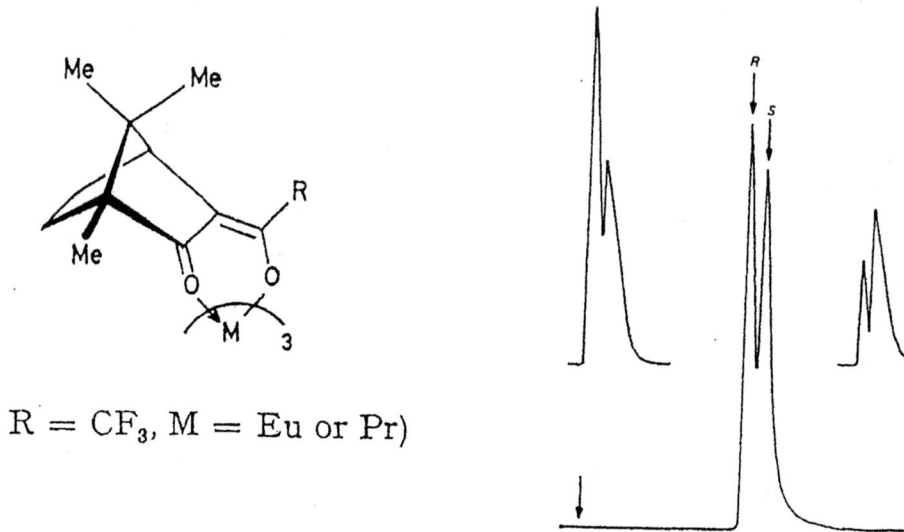

R = CF$_3$, M = Eu or Pr)

Figure 3. Analytical enantiomeric separation of methyloxirane on Eu-CAM$_3$ by complexation GC (center). Racemic sample spiked with *(R)*-enantiomer (left) and with *(S)*-enantiomer (right). Experimental conditions cf. text. *(R)*-methyloxirane elutes after 9 min (Golding et al., 1977).

Although the separation was performed on an analytical scale, the potential of the approach for preparative-scale resolutions by enantioselective complexation GC was immediately apparent. This was borne out by the semipreparative separation of an aromatic oxirane exhibiting the large separation factor of $\alpha \sim 2$ by complexation GC. Thus, 20 μl of neat 2,2-dimethyl-3-phenyloxirane were resolved on a 3 m x 9 mm (i.d.) glass column containing 12.5 % (w/w) of manganese(II)-*bis*[(3-heptafluorobutanoyl)-*(1R)*-camphorate] (Mn-CAM$_2$) (860 mg) in polysiloxane OV-101 (8.2 g) coated on Chromosorb W (AW-DMCS, 60-80 mesh) (65 g) in 20 h at 60°C and 1.5 bar N$_2$ as carrier gas (cf. Figure 4) (Schurig et al., 1990). The following merits of enantioselective preparative-scale GC in comparison to LC had been noted (Schurig et al., 1990):

  – the gaseous mobile phase N$_2$ is inexpensive and innocuous
  – the fractions are free of a liquid phase
  – solubility requirements in a liquid phase are absent

In Figure 5 an over-loading experiment is depicted (Schurig et al., 1990). The second fraction was eluted by increasing the temperature to 70°C. Incidentally, this leads to peak splitting due to temperature fluctuations in the oven (thermal 'christmas-tree'-effect).

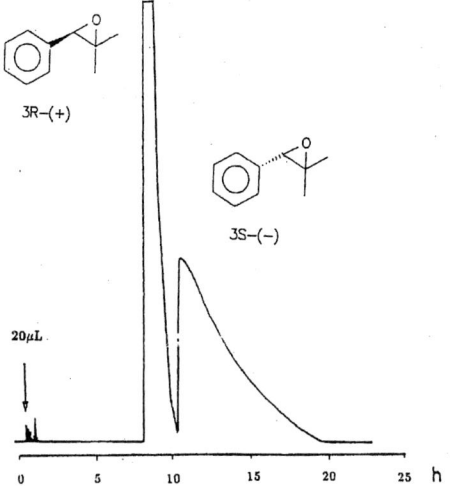

Figure 4. Semi-preparative-scale separation of 2,2-dimethyl-3-phenyloxirane on Mn-CAM₂ (Schurig et al., 1990).

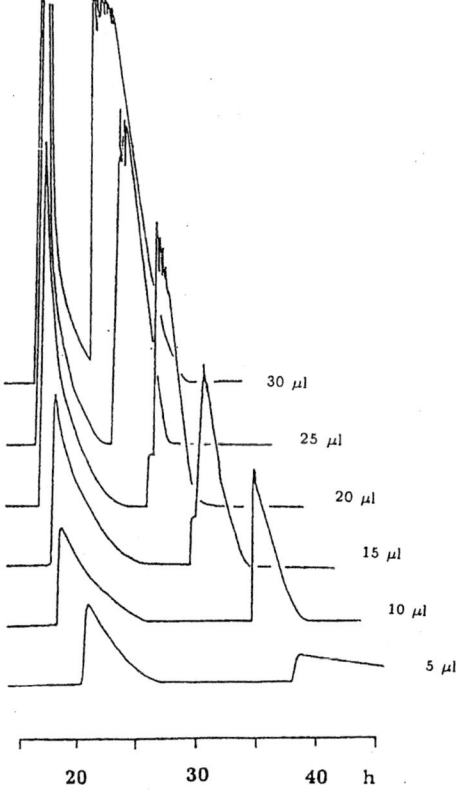

Figure 5. Over-loading experiment of 5 - 30 μl 2,2-dimethyl-3-phenyloxirane on Mn-CAM₂ (experimental conditions cf. text).

3.2.2.   1-Chloro-2,2-dimethylaziridine

Stereogenic three-coordinated nitrogen is conformationally labile due to a small energy barrier to pyramidal inversion proceeding without bond-breaking. Yet stable trisubstituted nitrogen invertomers may be isolated at room temperature if the inversion process is prevented or hindered. This is achieved by incorporation of the nonplanar nitrogen as bridge-head atom into a bicyclic structure as in Tröger's base. For more than a century Tröger's base has been one of the most fascinating and stimulating chiral molecules in organic chemistry. The separation of the enantiomers of Tröger's base was achieved by liquid chromatography on an 0.9 m column containing lactose hydrate followed by fractional crystallization (Prelog and Wieland, 1944). This separation represented the first reproducible chromatographic separation of an enantiomeric pair. Alternatively, the barrier to inversion can also be enhanced by constraining the nitrogen atom into a tight three-membered aziridine ring accompanied by halogeno-substitution at nitrogen (Felix and Eschenmoser, 1968; Brois, 1968a; Brois, 1968b; Lehn and Wagner, 1968).

The invertomers of the nitrogen pyramid 1-chloro-2,2-dimethylaziridine have been separated by enantioselective complexation GC on the chiral metal chelate nickel(II)-*bis*[(3-heptafluorobutanoyl)-*(1R)*-camphorate] (Ni-CAM$_2$) in squalane (Schurig et al., 1979). The resolution of the chiral aziridine into two sharp peaks with $\alpha = 1.5$ in 45 min at 60°C demonstrated the stereochemical integrity of the invertomers (cf. Figure 6).

Employing a stainless steel column (7 m x 3 mm (i.d.)) containing 250 mg Ni-CAM$_2$ in 6.25 g OV-101 coated on 43 g Chromosorb W (AW-DMCS, 60-80 mesh), enantiomerically pure invertomers of 1-chloro-2,2-dimethylaziridine were obtained in mg quantities at room temperature (cf. Figure 7). The high enantiomeric excess ee of the enantiomers was established via analytical complexation GC (cf. Figure 8) (Schurig and Leyrer, 1990). The amount of the isolated enantiomerically highly enriched invertomers (appr. 2 mg each) was sufficient for the determination of the sign of the specific rotation. Also the enantiomerization kinetics could be measured in the gas phase at 82°C by complexation GC employing the analytical column (cf. Figure 8) for screening the decrease of ee with time (cf. Figure 9). The rate of inversion at 338.6 K (the reaction vessel was placed in boiling methanol of 65.3°C) was $k = 6.9 \cdot 10^{-6}$ sec^{-1}, $t_{1/2} = 842$ min, $\Delta G^{\ddagger} = 115.5 \pm 1.2$ kJ/mol (mean value of four measurements) (Schurig and Leyrer, 1990).

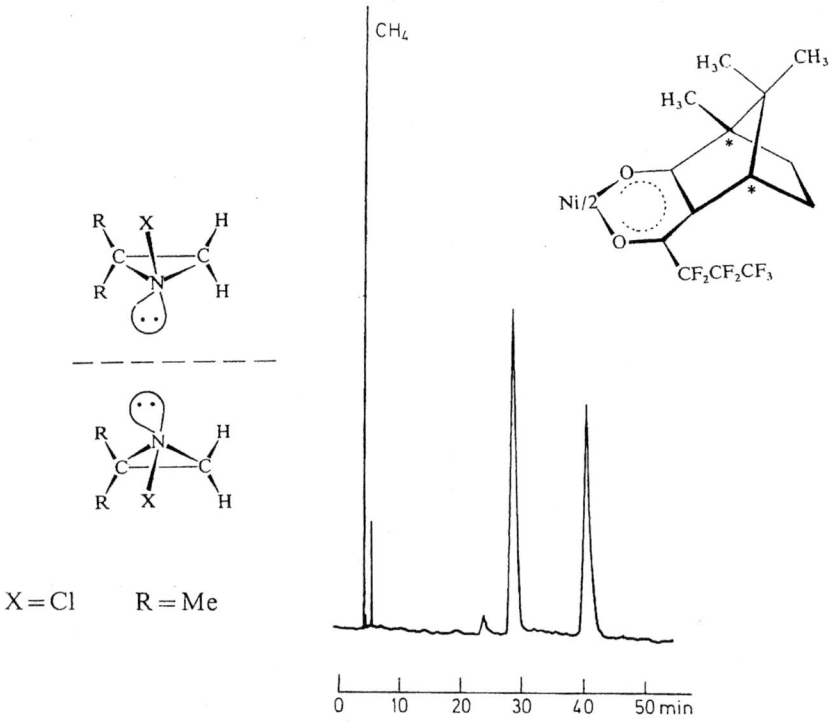

Figure 6. Analytical invertomer separation of 1-chloro-2,2-dimethylaziridine by complexation GC on Ni-CAM₂ (0.133 molal in squalane) at 60°C. Column: 100 m x 0.5 mm (i.d.) metallic nickel capillary. Carrier gas: N₂ (3.8 ml/min), split ratio 1:50 (Schurig et al., 1979).

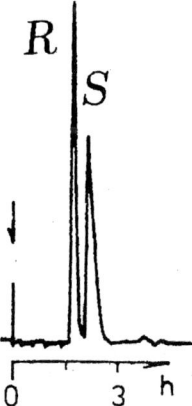

Figure 7. Semi-preparative-scale separation of the invertomers of 1-chloro-2,2-dimethylaziridine by complexation GC on Ni-CAM₂ at 22°C and 1.5 bar N₂ in 3 h at 22°C. Injected amount of racemate: 6 μl (further experimental conditions cf. text). The (S)-enantiomer is eluted before the (R)-enantiomer on Ni-CAM₂ (Schurig and Leyrer, 1990).

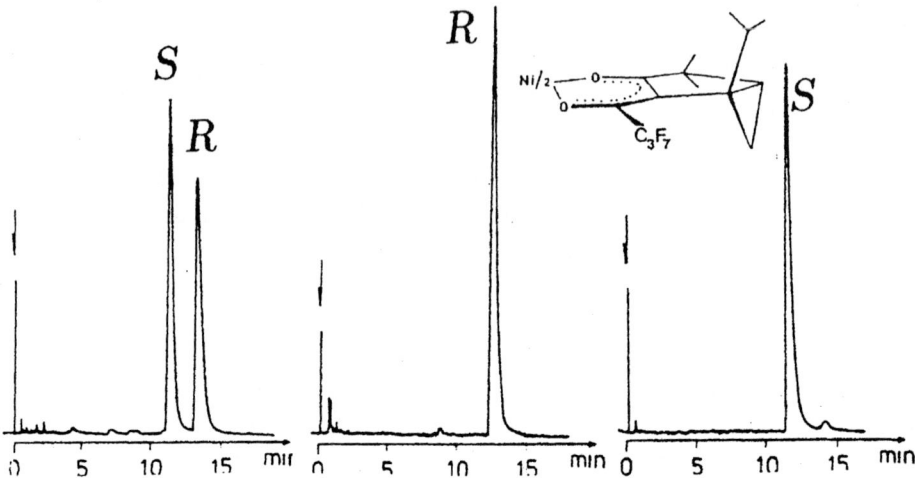

Figure 8. Enantiomeric purity of the isolated invertomers. Left: racemic sample; middle: pure *(R)*-enantiomer; right: *(S)*-enantiomer contaminated with *(R)*-enantiomer from incomplete preparative fractionation (*ee* = 98 %). Analytical column: 25 m x 0.25 mm glass capillary coated with 0.125 molal nickel(II) *bis*[(2-heptafluorobutanoyl)-*(1S,5S)*-4-methylthujonate] (Ni-METHU₂) in OV-101 at 50°C. The *(R)*-enantiomer is eluted before the *(S)*-enantiomer on *(1S,5S)*-Ni-METHU₂ (Schurig and Leyrer, 1990).

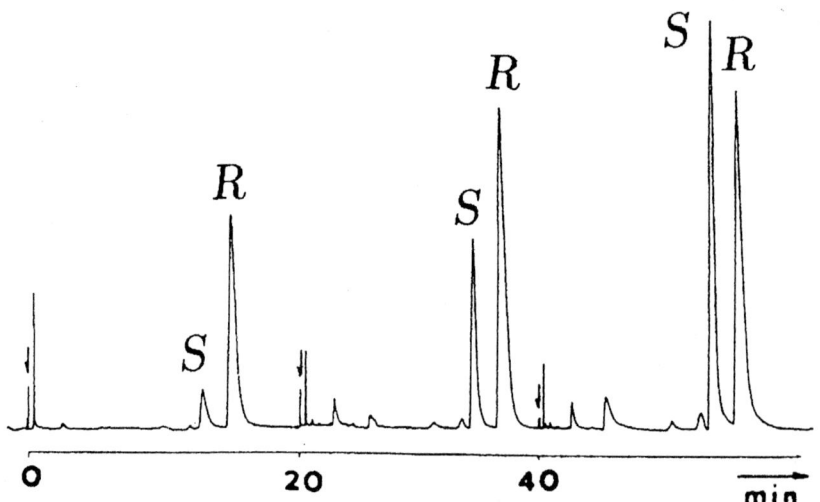

Figure 9. Racemization kinetics of *(R)*-1-chloro-2,2-dimethylaziridine at 82°C by monitoring the increase of the *(S)*-enantiomer with time by complexation GC (experimental conditions cf. Figure 8).

The invertomer of 1-chloro-2,2-dimethylaziridine, eluted as the second fraction on $(1R)$-Ni-CAM$_2$, showed a positive sign of the specific rotation at all wavelengths and a positive Cotton effect at 217 and 260 nm in n-pentane. By indirect gas-chromatographic evidence, the configuration (S) was assigned to the second eluted enantiomer of 1-chloro-2,2-dimethylaziridine on $(1R)$-Ni-CAM$_2$ because diastereomeric trans-$(1S,2S)$-1-chloro-2-methylaziridine, which differs from the former only by the absence of a methyl group in anti-position to the coordinating nitrogen lone pair, is also eluted as the second peak on $(1R)$-Ni-CAM$_2$ (cf. Figure 10).

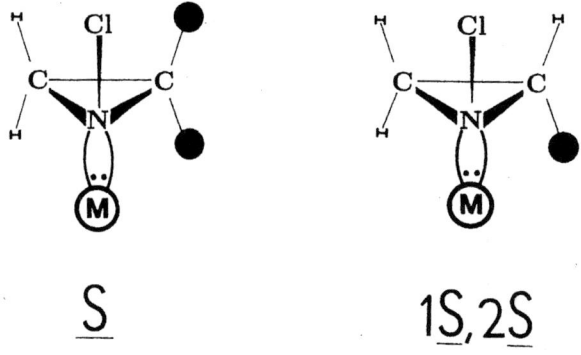

Figure 10. Absolute configuration of (S)-1-chloro-2,2-dimethylaziridine and of trans-$(1S,2S)$-1-chloro-2-methylaziridine (the filled circle represents a methyl group, M denotes the coordinating metal ion).

This assignment was at odds with that obtained by the octant rule but was later confirmed also by chiroptical evidence (cf. cited references in Schurig and Leyrer, 1990). Thus the absolute configuration of (+)-1-chloro-2,2-dimethylaziridine was assigned as (S). The results show that the assignment of absolute configuration can be aided by semipreparative GC via the determination of the sign of the specific rotation of isolated enantiomers.

3.2.3.   1.6-Dioxaspiro[4.4]nonanes

Spiroketals are widespread in nature. Alkyl-substituted 1.6-dioxaspiro-[4.4]nonanes, 1,6-dioxaspiro[4.5]decanes, 1.6-dioxaspiro[4.6]undecanes and 1,7-dioxaspiro[5.5]undecanes are important components of insect pheromones (cf. structures *79-87* in Mori, 1989). Complexation GC exhibits quite large separation factors for the stereoisomers of this type of compounds ($\alpha > 1.3$) (Weber and Schurig, 1984). The semipreparative-scale enantiomeric separation of 1 μl of the three chiral spiroketals 1.6-dioxaspiro[4.4]-nonane, *E,Z*-2-ethyl-1.6-dioxaspiro[4.4]nonane (*Chalcogran*, the aggregation pheromone of *Pityogenes chalcographus (L.)* (Mori, 1989)) and *EE,EZ, ZZ*-2,7-dimethyl-1.6-dioxaspiro[4.4]nonane were performed on an analytical 3.5 m x 4 mm (i.d.) stainless steel column containing 80 mg of the chiral metal chelate nickel(II)-*bis*[(6-heptafluorobutanoyl)-*(5S)*-carvonate] (Ni-CAR$_2$) in 2 g squalane coated on 11.6 g Chromosorb W (AW-DMCS, 60-80 mesh) (Schurig et al., 1989a) at 45°C and 1.4 bar N$_2$ (Schurig, 1987; Schurig, 1988) (cf. Figure 11). The split ratio (trap/detector) was set 60 : 1. The trap for the eluents was immersed into crushed dry ice. The exceedingly long separation times as a result of large retention factors could only be moderately alleviated by temperature-programming. The enantiomeric excess of the recovered stereoisomers was determined on an analytical column. It amounted to *ee* > 99 % for the first eluted enantiomer. Only trace amounts of enantiopure spiroketals were obtained. Whereas these minor quantities might suffice for biological trials, up-scaling, if required, may be performed by exchanging the analytical packed column for a column of larger dimensions.

In semipreparative chromatography, the second eluted enantiomer is prone to contamination by the first eluted enantiomer. The second eluted enantiomer can be reverted to the first eluted enantiomer when a CSP of opposite chirality is used. Since both enantiomers of carvone are readily available as starting material for the preparation of the CSP Ni-(CAR)$_2$ one of the two 'enantiomeric' columns can be selected in an effort to elute the enantiomer of interest as a first fraction.

On close inspection of the chromatograms, an elevated baseline can be discerned between the terminal peaks of the spiroketals. This peak distortion is caused by interconversion of the stereoisomers during separation. On-column enantiomerization leads to the formation of a plateau between the terminal peaks (Trapp et al., 2001). Such a kinetic phenomenon of molecular interconversion was first observed for 1.6-dioxaspiro[4.4]nonane by complexation GC (Schurig and Bürkle, 1982). This interesting pheno-menon sheds light on the inherent stereochemical lability of the spiroketal functionality. Consequently, the stereoisomeric separation of spiroketals should be performed preferentially at low temperatures.

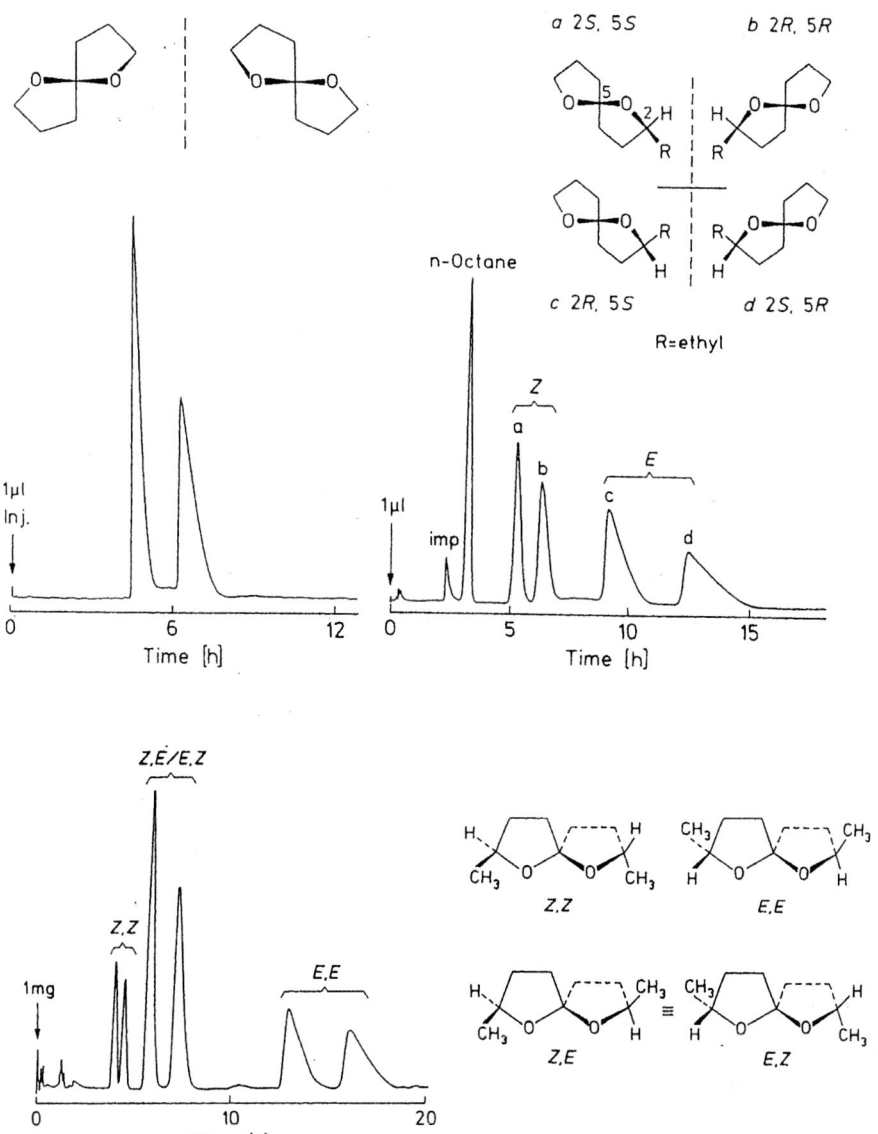

Figure 11. Semi-preparative-scale stereoisomeric separation of 1.6-dioxaspiro[4.4]nonanes on Ni-CAR$_2$ (experimental conditions cf. text) (Schurig, 1987).

3.3.   CYCLODEXTRINS AS CSPs

3.3.1.   *Terpenoic hydrocarbons*

Large separation factors (α up to 2.17) were observed for the terpenoic hydrocarbons α- and β-pinene, *cis*- and *trans*-pinane and 2-carene (cf. Figure 12) at 35 - 50°C on a 2 m x 4 mm (i.d.) glass column containing the solution of 0.65 mol % α-cyclodextrin hydrate in formamide (4.5 g) coated on Celite (20 g) (Kościelski et al., 1983; Kościelski et al., 1986). Unfortunately it was found that the peak efficiency of the column was low (950 - 1 250 theoretical plates), the column temperature was limited to 70°C and the life-time of the system was short due to extensive column bleeding of the solvent formamide and dehydration of the stationary phase by the dry carrier gas helium. This work nevertheless started an impressive development of enantioselective GC employing derivatized cyclodextrins as CSPs (Schurig and Nowotny, 1990; König, 1992)

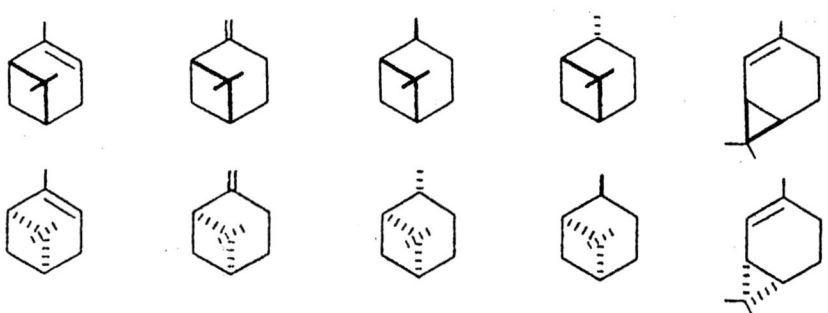

Figure 12. Structures of terpenoic hydrocarbons α-pinene, β-pinene, *cis*-pinane, *trans*-pinane and 2-carene resolved on α-cyclodextrin hydrate (Kościelski et al., 1986).

The presence of water was subsequently identified as being essential for the performance of the above set-up and for maintaining high enantioselectivity (Lindström et al., 1990). The preparative-scale enantiomeric separation of racemic camphene with a separation factor as high as $\alpha = 3.7$ (measured at 30°C on an analytical column) was subsequently achieved at 45°C on a 2.1 m x 4 mm (i.d.) column containing α-cyclodextrin hydrate in formamide (1 : 5, w/w) coated on Chromosorb W (AW, 45-60 mesh) by saturating the carrier gas (helium) with water vapour (Lindström et al., 1990).

3.3.2.   Flavour compounds and terpenoids

With the advent of alkylated/acylated cyclodextrins (CDs) as versatile CSPs for analytical separation of enantiomers by GC (Schurig and Nowotny, 1990; König, 1992), their potential for preparative-scale separations was immediately apparent. Thus, the enantiomers of 2 mg racemic methyl jasmonate were separated in 80 min on a 1.8 m x 4 mm (i.d.) stainless steel column containing heptakis(2,6-di-O-methyl-3-O-n-pentyl)-β-cyclodextrin in polysiloxane PS 86 (1:1, w/w) coated on Chromosorb W (HP, 100-120 mesh) (2.5 % w/w) at 120°C and 0.4 bar helium (Hardt and König, 1994). Recovery of the samples varied between 40 and 80 %. The separation of 5 mg racemic muscone with the low separation factor $\alpha = 1.025$ yielded only an incomplete resolution. 1 mg of the pharmaceutical hexobarbital could also be resolved (Hardt and König, 1994). In a number of later papers by König et al. (cf. references 11-14 in Bicchi et al., 1997; König et al., 1999), the semipreparative separations of terpenoids guided the elucidation of their stereochemistries.

Emerging from an industrial-related laboratory, the preparative gas-chromatographic separation of the 'mushroom odour' racemate, i.e. 1-octenyl 3-acetate, has been attempted employing two 1 m x 20 mm (i.d.) columns (connected in series) containing undiluted permethylated β-cyclo-dextrin coated on Chromosorb P (NAW) (10 %, w/w) in an automated THN 102 instrument (Chromatelf) with a katharometer as detector at 120°C and H_2 (2.6 l/min) (Fuchs and Perrut, 1994). Large amount of racemate (up to 500 μl of ester) were injected onto the column, however, enantiomeric separation was incomplete and the recovered enantiomers had a low enantiomeric excess *ee*. Columns containing undiluted permethyla-ted CD ought to be operated at high temperatures due to the crystalli-ne nature of CD. To overcome this restriction, dilution of the CD in the polysiloxane OV-1701 (cf. Schurig and Nowotny, 1990) might have been a viable alternative approach allowing a lower oven temperature thereby increasing enantioselectivity. The semipreparative-scale separation of the enantiomers of filbertone (E-5-methyl-hept-2-en-4-one), the principal fla-vour component of filberts (hazelnuts), has been achieved by GC on a glass column (2.3 m x 3 mm i.d.) containing heptakis(2,3-di-O-acetyl-6-O-tert-butyldimethylsilyl)-β-cyclodextrin (TBDMS-β-CD) in polysiloxane PS086 (20 %) coated on Chromosorb P (AW, DMCS 80-100 mesh) (Blanch and Schurig, 2004). Only 0.3 mg of racemic α-ionone could be resolved on a short glass column (0.2 m x 4 mm i.d.) containing Lipodex E in SE-54 coated on Chromosorb at 75°C and N_2 (Quattrini et al., 1999).

3.3.3.   All-trans-perhydrotriphenylene (PHTP)

All-*trans*-perhydrotriphenylene (PHTP) (cf. insert in Figure 14) is the product of exhaustive hydrogenation of triphenylene. It belongs to one of ten stereoisomers of PHTP. The chiral compound of high rotational symmetry $(D_3 = C_3 + 3\,C_2)$ forms inclusion complexes. The stereoselective polymerization via γ-radiation of the prochiral diolefin 1,3-pentadiene within the chiral nano channels of (R)-$(-)$-all-*trans*-PHTP led to an optically active 1,4-*trans*-isotactic polymer (Natta and Farina, 1976) (cf. Figure 13).

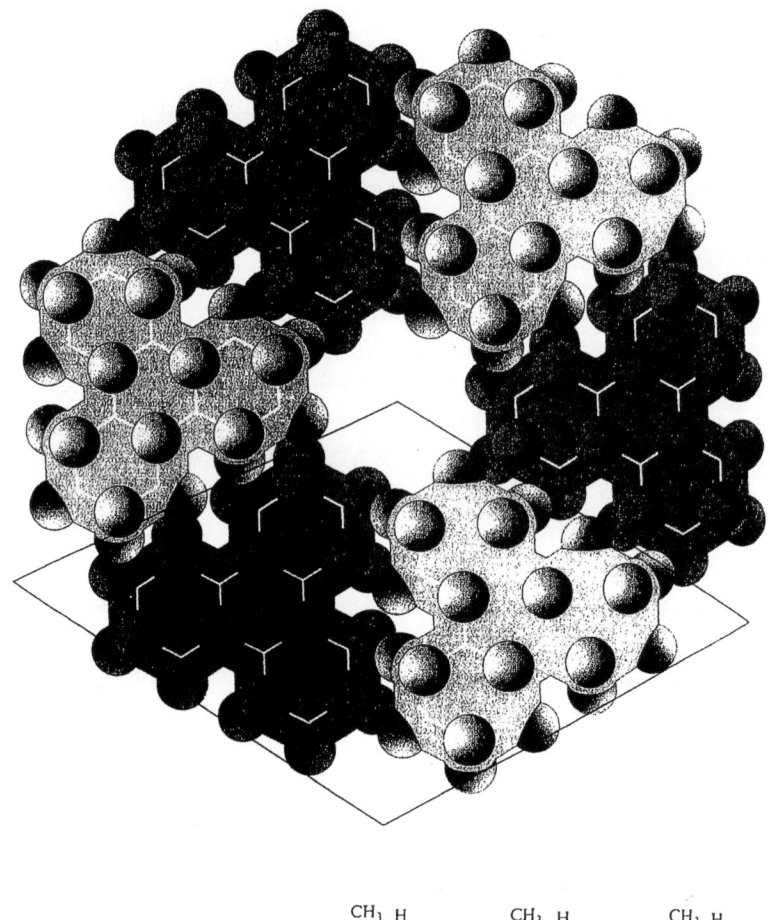

Figure 13. Top: Unit cell of all-*trans*-PHTP. Bottom: stereoselective polymerization of 1,3-pentadiene (Natta and Farina, 1976)

Optically active (R)-$(-)$-all-*trans*-PHTP ($[\alpha]_D$ -93, m.p. 142°C) was obtained by diastereomeric crystallization of the e-2-carboxylic acid of PHTP with dehydroabietylamine as resolving agent (Farina and Audisio, 1970). Because racemic unfunctionalized 1,2-dialkylcycloalkanes are amenable to resolution on permethylated-β-cyclodextrin by GC (Schurig et al., 1989b), it is not unexpected that PHTP is also prone to enantiomeric separation by inclusion GC although the cavity size of β-cyclodextrin is small in respect to the molecular size of PHTP. Indeed, the analytical and semipreparative-scale gas chromatographic separation of the enantiomers of all-*trans*-PHTP on heptakis(2,3-di-*O*-acetyl-6-*O*-*tert*-butyldimethylsilyl)β-cyclodextrin (TB-DMS-β-CD) has been achieved (Schürch et al., 2001). For semipreparative-scale separations, two columns (column A: 2 m x 2 mm (i.d.) glass; column B: 1.8 m x 4 mm (i.d.) stainless steel) were used. The ratio of TBDMS-β-CD to polysiloxane SE-54 was 89 : 11 (w/w). The coating of Chromosorb P (AW-DMCS 80-100 mesh) (column A) and Chromosorb P (AW-DMCS 100-120 mesh) (column B) with the CSP was 6 % each. The total amount of packing was 3.2 g for column A and 18.1 g for column B, respectively. N_2 was used as carrier gas. The amount of sample injected in column A was 1 μl of a 50 mg/ml PHTP solution in *n*-pentane, whereas in column B, 2 mg PHTP in 40 μl *n*-pentane were injected with a cycle time of 60 min. The isothermal column temperature was held between 140 - 190°C. Sample fractions were collected in glass traps held at $-$ 15°C. Although the analytical separation of PHTP on TBDMS-β-CD exhibited the low separation factor α of 1.04 and resolution factor R_s of 2.48 (cf. Figure 15), respectively, a semipreparative-scale separation of enantiomers was still feasible by carefully adjusting column temperature, carrier gas flow and column load. The time constraint, however, necessitated the use of an elevated temperature thereby sacrifying resolution and amount of sample to be resolved. A column load study is depicted in Figure 16. In five repetitive single-step injections of a total of 10 mg racemate per day by fractional collection into five traps, 16 % of the first eluted $(-)$-enantiomer with $ee = 99.2$ % (trap 1) and 20 % of the second eluted $(+)$-enantiomer with $ee = 97.6$ % (trap 4) were obtained. The total recovery including mixed fractions was 95 %. The results command interest as they show that a system which displays a low enantioselectivity is nonetheless amenable to a preparative-scale separation of the enantiomers. Moreover it is demonstrated that rather involatile enantiomers can be preparatively resolved by GC also at elevated column temperatures.

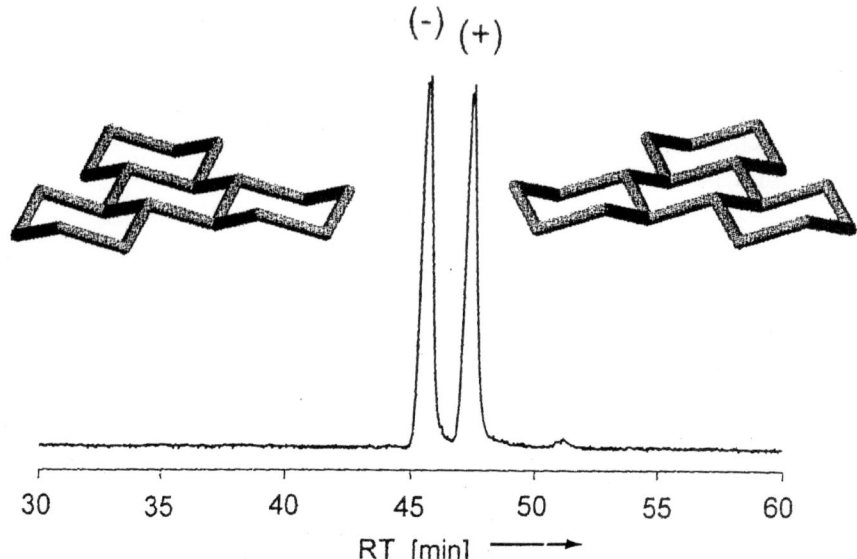

Figure 14. Analytical separation of racemic all-*trans*-PHTP on TBDMS-β-CD in OV-1701 (0.25 μm). Column: 25 m x 0.3 mm (i.d.) at 160°C, helium 95 kPa (Schürch et al., 2001).

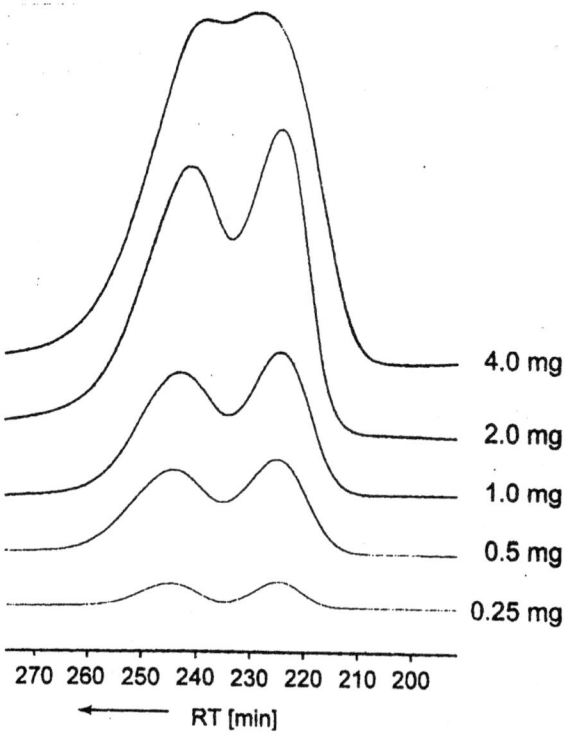

Figure 15. Over-loading experiment (column B) with incremental additions of racemic PHTP (Schürch et al., 2001).

3.3.4. Human inhalational anesthetics enflurane, isoflurane and desflurane

Diethyl ether was introduced into medicine as a human inhalational anesthetic in 1842. More recently, by substituting hydrogen for halogens, improved narcotic gases have been developed with reduced toxicity and flammability, improved metabolic integrity and increased volatility and potency. Contemporary anesthetics satisfy the six important requirements: muscle relaxation, analgesia, hypnosis, sedation, fast onset and rapid recovery. Substituting hydrogen by halogens, however, a stereogenic center was unintentionally introduced into the molecules. Thus, with the exception of sevoflurane, all contemporary inhalational anesthetics (ie, halothane, enflurane, isoflurane and desflurane) are chiral although they are currently used as racemic mixtures. Consequently, it became of great importance to provide sufficient amounts of single enantiomers of haloethers for biological and medical trials and for the determination of chiroptical properties. The absence of suitable functionalities precluded resolution of the racemates via diastereomers. Also enantioselective synthetic approaches for pure enantiomers appear not to be straightforward. The liquid phase chromatographic resolution on CSPs is hampered by the difficulty in separating the volatile compounds from the liquid mobile phase. Enantioselective GC is therefore the method of choice.

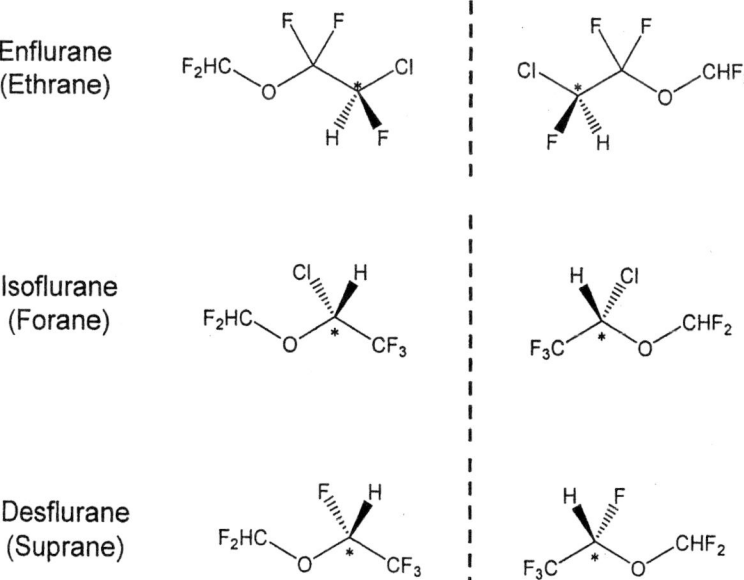

Figure 16. The structure of chiral inhalational anesthetics.

The preparative-scale enantiomeric separations of enflurane, isoflurane and desflurane (cf. Figure 16) by gas chromatography on modified cyclodextrins have been achieved. The separation was aided by unprecedentedly large separation factors α of the enantiomers. Two approaches were followed.

(i) The enantiomers of isoflurane and enflurane were preparatively separated on undiluted 3-*O*-trifluoroacetylated-*2,6-di-O-n*-pentylated γ-cyclodextrin, a commercially available mixture of regioisomers and homologues (Staerk et al., 1994a; Staerk et al., 1994b). After a careful optimization study in regard to sample load and enantiomeric purity, a 2 m x 10 mm (i.d.) stainless steel column containing 23.4 % (w/w) of the modified γ-cyclodextrin coated on Chromosorb W (AW, 80-100 mesh) was employed for the preparative-scale enantiomeric separation of isoflurane at 40°C with helium as carrier gas. Using a sampling interface between the exit of the column and the fraction collector, the enantiomeric composition was detected on-line by an analytical enantioselective column. This set-up allowed the calculation of enantiomeric purity, recovery and production rate. The latter could be increased up to a threshold of approx. 400 mg of isoflurane racemate (Staerk et al., 1994a). In a similar approach but utilizing another packing material, a 1 m x 10 mm (i.d.) stainless steel column containing 25 % (w/w) of the modified γ-cyclodextrin coated on Chromosorb A (AW, 60-80 mesh) was employed for the preparative-scale enantiomeric separation of enflurane at 40°C with hydrogen as carrier gas. It was found that up to 24 mg of the first eluted enantiomer with *ee* = 100 % could be obtained when 75 mg of racemate was injected whereas only 6 mg of the second eluted enantiomer with *ee* = 100 % was obtained upon injecting the amount of 20 mg of racemate in the optimized system. The production rate could also be increased to gram/day quantities of pure enantiomers by consecutive injections around-the-clock. Repetitive injections were timed such that the end of the second peak of the former injection coincided with the start of the first peak in the subsequent injection (Staerk et al., 1994b). The production rate could be further enhanced when sacrifying enantiomeric excess *ee*.

(ii) Whereas the above described endeavour utilized an undiluted cylodextrin-based selector, the dilution of modified cyclodextrins in polar polysiloxanes, previously applied successfully in analytical enantioselective GC (Schurig and Nowotny, 1990), also proved useful for packed columns in a conventional GC apparatus (cf. Figure 17). Thus, 30 μl (47 mg) racemic enflurane was resolved in 45 min at 26°C into the first eluted *(R)* enantiomer and at 50°C in the second eluted (S) enantiomer with high chemical purity (99.9 %) and enantiomeric excess ($ee = 99.8$ %). A 4 m x 7 mm (i.d.) glass column containing $\sim$ 95 g of octakis(3-O-butanoyl-2,6-di-O-n-pentyl)-γ-cyclodextrin, Lipodex E (König et al., 1989) in polysiloxane SE-54 (10 %, w/w) coated on Chromosorb P (AW, DMCS, 80-100 mesh) (20 %, w/w) was used (Schurig et al., 1993; Schurig and Grosenick, 1994) (cf. Figure 18, left). The eluates from the columns were split (1 : 1500) (cf. Figure 17), one leading to the flame ionization detector and the other to a U-shaped cold trap cooled with dry ice/acetone or liquid nitrogen. While the separation of the enantiomers from the carrier gas helium is considered straightforward, mist formation nevertheless reduced the amount of the enantiomers eventually recovered ($\sim$ 12 mg each). An over-loading study is depicted in Figure 18, right.

Figure 17. Instrumental set-up of preparative-scale gas chromatography in a conventional GC apparatus (Carlo-Erba, HRGC 5300). The home-made packed column (4 m x 7 mm (i.d.)) utilized the modified condenser of a discarded rotary evaporator.

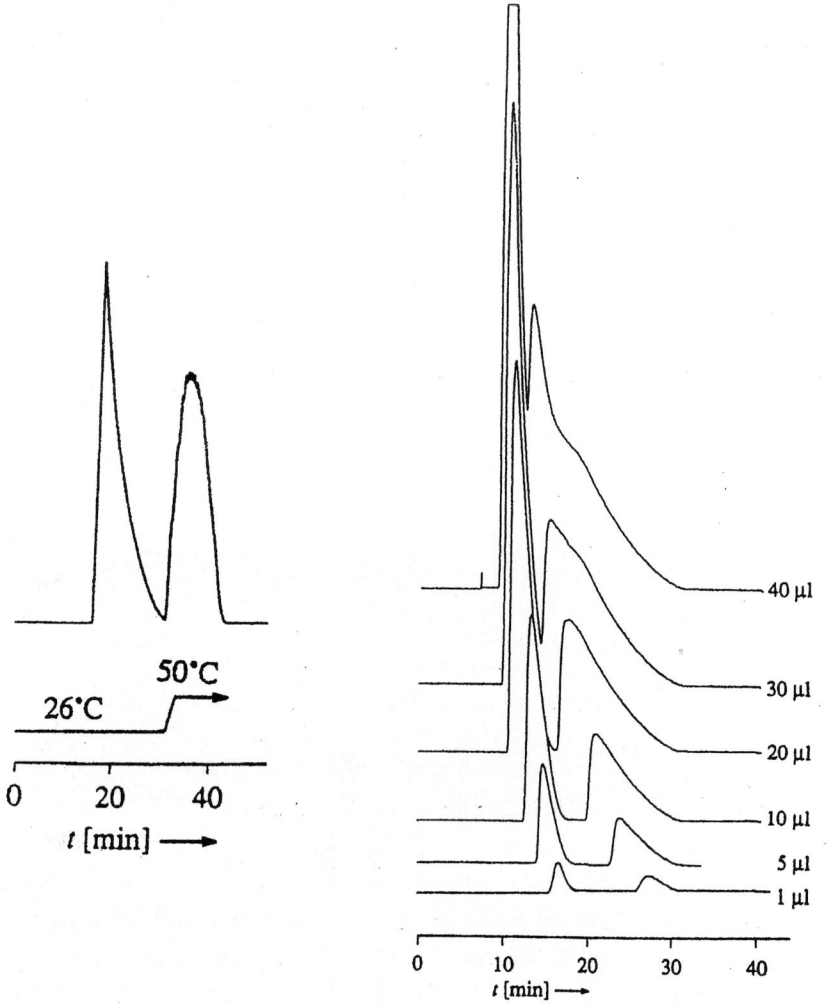

Figure 18. Left: Temperature-programmed preparative-scale separation of racemic enflurane by GC. Injected amount: 30 μl, 8.4 cm/s helium. Right: Over-loading experiment at 40°C and 8.4 cm/s helium (Schurig and Grosenick, 1994).

Repetitive injections allowed the collection of 250 mg each of enflurane enantiomer of ee = 99.9 % in one day suitable for chiroptical trials. The specific rotation was determined, i.e., $[\alpha]_D^{20}$ − 4.6 (first eluted enantiomer) and $[\alpha]_D^{20}$ + 4.6 (second eluted enantiomer) (c 1, n-hexane) (Schurig and Grosenick, 1994). The vibrational circular dichroism spectra were determined and the absolute configuration derived therefrom was assigned as (+)-*(S)* (Zhao et al., 2000).

By using the same procedure, 8 mg of isoflurane was separated into enantiomers while halothane could not be resolved (Schurig et al., 1993).

The above-described set-up was insufficient for the preparative-scale race-mate resolution of isoflurane and desflurane (chemical structures cf. Figure 17). Therefore an upscaled column design was employed. A 1 m x 24 mm (i.d.) stainless steel column (column A) and two 1 m x 6 mm (i.d.) co-lumns (combined to form column B) were tested in an automated Hupe & Busch preparative gas chromatograph unit built in the 1960s. Since large quantities of the chiral selector were required, unpurified octakis(3-O-butanoyl-2,6-di-O-n-pentyl)-γ-cyclodextrin, Lipodex E (König et al., 1989), consisting of a myriad of under- and over-pentylated species was used. Inte-restingly, the crude selector afforded a significantly higher separation factor α for isoflurane but reduced enantioselectivities for enflurane and desflu-rane. Approximately half a kilogram of stationary phase was prepared from 90.5 g polysiloxane SE-54 and 10.75 g crude Lipodex E (10.6 %, w/w) coa-ted on 398 g Chromosorb P (AW, DMCS, 80-100 mesh) (20.3 %, w/w). 300 mg of single enantiomers (ee = 99.9 %) of isoflurane were obtained per day with 130 automated repetitive injections using column A isothermally at room temperature. Desflurane could only be separated under overlapping conditions and recycling or discharging the middle fractions. 500 mg of the first eluted enantiomer (ee = 91 %) and 450 mg of the second eluted enan-tiomer (ee only 68 %) were obtained per day. This amount was drastically reduced when higher ee values were required.

The isolated enantiomers were investigated by cryogenic anomalous X-ray diffraction to determine their absolute configurations. The dextrorotatory enantiomers of isoflurane and desflurane possessed the (S) configuration. Thus the prior chiroptical assignment of dextrorotatory isoflurane as $(+)$-(S) was confirmed while the previous assignment of $(+)$-desflurane as (R) (Polavarapu et al., 1997) had to be revised to $(+)$-(S) with consequences in organic reaction mechanisms (Schurig et al., 1996). In regard to biological trials it could be established that the anesthetic action of single isoflurane enantiomers differed only marginally (17 %) in rats implying that interac-tion with a specific receptor is not important in the mechanisms of action of inhaled anesthetics (Eger et al., 1997).

3.3.5.   'Compound B', a degradation product of the human inhalational anesthetic sevoflurane

The perfluorodiether 'compound B' (1,1,1,3,3-pentafluoro-2-(fluoromethoxy)-3-methoxypropane) represents a chiral degradation product of the achiral inhalational anesthetics sevoflurane (cf. Figure 19). It is formed under alkaline conditions when sevoflurane is recycled during anesthesia and exhaled carbon dioxide is trapped by soda lime. 'Compound B' showed an unprecedentedly large separation factor of $\alpha = 4.1$ at 30°C on heptakis(2,3-di-O-acetyl-6-O-$tert$-butyldimethylsilyl)-β-cyclodextrin (TBDMS-β-CD) in the polysiloxane PS 86. The preparative scale resolution of 50 μl of racemic 'compound B' in one run was performed in 40 min on a 1 m x 18 mm (i.d.) stainless steel column containing unpurified TBDMS-β-CD in PS 86 (15,2 %, w/w) coated on Chromosorb P (AW, DMCS, 80-100 mesh) (19.2 %, w/w) at 70°C and 0.4 bar N_2 with a Hupe & Busch automated instrument (Schmidt et al., 2000) (cf. Figure 20, top). The high enantiomeric excess ($ee > 99.9$ %) was proved with an analytical column (cf. Figure 20, bottom). In ten automated runs, 275 mg of the first eluted enantiomer and 73 mg of the second eluted enantiomer were isolated. An intermediate fraction was stored for re-injection. Determination of the specific rotation and the absolute configuration by anomalous X-ray diffraction established that the first eluted fraction constitutes the (−)-(R) and the second eluted fraction the (+)-(S) enantiomer, respectively (Schmidt et al., 2000).

More recently, the highest separation factor ever reported in enantioselective GC was found for 'compound B' and Lipodex E by mere serendipity ($\alpha = 10$, Schurig and Schmidt, 2003). If 'compound B' was not a characteristic degradation product of sevoflurane, its extraordinary propensity for strong chiral bias would have escaped attention as such a structure would hardly be conceived by mere intuition. This highly enantioselective system constitutes an ideal target for future systematic studies of throughput in preparative-scale GC, including moving bed technology (*vide infra*).

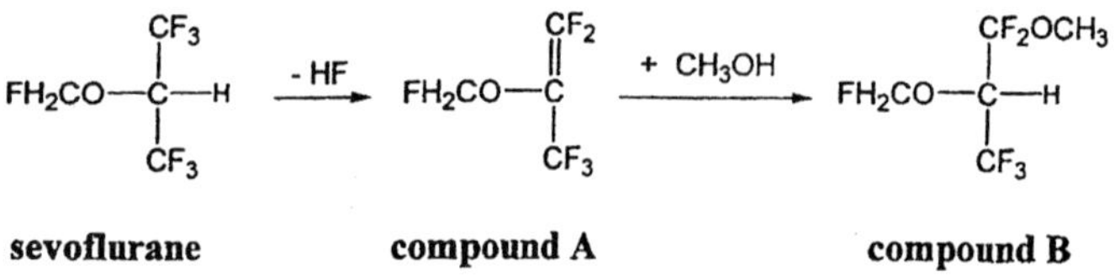

Figure 19. Decomposition products of desflurane in a rebreathing circuit during anesthesia (Schmidt et al., 2000).

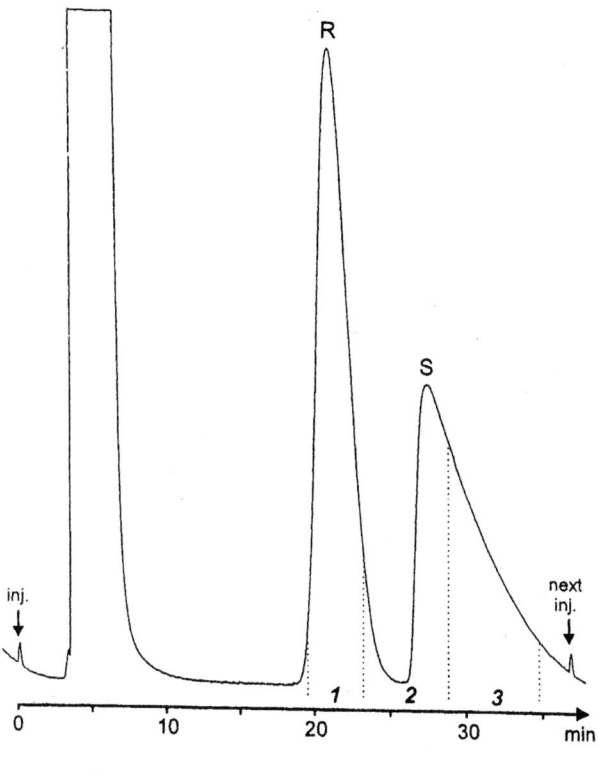

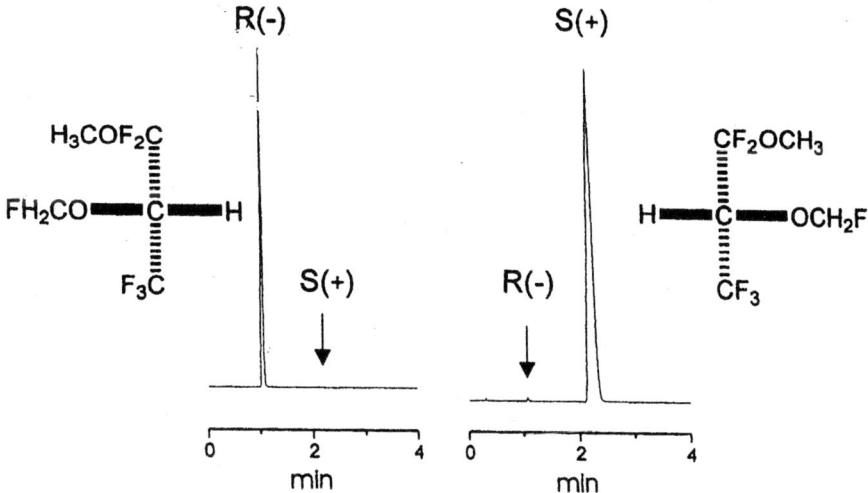

Figure 20. Top: Preparative separation of the enantiomers of 'compound B' (experimental conditions cf. text). Bottom: Evidence of the high enantiomeric excess *ee* by using an analytical column coated with TBDMS-β-CD in PS 86 (Schmidt et al., 2000).

4.   MOVING BED TECHNOLOGY IN PREPARATIVE-SCALE ENANTIOSELECTIVE GC

Conventional chromatography represents a discontinuous batch process and relies on periodic injections of feed. When the chromatographic bed is moved countercurrent to the inert mobile phase, a continuous separation process is feasible when one component moves in the direction of the mobile phase while the other component moves in the direction of the stationary phase (Barker, 1971). Such a scenario is realized in true moving bed chromatography (TMB) by the judicious choice of flow rates of bed and inert phase. Interestingly, the concept of continuous chromatography in a temperature-gradient was first discussed for the separation of enantiomers (Martin and Kuhn, 1941). The principle of an (achiral) TMB-GC-unit is depicted in Figure 21 (Nunnemann and Pauschmann, 1971).

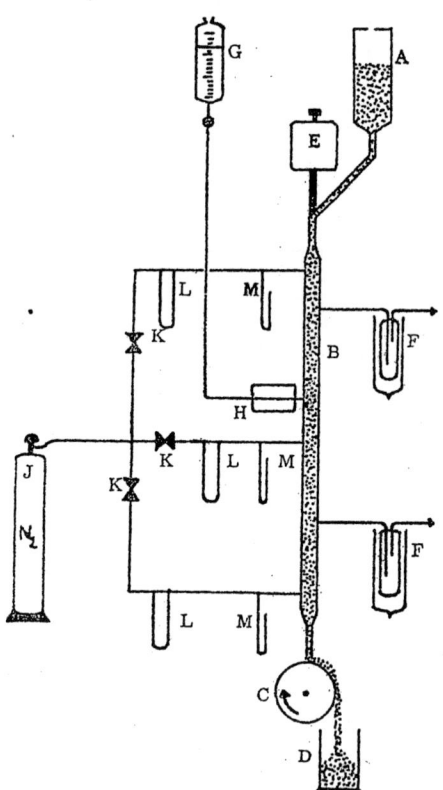

Figure 21. Set-up for TMB-GC. A: hopper. B: column. C: rotating cylinder. D: container. E: vibrator. F: top and bottom trap. G: feed. H: heated injector. J: gas supply. K: valve. L: flow meter. M: pressure gauge (Nunnemann and Pauschmann, 1971).

In this set-up ethanol and water ($\sim$ 1:1) of alcoholic beverages (Whisky, 43 % ethanol) were separated in fractions containing 7 % ethanol (bottom product) and 92 % ethanol (top product), respectively, using a 1 m x 16 mm (i.d.) glass column filled with saw dust passing by gravity through the column from a hopper against an up-stream of nitrogen gas (Nunnemann and Pauschmann, 1971).

The countercurrent movement of a stationary phase is cumbersome in practice but it can be circumvented by an array of short columns connected by multi-position valves connected with eluent, feed, extract and raffinate, a method referred to as simulated moving bed chromatography (SMB) (Schulte and Strube, 2001). In SMB chromatography, the continuous countercurrent flow of the fluid and of the solid adsorbent is simulated by periodically switching the different inlets and outlets in the multi-column unit. Enantioselective SMB-LC has first been demonstrated for racemic 1-phenylethanol resolved on the polysaccharide CSP Chiralcel OD. In this pioneering work the principle of the method and the set-up has been depicted in a lucid educational fashion (cf. Figures 22 & 23) (Negawa and Shoji, 1992).

Enantioselective SMB-GC has been realized adapting the results previously obtained in the batch processes described above. Preliminary studies involved racemic enflurane with a separation factor of $\alpha \sim 2$ (Juza et al., 1998a; Juza et al., 1998b; Biressi et al., 2000) and isoflurane (Biressi et al., 2002a) which were separated on unpurified octakis(3-O-butanoyl-2,6-di-O-n-pentyl)-γ-cyclodextrin, Lipodex E, in polysiloxane SE-54 and coated on Chromosorb P (AW, DMCS, 80-100 mesh).

An optimized version of the enantioselective SMB-GC unit was subsequently presented for enflurane enantiomers (chemical structure cf. insert in Figure 24) (Biressi et al., 2002b). It consisted of eight 80 cm x 15 mm (i.d.) stainless steel columns assembled in a home-made SMB-GC unit operated at 35°C (Scheme, cf. Figure 24). Each column with an adsorption bed volume of 140 ml each contained 20 % unpurified Lipodex E in the polysiloxane SE-54 and coated (17 %, w/w) on Chromosorb A (NAW, 20-30 mesh) $\sim$ 0.6 mm). This set-up represented the first gas-chromatographic SMB-GC unit for the preparative-scale separation of enantiomers.

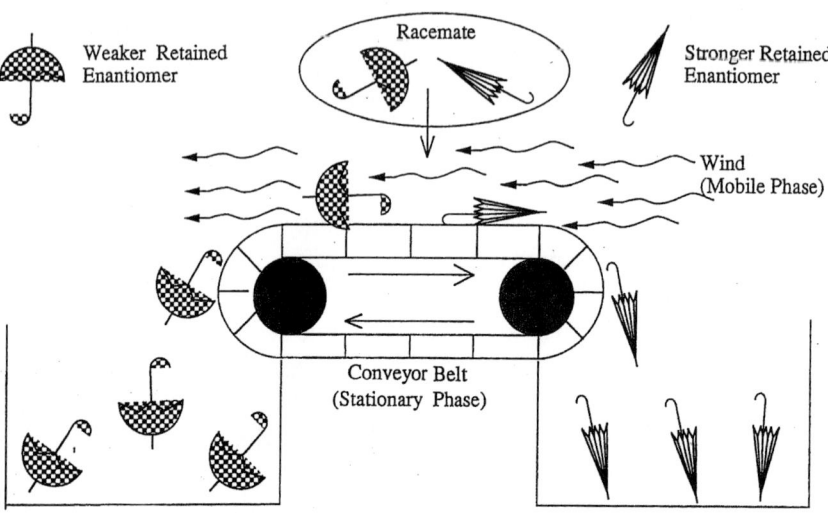

Figure 22. Visualization of the SMB approach in a binary separation system (the open and closed umbrellas do not represent real enantiomers) (Negawa and Shoji, 1992).

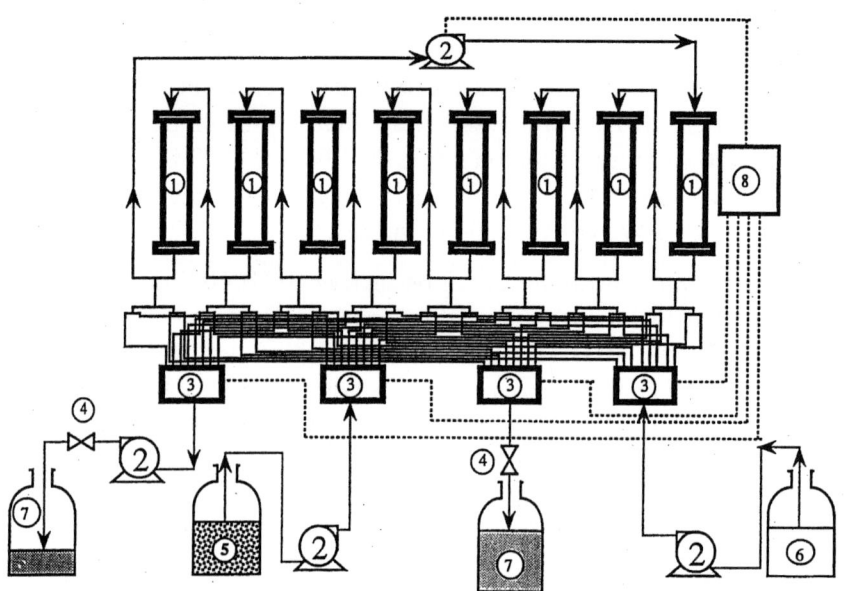

Figure 23. Realization of the enantioselective SMB liquid phase approach by using eight 15 cm x 20 mm (i.d.) columns containing Chiralcel OD. 1: columns. 2: LC-pump. 3: eight-port rotary valve. 4: back-pressure-valve. 5: feed reservoir (racemate). 6: desorbent reservoir. 7: extract (enantiomer 1) and raffinate (enantiomer 2) collector. 8: system controller (Negawa and Shoji, 1992).

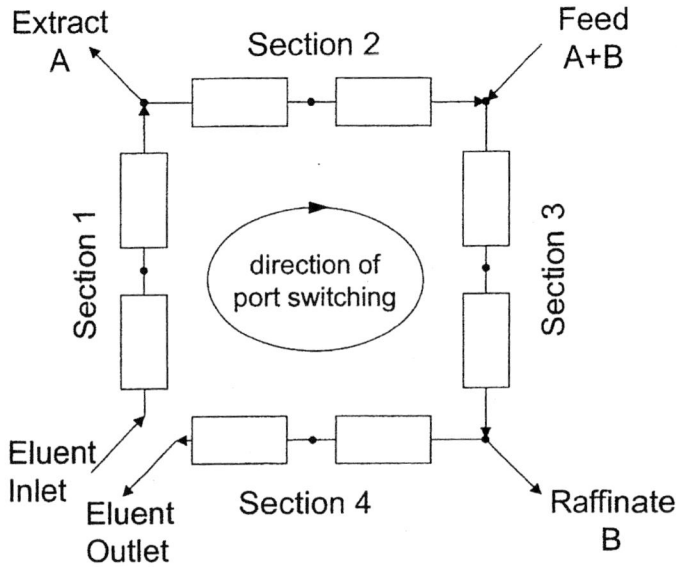

Figure 24. Scheme of an eight-column open-loop SMB-GC unit with 2-2-2-2 configuration (Biressi et al., 2002b).

Under carefully optimized conditions (internal flow rates, temperature, switch time, feed concentration and enantiomeric excess *ee*) the enantio-selective SMB-GC pilot unit furnished a total of 20 g of each enantiomer of enflurane with an enantiomeric excess of 96.6 % (time requirement not known) with N_2 as carrier gas (Biressi et al., 2002b). Noteworthy is the low particle size of the non-acid-washed Chromosorb A (NAW) used ($\sim$ 0.6 mm). Enflurane was introduced in the gaseous form (saturated in N_2). The SMB-GC unit produced the largest amounts of single enflurane enan-tiomers presently available for testing of potential differences in biological activity. No comparison of throughput between the batchwise *vs.* the SMB process is as yet available.

REFERENCES

Bayer, E. (1958) in D. H. Desty (ed.), *Gas Chromatography 1958*, Butterworth, London, p. 340.

Bayer, E., Hupe, K. P., and Mack H. (1963) Filling of analytical and preparative columns for gas chromatography. *Anal. Chem.* **35**, 492-496.

Barker, P. E. (1971) Continuous chromatographic techniques, in A. Zlatkis and V. Pretorius (eds) *Preparative Gas Chromatography*, Wiley-Interscience, New York. pp. 325-394.

Bicchi, C., D'Amato, A., Manzin, V., Galli, A., and Galli, M. (1997) Cyclodextrin derivatives in GC separation of racemic mixtures of volatiles - Part XII: thick-film wide-bore columns for enantiomer GC preparation. *J. High Resol. Chromatogr.* **20**, 493-498.

Bicchi, C., Balbo, C., D'Amato, A., Manzin, V., Schreier, P., Rozenblum, A., and Brunerie, P. (1998) Cyclodextrin derivatives in GC separation of racemic mixtures of volatiles - Part XIV: Some applications of thick-film wide-bore columns for enantiomer GC micropreparation. *J. High Resol. Chromatogr.* **21**, 103-106.

Biressi, G., Quattrini, F., Juza, M., Mazzotti, M., Schurig, V., and Morbidelli, M. (2000) Gas chromatographic simulated moving bed separation of the enantiomers of the inhalation anesthetic enflurane. *Chem. Engineer. Sci.* **55**, 4537-4547.

Biressi, G., Rajendran, G., Mazzotti, M., and Morbidelli, M. (2002a) The GC-SMB separation of the enantiomers of isoflurane. *Sep. Sci. Techn.* **37**, 2529-2543.

Biressi, G., Mazzotti, M., and Morbidelli, M. (2002b) Experimental investigation of the behavior of gas phase simulated moving beds. *J. Chromatogr. A* **957**, 211-225.

Blanch, G. P. and Schurig, V. (2004) Unpublished results.

Bonmati, R., Chapelet-Letourneux, G., and Guiochon, G. (1984) Gas chromatography: a new industrial process of separation. Application to essential oils. *Sep. Sci. & Techn.* **19**, 113-155.

Brois, S. J. (1968a) Aziridines. XI. Nitrogen inversion in N-haloaziridines. *J. Amer. Chem. Soc.* **90**, 506-508.

Brois, S. J. (1968b) Aziridines. XII. Isolation of a stable nitrogen pyramid. *J. Amer. Chem. Soc.* **90**, 508-509.

Dingenen, J. and Kinkel, J. N. (1994) Preparative chromatographic resolution of racemates on chiral stationary phases on laboratory and production scales by closed-loop recycling chromatography. *J. Chromatogr. A* **666**, 627-650.

Eger II, E. I., Koblin, D. D., Laster, M. J., Schurig, V., Juza, M., Ionescu, P., and Gong. D. (1997) Minimum alveolar anesthetic concentration values for the enantiomers of isoflurane differ minimally. *Anesth. Analg.* **85**, 188-192.

Farina, M. and Audisio, G. (1970) Stereochemistry of perhydrotriphenylene II. Absolute rotation and configuration of optically active *anti-trans-anti-trans-anti-trans*-perhydrotriphenylene, *Tetrahedron* **26**, 1839-1844.

Felix, D. and Eschenmoser, A. (1968) Langsame Inversion am pyramidal gebundenen Stickstoff: Isolierung von diastereomeren 7-Chlor-7-azabicyclo[4.1.0]heptanen bei Raumtemperatur. *Angew. Chem.* **80**, 197-199.

Francotte, E. (1994) Contribution of preparative chromatographic resolution to the investigation of chiral phenomena. *J. Chromatogr. A* **666**, 565-601.

Francotte, E. (1996) Chromatography as a separation tool for the preparative resolution of racemic compounds, in S. Ahuja (ed.) *Chiral Separations: Applications and Technology*, ACS, New York, pp. 271-308.

Francotte, E. (2001) Enantioselective chromatography as a powerful alternative for the preparation of drug enantiomers. *J. Chromatogr. A* **906**, 379-397.

Fuchs, G. and Perrut, M. (1994) Enantiomer fractionation by preparative gas chromatography. *J. Chromatogr. A.* **658**, 437-443.

Gil-Av, E., Feibush, B., and Charles-Sigler, R. (1966) Separation of enantiomers by gas liquid chromatography with an optically active stationary phase. *Tetrahedr. Lett.* 1009-1015.

Gil-Av, E. and Feibush B. (1967) Resolution of enantiomers by gas liquid chromatography with optically active stationary phases. Separation on packed columns. *Tetrahedr. Lett.* 3345-3347.

Gil-Av, E. and Schurig, V. (1994) Resolution of non-racemic mixtures in achiral chromatographic systems: a model for the enantioselective effects observed. *J. Chromatogr. A*, **666**, 519-525.

Golding, P. T., Sellars, P. J., and Wong, A. K. (1977) Resolution of racemic epoxides on g.l.c. columns containing optically active lanthanoid complexes. *J. Chem. Soc., Chem. Comm.*, 570-571.

Hardt, I. and König, W. A. (1994) Preparative enantiomer separation with modified cyclodextrins as chiral stationary phases. *J. Chromatogr. A*, **666**, 611-615.

Hupe, K.-P. (1971) Outlet system, in A. Zlatkis and V. Pretorius (eds) *Preparative Gas Chromatography*, Wiley-Interscience, New York. pp. 143-162.

Juza, M., Braun, M., and Schurig, V. (1997) Preparative enantiomer separation of the chiral inhalation anesthetics enflurane, isoflurane and desflurane by gas chromatography on a derivatized γ-cyclodextrin stationary phase. *J. Chromatogr. A* **769**, 119-127.

Juza, M., Di Giovanni, O., Biressi, G., Schurig, V., Mazzotti, M., and Morbidelli, M. (1998a) Continuous enantiomer separation of the volatile inhalation anesthetic enflurane with a gas chromatographic simulated moving bed unit. *J. Chromatogr. A* **813**, 333-347.

Juza, M., Biressi, G., Di Giovanni, O., Mazzotti, M., Schurig, V., and Morbidelli, M. (1998b) Resolution of the inhalation anesthetic enflurane on a cyclodextrin-based chiral stationary phase: development of the GC-SMB separation, in F. Mennier (ed.), *Fundamentals of Adsorption*, Elsevier, Amsterdam, pp. 455-460.

König, W. A. (1987) *The Practice of Enantiomer Separation by Capillary Gas Chromatography*, Hüthig, Heidelberg.

König, W. A., Krebber, R., and Mischnick, P. (1989) Cyclodextrins as chiral stationary phases in capillary gas chromatography. Part V: octakis(3-O-butyryl-2,6-di-O-pentyl)-γ-cyclodextrin. *J. High Resol. Chromatogr.* **12**, 732-738.

König, W. A. (1992) *Enantioselective Gas Chromatography with Modified Cyclodextrins*, Hüthig, Heidelberg.

König, W. A., Bülow, N., and Saritas, Y. (1999) Identification of sesquiterpene hydrocarbons by gas phase analytical methods. *Flavour & Fragrance J.* **14**, 367-378.

Kościelski, T., Sybilska, D., and Jurczak, J. (1983) Separation of α- and β-pinene into enantiomers in gas-liquid chromatography systems via α-cyclodextrin inclusion complexes. *J. Chromatogr.* **280**, 131-134.

Kościelski, T., Sybilska, D., and Jurczak, J. (1986) New chromatographic method for the determination of the enantiomeric purity of terpenoic hydrocarbons. *J. Chromatogr.* **364**, 299-303.

Lehn, J. M. and Wagner, J. (1968) Hindered nitrogen inversion in *N*-halogenoaziridines and in *N*-halogenoazetidines. *J. Chem. Soc., Chem. Commun.*, 148-150.

Lindström, M., Norin, T., and Roeraade, J. (1990) Gas chromatographic separation of monoterpene hydrocarbon enantiomers on α-cyclodextrin. *J. Chromatogr.* **513**, 315-320.

Martin H. and Kuhn W. (1941) Multiplikationsverfahren zur Spaltung von Racematen. *Z. Elektrochem.* **47**, 216-220.

Mori, K. (1989) Synthesis of optically active pheromones. *Tetrahedron* **45**, 3233-3298.

Natta, G. and Farina, M. (1976) *Struktur und Verhalten von Molekülen im Raum*, Verlag Chemie, Weinheim, New York.

Negawa, M. and Shoji, F. (1992) Optical resolution by simulated moving-bed adsorption technology. *J. Chromatogr.* **590**, 113-117.

Nicoud, R.-M., Jaubert, J.-N., Rupprecht, I., and Kinkel, J. (1996) Enantiomeric enrichment of non-racemic mixtures of binaphthol with non-chiral packings. *Chirality* **8**, 234-243.

Nunnemann, F. and Pauschmann, H. (1971) private communication (cf. F. Nunnemannn, Diploma Thesis, University of Tübingen, Germany 1971).

Pescar, R. E. (1971) Preparative Column Technology, in A. Zlatkis and V. Pretorius (eds) *Preparative Gas Chromatography*, Wiley-Interscience, New York. pp. 73-141.

Polavarapu, P. L, Cholli, A., and Vernice, G. G. (1997) Determination of absolute configurations and predominant conformations of general anesthetics: desflurane (vol 82, pg 791, 1993). *J. Pharm. Sci.* **86**, 267.

Prelog V. and Wieland P. (1944) Über die Spaltung von *Trögerscher* Base in optische Antipoden, ein Beitrag zur Stereochemie des dreiwertigen Stickstoffs. *Helv. Chim. Acta* **27**, 1127-1134.

Quattrini, F., Biressi, G., Juza, M., Mazzotti, M., Fuganti, C., and Morbidelli, M. (1999) Enantiomer separation of alpha-ionone using gas chromatography with cyclodextrin derivatives as chiral stationary phases. *J. Chromatogr. A* **865**, 201-210.

Schmidt, R., Roeder, M., Oeckler, O., Simon, A., and Schurig, V. (2000) Separation and absolute configuration of the enantiomers of a degradation product of the new inhalation anesthetic sevoflurane. *Chirality* **12**, 751-755.

Schulte, M. and Strube, J. (2001) Preparative enantioseparation by simulated moving bed chromatography. *J. Chromatogr. A* **906**, 399-416.

Schürch, S., Saxer, A., Claude, S., Tabacchi, R., Trusch, B., and Hulliger, J. (2001) Semipreparative gas chromatographic separation of *all-trans*-perhydrotriphenylene enantiomers on a chiral cyclodextrin stationary phase, *J. Chromatogr. A* **905**, 175-182.

Schurig, V., Bürkle, W., Zlatkis, A., and Poole C. F. (1979) Quantitative resolution of pyramidal nitrogen invertomers by complexation chromatography, *Naturwissenschaften* **66**, 423.

Schurig, V. and Bürkle, W. (1982) Extending the scope of enantiomer resolution by complexation gas chromatography, *J. Amer. Chem. Soc.* **104**, 7573-7580.

Schurig, V. (1984) Gas chromatographic separation of enantiomers on optically active metal-complex-free stationary phases. *Angew. Chem. Int. Ed.* **23**, 747-765.

Schurig, V. (1987) Semi-preparative enantiomer separation of 1,6-dioxaspiro[4.4]nonanes by complexation gas chromatography, *Naturwissenschaften* **74**, 190-191.

Schurig, V. (1988) Enantiomer separation by complexation gas chromatography – applications in chiral analysis of pheromones and flavours, in P. Schreier (ed.), *Bioflavour '87*, Walter de Gruyter, Berlin, New York, pp. 35-54.

Schurig, V., Bürkle, W., Hintzer, K., and Weber R. (1989a) Evaluation of nickel(II) *bis*[α-(heptafluorobutanoyl)-terpeneketonates] as chiral stationary phases for the enantiomer separation of alkyl-substituted cyclic ethers by complexation gas chromatography, *J. Chromatogr.* **475**, 23-44.

Schurig, V., Nowotny, H.-P., and Schmalzing, D. (1989b) Gas-chromatographic enantiomer separation of unfunctionalized cycloalkanes on permethylated β-cyclodextrin, *Angew. Chem. Int. Ed.* **28**, 736-737.

Schurig, V. and Leyrer U. (1990) Semi-preparative enantiomer separation of 1-chloro-2,2-dimethylaziridine by complexation gas chromatography - absolute configuration and barrier of inversion, *Tetrahedr. Asymm.* **1**, 865-868.

Schurig, V. and Nowotny, H.-P. (1990) Gas chromatographic separation of enantiomers on cyclodextrin derivatives. *Angew. Chem. Int. Ed.* **29**, 939-957.

Schurig, V., Schleimer, M., and Wistuba, D. (1990) Semipräparative Enantiomerentrennung durch Komplexierungs-Gaschromatographie. Semipreparative enantiomer separation by complexation gas chromatography, in W. Günther, J. P. Matthes and H.-H. Perkampus (eds.), *Instrumentalized Analytical Chemistry and Computer Technology 1990*, GIT Verlag, Darmstadt, Germany, pp. 366-370.

Schurig, V., Grosenick, H., and Green, B. S. (1993) Preparative enantiomer separation of the anesthetic enflurane by gas inclusion chromatography. *Angew. Chem. Int. Ed.* **32**, 1662-1663.

Schurig, V. and Grosenick, H. (1994) Preparative enantiomer separation of enflurane and isoflurane by inclusion chromatography. *J. Chromatogr. A* **666** 617-625.

Schurig, V., Juza, M., Green, B. S., Horakh, J., and Simon, A. (1996) Absolute configuration of the inhalation anesthetics isoflurane and desflurane. *Angew. Chem. Int. Ed.* **35**, 1680-1682.

Schurig, V. (2001) Separation of enantiomers by gas chromatography - Review. *J. Chromatogr. A* **906**, 275-299.

Schurig, V. (2002) Practice and theory of enantioselective complexation gas chromatography. *J. Chromatogr. A* **965**, 315-356.

Schurig, V. and Schmidt, R. (2003) Extraordinary chiral discrimination in inclusion gas chromatography. Thermodynamics of enantioselectivity between a racemic perfluorodiether and a modified γ-cyclodextrin. *J. Chromatogr.* **1000**, 311-324.

Snopek, J., Smolková-Keulemansová, E., Cserháti, T., Gahm, K., and Stalcup, A. (1996) Cyclodextrins in analytical separation methods, in J. Szejtli and T. Osa (eds.), *Comprehensive Supramolecular Chemistry*, Pergamon, Vol 3 Chapter 18, pp. 515-571.

Schreier, P., Bernreuther, A., and Huffer, M. (1995) *Analysis of Chiral Organic Molecules*, Walter de Gruyter, Berlin, 1995.

Staerk, D. U., Shitangkoon, A., and Vigh, G. (1994a) Gas chromatographic separation of the enantiomers of volatile fluoroether anesthetics by derivatized cyclodextrins II. Preparative-scale separations for isoflurane. *J. Chromatogr. A* **663**, 79-85.

Staerk, D. U., Shitangkoon, A., and Vigh, G. (1994b) Gas chromatographic separation of the enantiomers of volatile fluoroether anesthetics by derivatized cyclodextrins III. Preparative-scale separations for enflurane. *J. Chromatogr. A* **677**, 133-140.

Staerk, D. U., Shitangkoon, A., and Vigh, G. (1995) Preparative gas chromatographic separation of the enantiomers of methyl-chloropropionate using a cyclodextrin-based stationary phase. *J. Chromatogr. A* **702**, 251-257.

Trapp, O., Schoetz, G., and Schurig, V. (2001) Determination of enantiomerization barriers by dynamic and stopped-flow chromatographic methods - Review article. *Chirality*, **13** 403-414.

Wang, F., Polavarapu, P. L., Schurig. V., and Schmidt, R. (2002) Absolute configuration and conformational analysis of a degradation product of the inhalation anesthetic sevoflurane: a vibrational circular dichroism study. *Chirality*, **14**, 618-624.

Weber, R. and Schurig V. (1984) Complexation gas chromatography - a valuable tool for the stereochemical analysis of pheromones. *Naturwissenschaften* **71**, 408-413.

Zhao, C., Polavarapu, P. L., Grosenick, H., and Schurig, V. (2000) Vibrational circular dichroism, absolute configuration and predominant conformations of volatile anesthetics: enflurane. *J. Mol. Struct.* **550**, 105-115.

PRACTICAL RESOLUTION OF ENANTIOMERS BY HIGH-PERFORMANCE LIQUID CHROMATOGRAPHY

Chiyo Yamamoto and Yoshio Okamoto
Department of Applied Chemistry, Graduate School of Engineering, Nagoya University
Furo-cho, Chikusa-ku, Nagoya 464-8603, Japan

1. Introduction

Resolution by liquid chromatography is a rather new method for separating racemates, which had not been mentioned by Pasteur. The first baseline separation of enantiomers by liquid chromatography was reported for the resolution of α-amino acids by ligand-exchange chromatography using a copper complex by Davankov in 1971,[1] and in the 1980s, the instrumentation for high-performance liquid chromatography (HPLC) remarkably advanced. This encouraged the development of efficient chiral stationary phases (CSP) for HPLC, and many CSPs have been since reported. Today, nearly one hundred CSPs are commercially available. The CSPs have been prepared from both optically active small molecules and polymers, which are usually supported on silica gel.[2]

Thirty years ago, only one method existed for estimating optical purity or enantiomeric excess (ee); that is the measurement of optical rotation by a polarimeter. This method usually requires an enantiomerically pure isomer to obtain a standard specific rotation. However, the purity of the pure isomer is often difficult to prove. Therefore, the specific rotation for an isolated compound sometimes became greater than the standard value for the enantiomer that was expected to be enantiomerically pure. In recent years, however, this situation has completely changed. Figure 1(a) indicates how organic chemists estimated ee in the journal, *"Tetrahedron Asymmetry"*, during 1995-2001. The three major methods are NMR, gas chromatography (GC), and HPLC, while the polarimetric method has lost its position. Of these three, the NMR method was first introduced, but its role is gradually decreasing probably because of its lower accuracy. The GC using a chiral stationary phase (CSP) is particularly useful for volatile compounds of relatively low-molecular weight and has been mostly used only for analytical purposes. On the other hand, the HPLC method using a CSP is suitable for compounds with a relatively high molecular weight for both analytical and preparative purposes.

Figure 1(b) shows what kinds of CSPs has been used in HPLC for analyzing chiral compounds. There are two types of CSPs; one is prepared from small molecules with a chiral recognition ability and the other from optically active polymers. It is clear that the

301

F. Toda (ed.), Enantiomer Separation, 301-322.

most popular CSPs are the derivatives (carbamates and esters) of polysaccharides, cellulose and amylose. According to our data on the resolution of about 500 racemates using the two most useful derivatives,[3] 3,5-dimethylphenylcarbamates (commercial name: Chiralcel OD and Chiralpak AD) of cellulose and amylose, respectively, nearly 80% of the racemates have been resolved by at least one of the CSPs, and a recent report from the Daicel Co. indicates that nearly 90 % of the chiral compounds can be resolved by the polysaccharide-based CSPs. Besides analytical use, these phases can also be used for a large scale preparative separation. These are also useable for supercritical fluid chromatographic[4] and capillary electrochromatographic conditions.[5] In this chapter, we mainly explain the HPLC resolution of various types of compounds using these CSPs.[6]

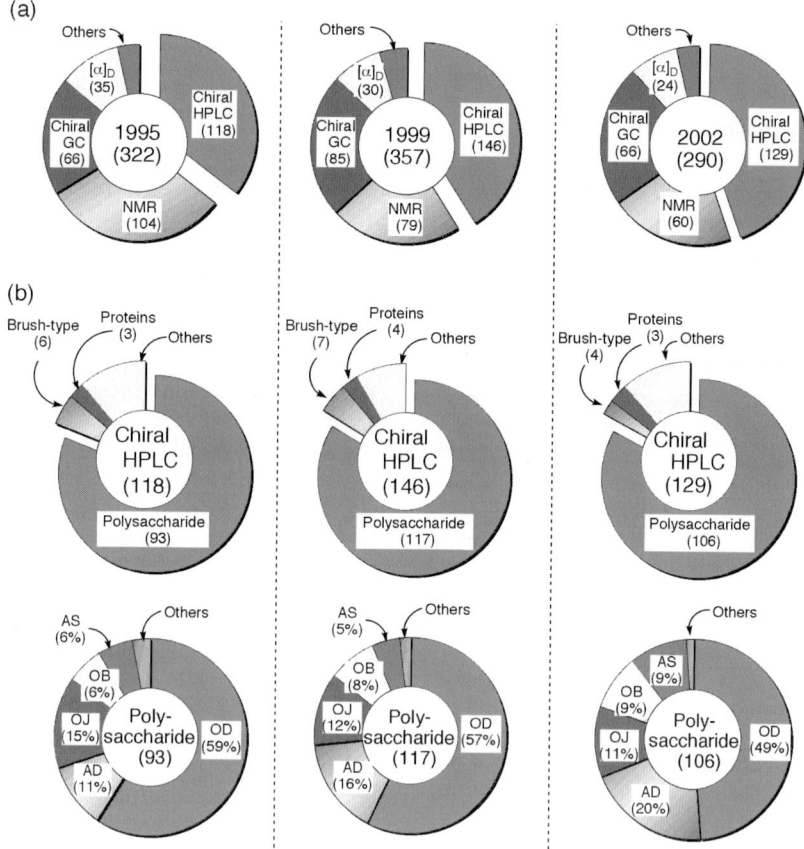

Figure 1. Distribution of the methods for the determination of enantiomer composition (a) and items of CSPs used in chiral HPLC (b) appeared in *Tetrahedron Asymmetry* in 1995-2002.

2. Chiral Stationary Phases (CSPs)

As already explained, CSPs have been prepared from both optically active small molecules and polymers. Some of the small molecule CSPs are shown in Figure 2. In 1970, Davankov reported the first baseline separation of enantiomers by CSP **1** consisting of an L-proline residue. This can separate various α-amino acids in the presence of copper ions.[7] Analogous phases are **2** and **3**. CSP **4** found by Pirkle contains polar amide groups that can interact through a hydrogen bond and a 3,5-dinitrophenyl group that can interact through a π•π interaction.[8] Many similar phases including **5-10** have been prepared.[9] CSP **11** separates acidic compounds through an ionic interaction.[10] The crown ether-containing CSP **12** exhibits a high chiral recognition to α-amino acids.[11] Cyclodextrin-based phase **13** resolves enantiomers through the formation of inclusion complexes.[12]

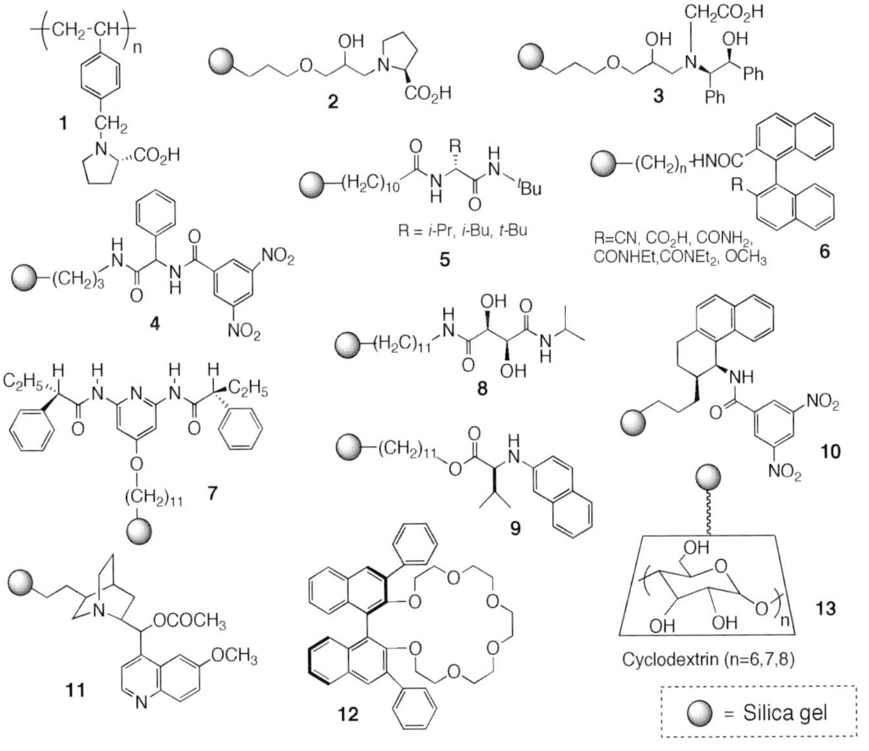

Figure 2. Small molecule chiral stationary phases for HPLC.

Various optically active polymers have also been used as CSPs (Figure 3). We can produce the one-handed helical polymer **14** by helix-sense-selective polymerization using a chiral anionic initiator.[13] The polymer is totally isotactic and optically active due to its stable helical structure, and shows a high chiral recognition when coated on a macroporus silica gel.[14]   Blaschke prepared various cross-linked gels consisting of poly(acrylamides) and poly(methacrylamides) with chiral side groups. Using these gels, various drug enantiomers were separated.[15] He pointed out that the teratogenic effect of thalidomide is due to one of the enantiomers.[16] Poly(acetylene) **20** exhibits a high chiral recognition only when the polymer has a stereoregular structure.[17,18]   Saigo prepared polyamides **25** from chiral cumarine derivatives and diamine (HNR-(CH₂)ₙ-RNH). The chiral recognition ability of **25** depends on the number of methylene groups and an odd-even effect has been observed.[19]

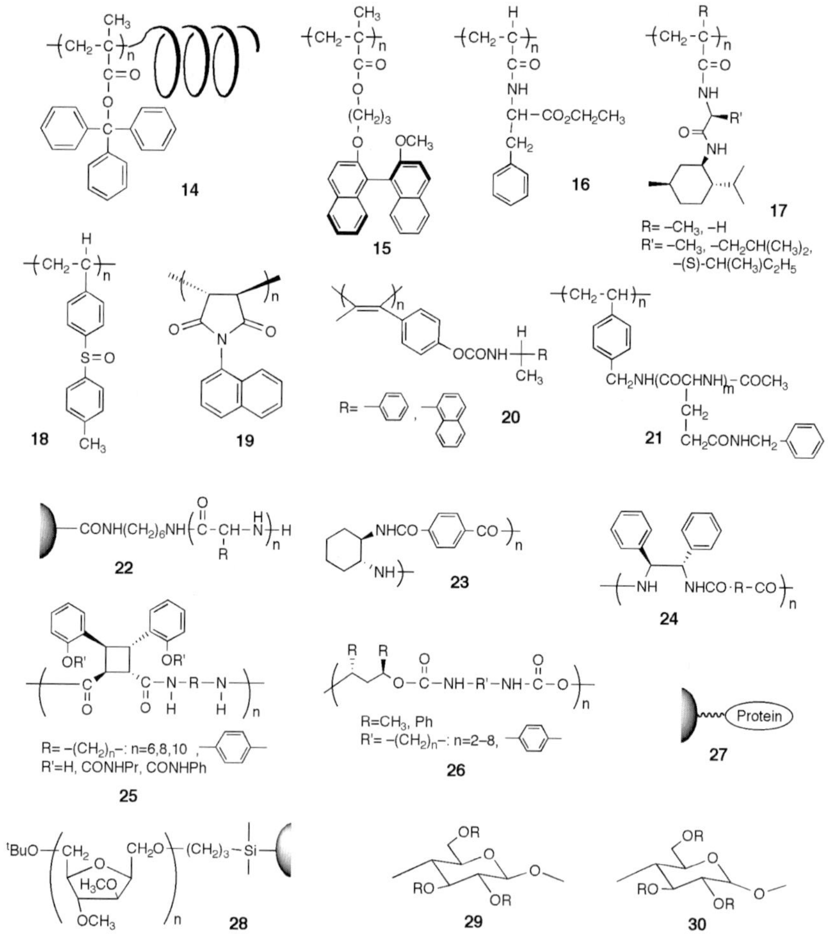

Figure 3. Chiral polymer stationary phases for HPLC.

Chiral recognition by a polymer-based CSP is often influenced by the higher-order structure of the polymer. Therefore, it is difficult to predict the chiral recognition ability of a CSP from the monomer structure. To achieve a high chiral recognition, it is usually preferable for polymers to have a regular conformation, particularly a chiral helical conformation arising from a stereoregular main chain.[20]

Besides the synthetic polymers, we can also use naturally occurring chiral polymers, proteins (27) and polysaccharides (29, 30). Fibrous proteins, such as wool and silk, do not show high chiral recognition probably due to the existence of too many different chiral sites along a polymer chain. On the other hand, some globular proteins, such as albumin and glycoproteins, show a high chiral recognition that can be used for chiral separation.[21] However, these proteins are usually conformationally unstable, and therefore, it is difficult to maintain the same chiral recognition ability. The protein-based CSPs are often not useful for preparative separation because of the limited number of chiral recognition sites. Polysaccharides, particularly cellulose and amylose, are the most abundant polymers on the earth. These polymers are completely stereoregular and easily modified to esters and carbamates, which exhibit a high chiral recognition as described above.[20,22-25]

3. Evaluation of Chiral Recognition Ability of a CSP

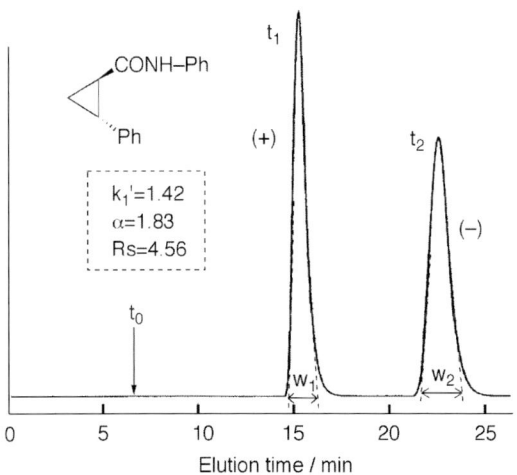

Figure 4. Optical resolution of *N*,2-diphenylcyclopropanecarboxamide on amylose. tris(3,5-dimethylphenylcarbamate) column: 25 x 0.46 (i.d.) cm; eluent: hexane/2-propanol (90/10); flow rate: 0.5 ml/min.

The chiral recognition ability of a CSP for a compound can be quantitatively evaluated from chromatographic resolution data (Figure 4).   For the HPLC resolution of the enantiomers of N,2-diphenylcyclopropanecarboxamide using Chiralpak AD as a chiral stationary phase, the enantiomers elutes at t_1 and t_2 and are completely resolved. T_0 is the elution time for a non-retained compound. The capacity factors (k_1 and k_2) for each enantiomer, the separation factor (α• that represents the chiral recognition ability of the CSP, and the resolution factor (Rs) that shows the degree of resolution of two peaks, are then defined as follows:

$$k_1 = (t_1 - t_0)/t_0, \ k_2 = (t_2 - t_0)/t_0 \tag{1}$$

$$\alpha = k_2/k_1 = (t_2 - t_0)/(t_1 - t_0) \tag{2}$$

$$Rs = 2(t_2 - t_1)/(w_1 + w_2) \tag{3}$$

From the • value, we can estimate the energy difference ($\Delta\Delta G$) between the interactions of a CSP with a pair of enantiomers using the equation, $\Delta\Delta G = -RT \ln \alpha$. We can usually observe the baseline separation of enantiomers for $\alpha = 1.2$, which corresponds to $\Delta\Delta G = -0.11$ kcal/mol. A very small energy difference in the interaction leads to the baseline separation.

4. Polysaccharide-based CSPs

4.1 CELLULOSE ESTERS

A large number of polysaccharide derivatives, such as esters and carbamates of cellulose and amylose, have been evaluated as CSPs, and at present, more than ten derivative-based CSPs are on commercially available. Their structures and suppliers are shown in Figure 5. Cellulose triacetate has been utilized in two different forms. One is prepared by the heterogeneous acetylation of native microcrystalline cellulose (Avicel) in benzene and is used as a CSP after being ground and sieved into small particles. It was the first practical CSP derived from polysaccharides, and has been called "microcrystalline cellulose triacetate" (**CTA-I**) as its structure is closely related to native cellulose (form I).[26,27] The other one (**OA**) is prepared by coating **CTA-I** dissolved in a solvent on silica gel and exhibits a completely different chiral recognition ability from that of **CTA-I**.[28-30] This may be because the higher order structure of **OA** is different from the original one (**CTA-I**). On these two CSPs, a reverse elution order of enantiomers has been observed for several racemates. The **CTA-I** is useful for the large-scale separation due to its high loading capacity.[31,32] Some stereochemically interesting racemates resolved on **CTA-I** are shown in Figure 6.[33-40] On the other hand, **OA** has advantages in column efficiency and durability. Other polysaccharide-based CSPs, except for **CTA-I**, have been prepared by coating them on silica gel.

Cellulose benzoate (**OB**),[41] 4-methybenzoate (**OJ**)[42] and cinnamate (**OK**)[29] are commercially available besides the acetates. These three show a characteristic chiral recognition. In particular, **OB** and **OJ** are very different in chiral recognition, although the latter **OJ** has only an additional 4-methyl group to **OB**. Empirically, **OJ** seems to be more suitable for the separation of large size compounds than **OB** as shown in Table 1. As the molecular size increases, separation factors on **OB** and **OJ** tend to decrease and increase, respectively. **OB** is useful for the enantioseparation of aliphatic carbonyl compounds, such as ketones, esters, imides, and sulfoxides. On the other hand, **OJ** exhibits a characteristic ability for the resolution of aryl propionic acids. **OK** performs a chiral recognition similar to **OJ**. The main chiral recognition sites of the cellulose benzoates are considered to be the polar carbonyl groups of esters, which can interact with analytes via hydrogen bonding and a dipole-dipole interaction. Amylose benzoates show much lower chiral recognition abilities.[42]

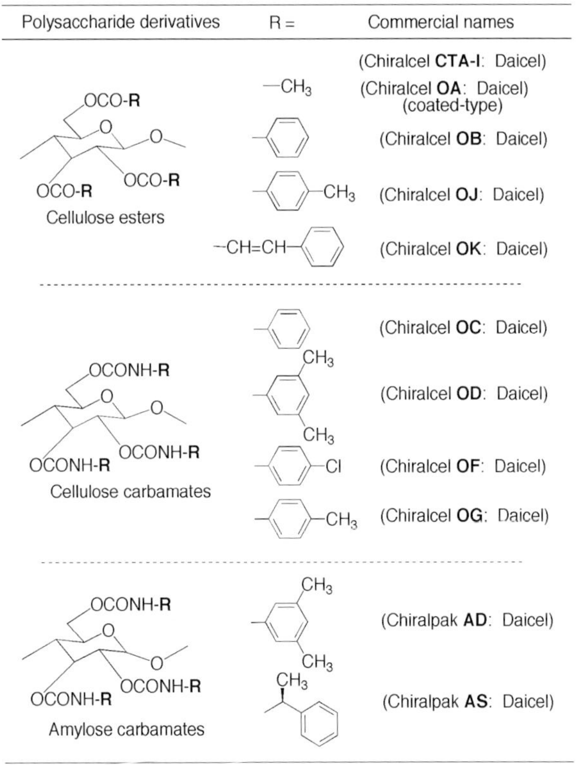

Figure 5.   Structures and suppliers of commercially available polysaccharide-based CSPs.   **CTA-I** is
also available from Merck and Macherey-Nagel.

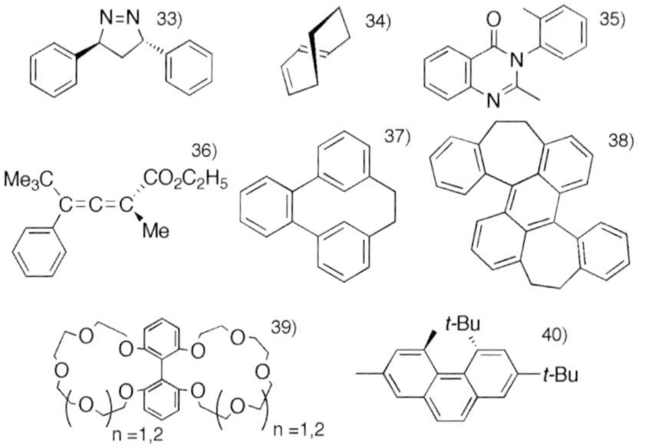

Figure 6. Stereochemically interesting compounds resolved on **CTA-I**.

TABLE 1. Difference of the chiral recognition ability (α) between **OB** and **OJ**

	OB	OJ		OB	OJ		OB	OJ
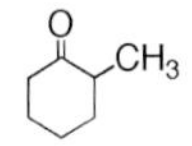 cyclohexyl methyl ketone	1.15	1.0	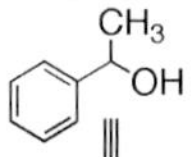 1-phenylethanol (CH₃, OH)	1.57	1.17	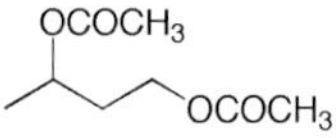 diacetate (OCOCH₃, OCOCH₃)	1.73	1.0
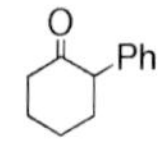 cyclohexyl phenyl ketone	1.47	1.18	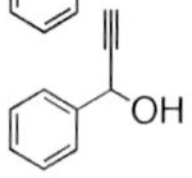 1-phenyl-2-propyn-1-ol (OH)	1.20	1.58	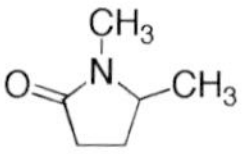 N,N-dimethyl lactam (CH₃, CH₃)	1.21	1.0
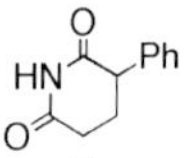 N-phenyl lactam (HN, Ph)	1.19	2.15	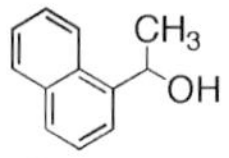 1-(1-naphthyl)ethanol (CH₃, OH)	1.37	1.54	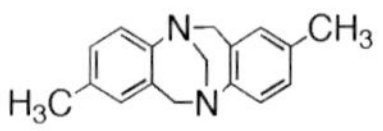 dimethyl diamine (CH₃)	1.0	6.05
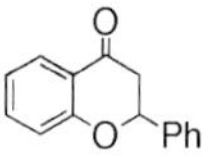 chroman (O, Ph)	1.0	1.17	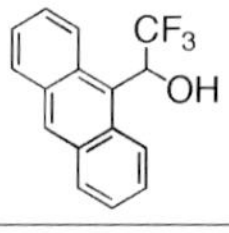 anthryl-CF₃-carbinol (CF₃, OH)	1.0	1.22			

4.2 PHENYLCARBAMATES OF CELLULOSE AND AMYLOSE

A variety of phenylcarbamates of cellulose and amylose have been prepared and evaluated as the CSPs for HPLC. Phenylcarbamates have the amide groups CONH capable of hydrogen bonding as a hydrogen accepter and donor with polar racemates under normal phase conditions. The substituents on the phenyl groups influence the polarity of the carbamate residues; that is, the electron density of the C=O oxygen is increased by an electron-donating group like a methyl and the acidity of the NH proton is increased by an electron-withdrawing group like chlorine. Generally, the introduction of an electron-donating alkyl group or an electron-withdrawing halogen at the *m*- and/or *p*-position on the phenyl ring improves the chiral recognition ability for many racemates.[43] In particular, the tris(3,5-dimethylphenylcarbamate) derivatives of both cellulose and amylose (**OD**, **AD**) exhibit interesting and excellent resolving abilities for a variety of racemates.[43-45] For instance, **OD** resolves aromatic hydrocarbons, amines, carboxylic acids,[46] alcohols, amino acid derivatives,[47] and many drugs,[48] including β-adrenergic blocking agents (β-blockers)[49] with high α values (Figure 7). These two CSPs often show a complementary recognition, and many enantiomers unresolved on **OD** can be resolved on **AD**, and *vice versa*. Enantiomers often elute in the reverse order on **OD** and **AD**. Figure 8 shows the probability of the resolution of about 500 racemates on **OD** and **AD**. **OD** resolved about 60% of the compounds and **AD** about 55%. A total of about 80% can be resolved at least on one of the two.[3] The resolution of a compound can be attained with a high possibility.

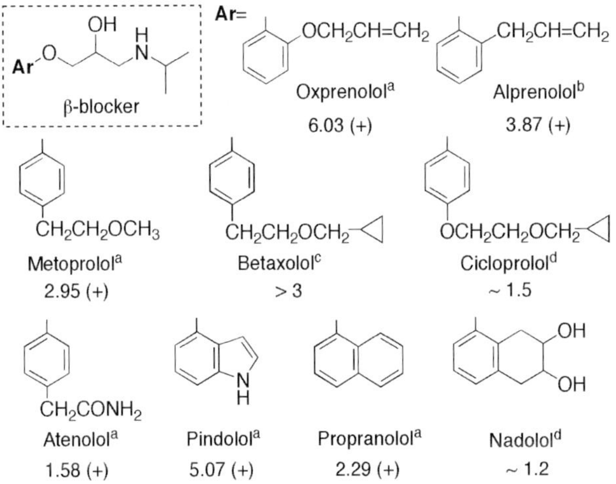

Figure 7. Resolution of β-blockers on **OD**.

Resolution of racemates on OD and AD		
	OD	**AD**
Numbers of racemates	505	384
Resolved	314	211
-completely	228	109
-partially	86	102

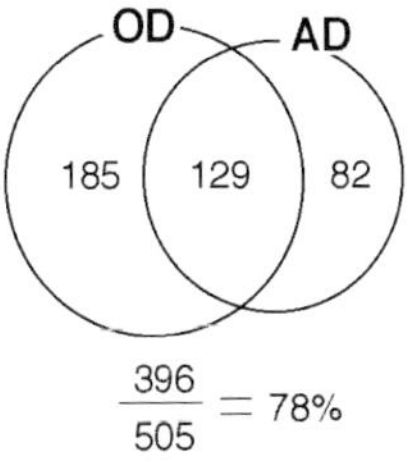

$$\frac{396}{505} = 78\%$$

Figure 8. The probability of the resolution of racemates on **OD** and **AD**.

4.3 BENZYLCARBAMATES OF CELLULOSE AND AMYLOSE

Amylose benzylcarbamates show an interesting chiral recognition different from those of the phenylcarbamates.[50,51] Especially, (S)-1-phenylethylcarbamate (**AS**) has an high resolving ability and the CSP is commercially available. Other less or more bulky benzylcarbamates of amylose and of cellulose exhibit very low recognition abilities, and the (R)-1-phenylethylcarbamate shows lower different recognitions. **AS** is useful for the resolution of β-lactams[52] and 3-hydroxy-2-cyclopentanone derivatives[53] (Figure 9).

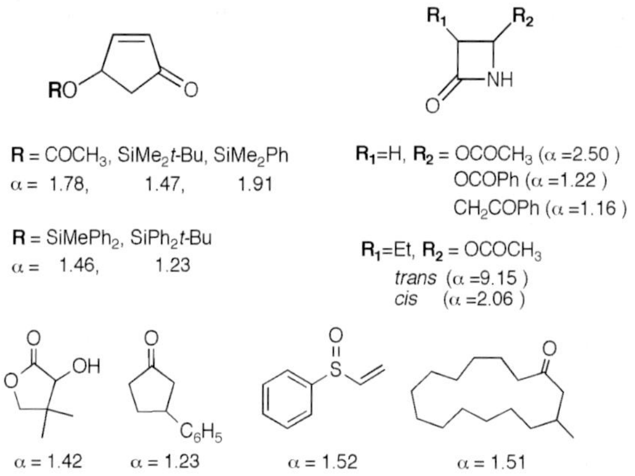

R = COCH₃, SiMe₂t-Bu, SiMe₂Ph
α = 1.78,   1.47,   1.91

R = SiMePh₂, SiPh₂t-Bu
α =   1.46,   1.23

R₁=H, R₂ = OCOCH₃ (α =2.50)
              OCOPh (α =1.22)
              CH₂COPh (α =1.16)

R₁=Et, R₂ = OCOCH₃
  trans (α =9.15)
  cis   (α =2.06)

α = 1.42   α = 1.23   α = 1.52   α = 1.51

Figure 9. Compound efficiently resolved on **AS**.

4.4 CYCLOALKYLCARBAMATES OF CELLULOSE AND AMYLOSE

The most simple alkylcarbamates such as methyl and *tert*-butyl show a low chiral recognition.[43] However, cellulose and amylose cycloalkylcarbamates such as cyclohexyl and norbornylcarbamates (Figure 10) have a wide applicability for the enantioseparation of many racemates, although these CSPs are not commercially available at this moment.[54,55] Cycloalkylcarbamates can be used as the CSP not only for HPLC but also for thin-layer chromatography (TLC) because of the absence of any UV absorption above 220 nm. The TLC results can be compared with those obtained in HPLC with the same CSP, showing a good correlation between the α values.[54] Therefore, the TLC enables the rapid set up of the conditions for the HPLC resolution.

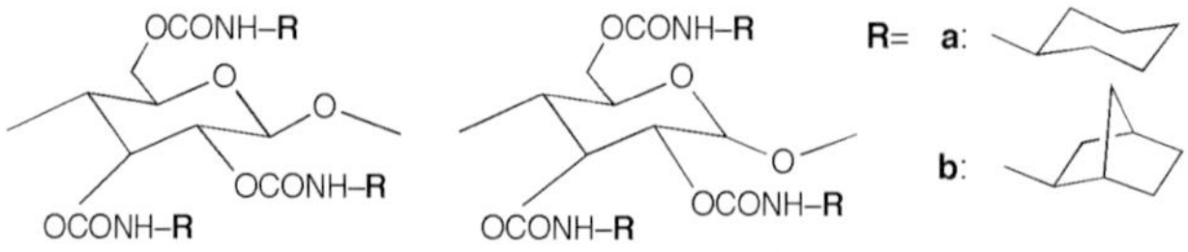

OCONH–R   OCONH–R   R= a:

              OCONH–R   b:
OCONH–R   OCONH–R   OCONH–R

Figure 10. Cyclohexyl (a) and norbornylcarbamates (b) of cellulose and amylose.

4.5 OTHER POLYSACCHARIDE PHENYLCARBAMATES

3,5-Dimethyl- and 3,5-dichlorophenylcarbamates of other polysaccharides such as chitin, chitosan, xylan, curdlan, dextran, galactosamine, and inulin (Figure 11) have been prepared and their chiral recognition abilities as CSPs for HPLC have been evaluated.[56-59] These polysaccharides consist of different sugar units or linkage positions, and therefore, the enantioselectivity and the elution order of enantiomers on these polysaccharides are significantly different. The 3,5-dimethylphenylcarbamates of xylan, chitosan, and chitin and the 3,5-dichlorophenylcarbamates of galactosamine and chitin showed relatively high chiral recognition abilities, and better resolve some compounds than **OD** and **AD**. Especially, chitin phenylcarbamates are effective for the resolution of several acidic drugs such as 2-arylpropionic acids including ketoprofen and ibuprofen (Table 2).[58]

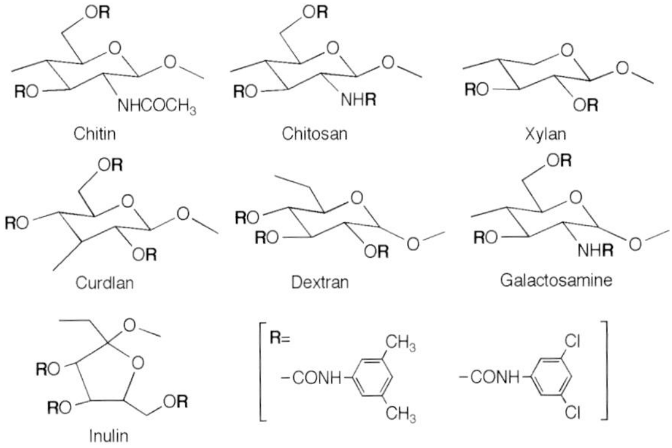

Figure 11. Structures of phenylcarbamates of various polysaccharides.

TABLE 2. Resolution of 2-arylpropionic acids on chitin carbamates[a]

R—CHCOOH CH₃	3,5-(CH₃)₂		3,5-(Cl)₂	
R=	k_1'	α	k_1'	α
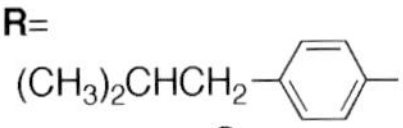 (CH₃)₂CHCH₂— (ibuprofen)	0.67 (+)	~ 1	0.54 (+)	1.11
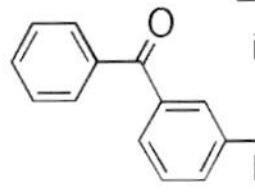 ketoprofen	3.81 (+)	1.21	8.29 (+)[b]	1.72
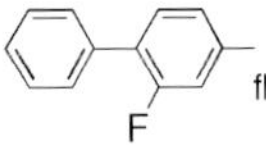 flurbiprofen	1.33 (−)	1.08	0.29 (+)[c]	1.10
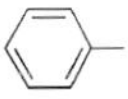	1.25 (+)	1.41	0.87 (+)	1.39

[a]The signs in parentheses represent the optical rotation of the first-eluted enantiomer. Flow rate: 0.5 ml min⁻¹. Eluent: hexane–2-propanol–CF₃COOH (95/5/1, v/v/v). [b]Flow rate: 1.0 ml min⁻¹. [c]Eluent: hexane–2-propanol–CF₃COOH (90/10/1, v/v/v).

5. Selection of a suitable column and an eluent for enantioseparation

Every polysaccharide-based CSP has a characteristic chiral resolution ability, and eluents significantly influence the selectivity of the CSPs. In order to achieve the efficient resolution of chiral compounds, it is very important to select the most suitable CSP and eluent. The polysaccharide-based CSPs can resolve a variety of chiral compounds including aromatic hydrocarbons, axially and planar dissymmetric compounds, metal-containing compounds, chiral sulfur or phosphorus compounds, cyano compounds, carbonyl compounds, amines, carboxylic acids, alcohols, amino acid derivatives, and ethers.[3,6] As examples, the chromatograms for the resolution of several racemates bearing different functional groups are shown in Figure 12.[43,44,60-62] Among the CSPs, **OD**, **AD**, **OJ**, and **AS** are particularly useful CSPs with high enantioselectivities. These different types of CSPs are somehow complementary to each other. Therefore, in order to efficiently select a suitable CSP, the following order of CSP selection may be recommended: Chiralcel OD, Chiralpak AD, Chiralcel OJ, and Chiralpak AS. By using these CSPs, 80-90% of compounds may be resolved.

As an eluent, both normal-phase and reversed-phase[63] eluents can be used for the polysaccharide-based CSPs; the former consists of nonpolar solvents like hexane containing an alcohol and the latter polar solvents like an alcohol or acetonitrile containing water. When an analyte is a polar compound, a mixture of hexane and 2-propanol or ethanol is often a suitable eluent. The structures of alcohols as an additive influence the enantioselectivity, and the change of the alcohol from 2-propanol to ethanol brings about a shorter elution time and a better separation in some cases.[64] When the retention times are rather short, methyl *tert*-butyl ether is effective as an additive instead of alcohols. On the other hand, for the nonpolar analytes, a reversed phase condition with a polar eluent is effective because the hydrophobic interaction between the hydrophobic outside of the polysaccharide derivatives and analytes is available. Aqueous eluents are also valuable for investigating pharmacokinetics, physiological, toxicological, and metabolic activities of drug enantiomers in living systems.[65]

The polysaccharide-based CSPs have been prepared by coating the polysaccharide derivatives on macroporous silica gel, and therefore, the solvents such as tetrahydrofuran, chloroform, and acetone, which dissolve or swell the polysaccharides, cannot be used as the main mobile phase. However, it is sometimes highly desirable to use these solvents as the eluent, because their addition in an eluent may improve the separation and also solve the problem of the low solubility of a sample in a preparative separation. So far, several immobilization methods of the polysaccharide derivatives onto silica gel have been reported.[66-73] Because the chiral recognition of the polysaccharide derivatives is significantly influenced by its high-order structure, the immobilization at the chain end is ideal. This has been realized for **AD**.[74] This CSP successfully resolved topologically interesting catenanes and molecular knots using a hexane/chloroform/2-propanol mixture (Figure 13).[75-77] In addition, the first direct HPLC resolution of the smallest chiral fullerene C_{76} was also achieved on this CSP using a hexane/chloroform mixture (80:20) as the eluent.[78]

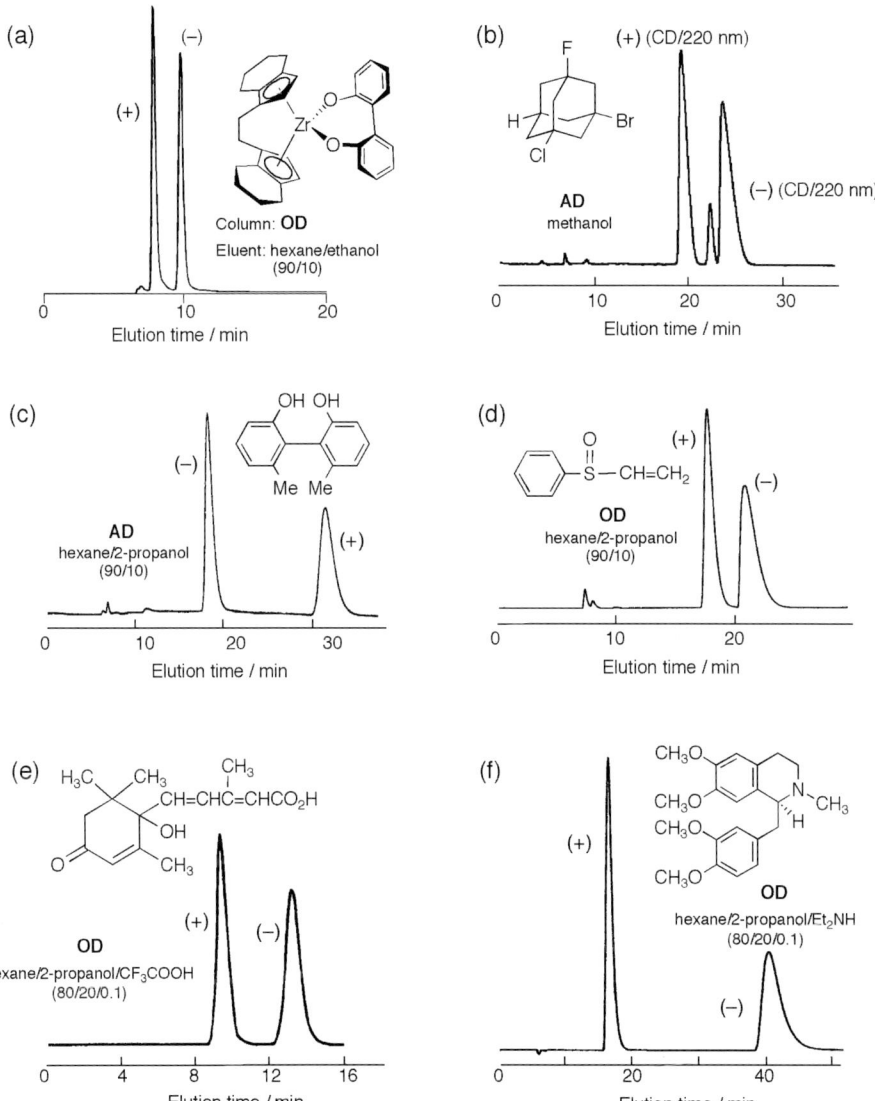

Figure 12. Chromatograms of the resolution of (a) the ansa-zirconocene derivative, (b) the adamantan derivative, (c) 2,2'-dihydroxy-6,6'-dimethylbiphenyl, (d) phenyl vinyl sulfoxide, (e) abscisic acid, and (f) laudanosine. Column, 25 x 0.46 cm (i.d.); flow rate, 0.5 ml/min. Chromatograms (a) and (e) are reproduced, with permission, from Ref. 61 (Copyright 1996, Chemical Society of Japan) and Ref. 62 (Copyright 1988, Elsevier Science B.V.), respectively.

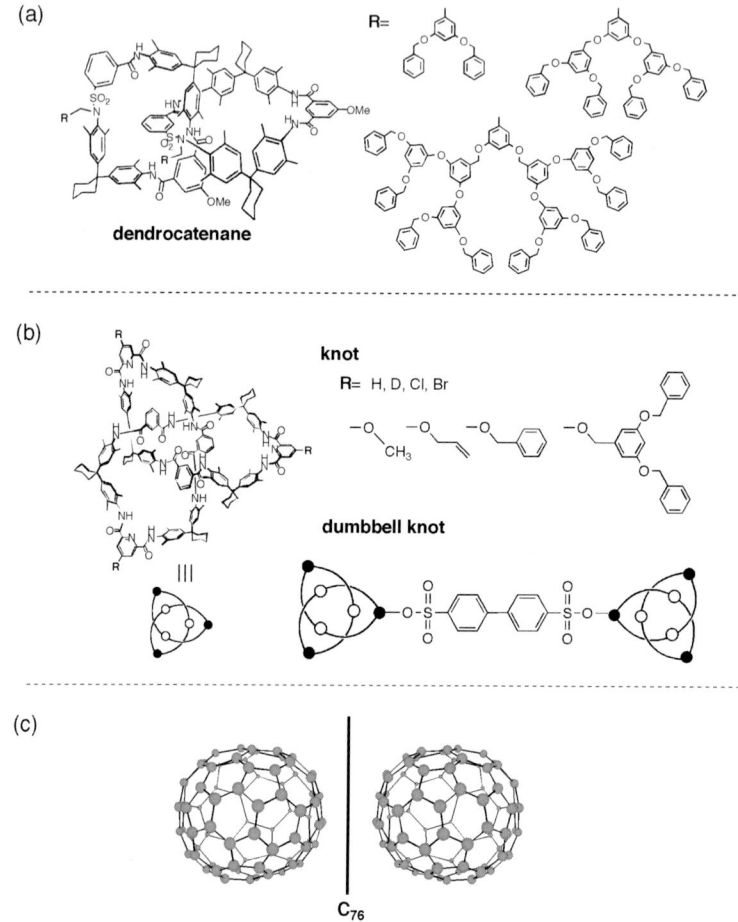

Figure 13. Structures of dendrocatenanes (a), molecular knot, dumbbell knot (b), and C$_{76}$ (c) resolved on chemically bonded-type CSP of **AD**.

6. Examples of the resolution of several typical racemates on the polysaccharide-based CSPs

6.1 CHIRAL AROMATIC HYDROCARBONS

Polysaccharide-based CSPs are effective for chiral aromatic hydrocarbons without polar substituents (Figure 14). For example, racemates **31** were resolved on **OD** using hexane-2-propanol as the eluent.[79] Compound **32** was not resolved with the mixture of hexane-alcohol and pure hexane, but it was completely separated on **OD** using isooctane.[80,81]   On the other hand, **33** was resolved on **AD** under the reversed phase condition, ethanol-water (9/1).[82]   As already described, hydrophobic interactions like a π-π interaction between the polysaccharide derivatives and analytes in polar solvents play a role in the recognition of chirality.   In this case, the addition of water into alcohol delays the elution times of the enantiomers.   Besides the polysaccharide-based CSPs, a one-handed helical poly(triphenylmethyl methacrylate) (CSP **14**) exhibits a high chiral recognition ability for axially or planar chiral compounds under reversed-phase conditions.[14,20,25]   However, this separation must be performed at a low temperature due to the instability of **14** against the hydrolysis of ester bonds.

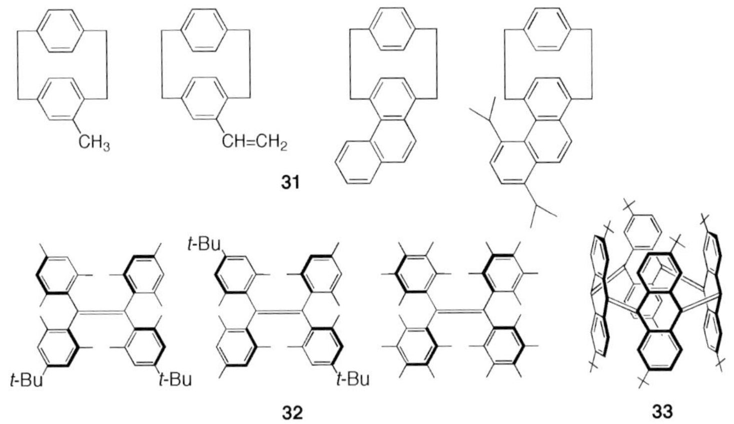

Figure 14. Chiral aromatic hydrocarbons without polar substituents resolved on polysaccharide-based CSPs.

6.2 NON-AROMATIC CHIRAL COMPOUNDS

Non-aromatic chiral compounds are also resolved on the polysaccharide-based CSPs. Several non-aromatic compounds resolved on cellulose tribenzoate (**OB**) are shown in Figure 15.[83]   The detection by a UV detector is often difficult and an RI detector is more useful to detect the enantiomer peaks.

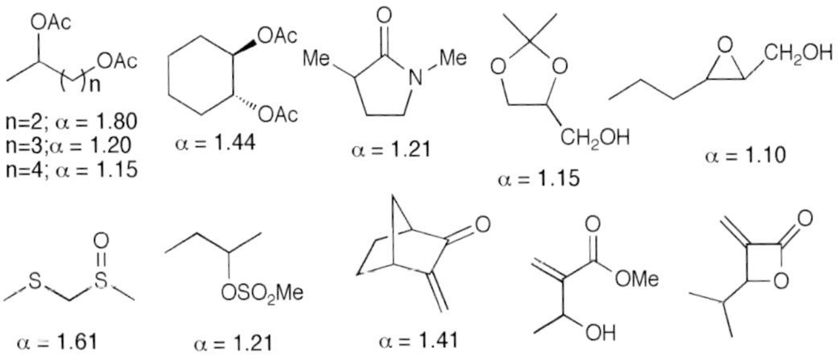

Figure 15. Non-aromatic compounds resolved on **OB**.

6.3 CHIRAL AMINES

For the resolution of basic compounds with a primary, secondary, or tertiary amino group including many basic drugs under normal phase conditions, the addition of a small amount of an amine, such as diethylamine or isopropylamine (ca. 0.1%), may bring about a better resolution without the tailing of peaks.[47,49] The separation of β-adrenergic blocking agents (β-blockers)[49] on **OD** has been carried out using hexane-2-propanol-diethylamine as the eluent. Under reversed-phase conditions, a basic mobile phase is not recommended for the column due to the low stability of the silica supports above pH 7. Therefore, it is important to use a suitable buffer with a proper pH. Under neutral and acidic mobile phase conditions, basic analytes are positively charged and cannot efficiently interact with the CSP. Therefore, it is effective to add a considerable amount of anions, for instance, PF_6^-, BF_4^-, and ClO_4^-, in the mobile phase to form an ion pair with the positively charged analytes. First, the use of 0.5 M $NaClO_4$ aq. /CH_3CN (60/40) is recommended.[63]

6.4 CHIRAL ACIDS

To resolve acidic compounds under normal phase conditions, the addition of a small amount of a strong acid such as trifluoroacetic acid or formic acid (ca. 0.5%) to an eluent is recommended (Figure 16).[46,62,84] Under reversed-phase conditions, an acidic mobile phase is useful in order to suppress the dissociation of an analyte. An aqueous solution or buffer of pH 2 containing an organic modifier (alcohol or acetonitrile), such as $HClO_4$ aq. (pH 2)/CH_3CN (60/40) and 0.5 M $NaClO_4$-$HClO_4$ aq. (pH 2)/CH_3CN (60/40), often results in good resolution.[63]

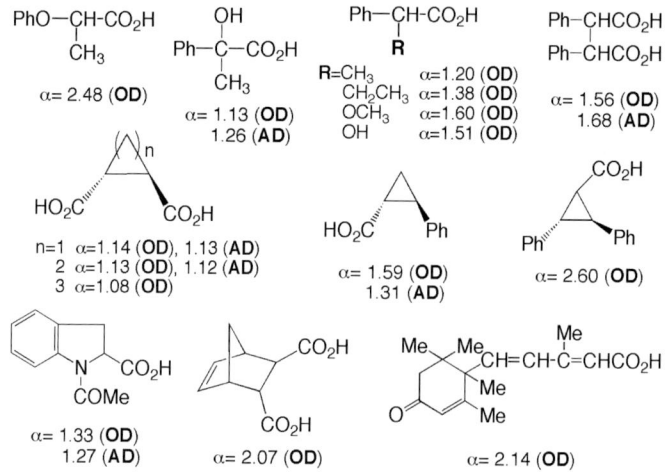

Figure 16. Carboxylic acids resolved on **OD** or **AD**.

6.5 CHIRAL ALCOHOLS

Polysaccharide-based CSPs also exhibit a chiral recognition for alcohols and a large number of resolutions have been reported. Chiral alcohols can usually be directly resolved with hexane containing a small amount of an alcohol as the eluent. For aliphatic alcohols, which cannot be directly resolved, their resolution is often efficiently attained as phenylcarbamate or benzoate derivatives on **OD** (Figure 17).[85] For example, 2-butanol and 2-pentanol are completely resolved with a very high selectivity on **OD** as their phenylcarbamates. The derivatization of alcohols to phenylcarbamates and benzoates can be easily achieved by the reaction with phenyl isocyanates and benzoyl chlorides, respectively. In most cases, the phenylcarbamates are better resolved than the benzoates. For chiral compounds bearing phenolic hydroxy groups, the addition of a small amount of an acid to an eluent is recommended to depress its dissociation.

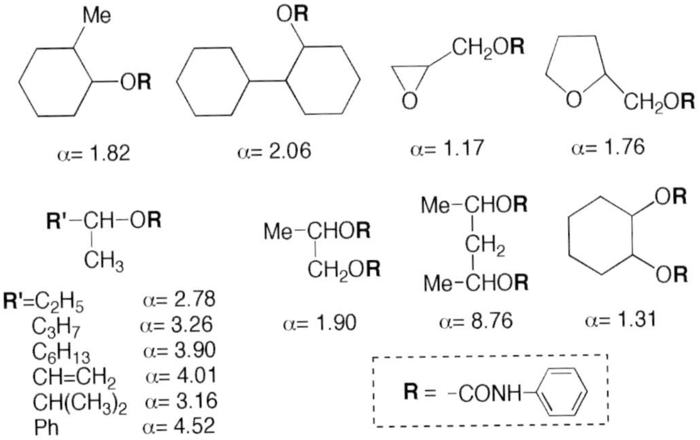

Figure 17. Resolution of alcohols as phenylcarbamates.

7. Conclusions

Chiral separation by HPLC is a practically useful method not only for determining optical purity but also for obtaining optical isomers, and numerous CSPs are presently on the market. In order to achieve the efficient resolution of chiral compounds, we have to choose a suitable chiral column and eluent. The polysaccharide-based CSPs have a high chiral recognition ability and offer a high possibility for the successful resolution of racemates including aliphatic and aromatic compounds with or without functional groups under normal and reversed-phase conditions.

The polysaccharide-based CSPs have already been utilized for the preparative separation of pharmaceutical compounds by simulated moving bed (SMB) chromatography, which is a powerful method for the separation of a two-component mixture.[86] On the other hand, miniaturized separation techniques such as capillary liquid chromatography (CLC), capillary electrophoresis (CE), and capillary electrochromatography (CEC) using polysaccharide-based CSPs have also been investigated.[87] A capillary column needs much lower amounts of stationary and mobile phases than the normal-size HPLC column, and therefore, is an environmentally friendly system. Recently, monolithic silica material was used as an alternative to a macroporous silica support for a CSP. This may allow a short time analysis and more efficient preparative separation mainly due to the reduction in the pressure drop and peak broadening.[88]

The chiral recognition mechanism of polysaccharide derivatives at a molecular level has been solved to some extent by chromatography, NMR, and computational methods.[89-93] This will be helpful in selecting suitable resolution conditions and in developing more effective CSPs.

8. References

1   Rogozhin, S.V. and Davankov, V.A. (1971) Ligand chromatography on asymmetric complex-forming sorbents as a new method for resolution of racemates, *J. Chem. Soc., Chem. Commun.* **192**, 490.

2   Ahuja, S. (1991) *Chiral Separation by Liquid Chromatography (ACS symposium Series, No. 471)*, American Chemical Society, Washington, D.C.

3   Kaida, Y. and Okamoto, Y. (1993) Optical resolution by high-performance liquid chromatography on benzylcarbamates of cellulose and amylose, *J. Chromatogr.* **641**, 267-278.

4   Kaida, Y. and Okamoto, Y. (1992) Optical resolution by supercritical fluid chromatography using polysaccharide derivatives as chiral stationary phases, *Bull. Chem. Soc. Jpn.* **65**, 2286-2288.

5   Girod, M., Chankvetadze, B., and Blaschke, G. (2000) Enantioseparations in non-aqueous capillary electrochromatography using polysaccharide type chiral stationary phases, *J. Chromatogr. A* **887**, 439-455.

6   Yashima, E., Yamamoto, C., and Okamoto, Y. (1998) Polysaccharide-based chiral LC columns, *Synlett*, 344-360.

7   Davankov, V.A. (1980) Resolution of racemates by ligand exchanges chromatography, *Adv. Chromatogr.* **18**, 139-195.

8   Pirkle, W.H., Finn, J.M., Schreiner, J.L., and Hamper, B.C. (1981) A widely useful chiral stationary phase for the high-performance liquid chromatography separation of enantiomers, *J. Am. Chem. Soc.* **103**, 3964-3966.

9   Pirkle, W.H. and Pochapsky, T.C. (1989) Considerations of chiral recognition relevant to the liquid chromatographic separation of enantiomers, *Chem. Rev.* **89**, 347-362.

10 Rosini, C., Altemura, P., Pini, D., Bertucci, C., Zullino, G., and Salvadori, P. (1985) Cinchona alkaloids for preparing new, easily accessible chiral stationary phases, *J. Chromatogr.* **348**, 79-87.

11 Shinbo, T., Yamaguchi, T., Nishimura, K., and Sugiura, M. (1987) Chromatographic separation of racemic amino acids by use of chiral crown ether-coated reversed-phase packings, *J. Chromatogr.* **405**, 145-153.

12 Chang, S.C., Reid III, G.L., Chen, S., Chang, C.D., and Armstrong, D.W. (1993) Evaluation of a new polar-organic high-performance liquid chromatographic mobile phase for cyclodextrin-bonded chiral stationary phases, *Trends in Anal. Chem.* **12**, 144-153.

13 Okamoto, Y., Suzuki, K., Ohta, K., Hatada, K., and Yuki, H. (1979) Optically active poly(triphenylmethyl methacrylate) with one-handed helical conformation, *J. Am. Chem. Soc.* **101**, 4763-4765.

14 Okamoto, Y. and Hatada, K. (1986), Resolution of enantiomers by HPLC optically active poly(triphenylmethyl methacrylate), *J. Liq. Chromatogr.* **9**, 369-384.

15 Blaschke, G. (1980), Chromatographic resolution on racemates, *Angew. Chem. Int. Ed. Engl.* **19**, 13-24.

16 Blaschke, G., Kraft, H.P., Fickentscher, K., and Kohler, F. (1979) Chromatographic separation of racemic thalidomide and teratogenic activity of its enantiomers, *Arzneim. Forsch.* **29**, 1640-1642.

17 Yashima, E., Huang, S., and Okamoto, Y. (1994) An optically active stereoregular polyphenylacetylene derivative as a novel chiral stationary phase for HPLC, *J. Chem. Soc., Chem. Commun.* 1811-1812.

18 Yashima, E., Matsushima, T., Nimura, T., and Okamoto, Y. (1996) Enantioseparation on optically active stereoregular polyphenylacetylene derivatives as chiral stationary phase for HPLC, *Korea Polym. J.* **4**, 139-146.

19. Saigo, K. (1992) Synthesis and properties of polyamides having a cyclobutanedicarboxylic acid derivative as a component, *Prog. Polym. Sci.* **17**, 35-86.

20 Okamoto, Y. Yashima, E. and Yamamoto, C. (2003) Optically active polymers with chiral recognition ability, in Green, M.M., Nolte, R.J.M., and Meijer, E.W. (eds.), *Materials Chirality: Volume 24 of Topics in Stereochemistry*, John Wiley & Sons Inc., New Jersey, pp. 157-208.

21 Haginaka, J. (2001) Protein-based chiral stationary phases for high-performance liquid chromatography enantioseparations, *J. Chromatogr. A* **906**, 253-273.

22 Yashima, E. and Okamoto, Y. (1995) Chiral discrimination on polysaccharides derivatives, *Bull. Chem. Soc. Jpn.* **68**, 3289-3307.

23 Okamoto, Y. and Yashima, E. (1998) Polysaccharide derivatives for chromatographic separation of enantiomers, *Angew. Chem. Int. Ed.* **37**, 1020-1043.

24 Yashima, E. (2001) Polysaccharide-based chiral stationary phases for high-performance liquid chromatographic enantioseparation, *J. Chromatogr. A* **906**, 105-125.

25 Yamamoto, C. and Okamoto, Y. (2004) Optically active polymers for chiral separation, *Bull. Chem. Soc. Jpn.* **77**, 227-257.

26 Hesse, G. and Hagel, R. (1973) Complete separation of a racemic mixture by elution chromatography on cellulose triacetate, *Chromatographia* **6**, 277-280.

27 Hesse, G. and Hagel, R. (1976) Chromatographic resolution or racemates, *Liebigs Ann. Chem.*, 966-1008.

28 Okamoto, Y., Kawashima, M., Yamamoto, K., and Hatada, K. (1984) Useful chiral packing materials for high-performance liquid chromatographic resolution. Cellulose triacetate and tribenzoate coated on macroporous silica gel, *Chem. Lett.*, 739-742.

29 Ichida, A., Shibata, T., Okamoto, I., Yuki, Y., Namikoshi, H., and Toga, Y. (1984) Resolution of enantiomers by HPLC on cellulose derivatives, *Chromatographia* **19**, 280-284.

30 Oguni, K., Oda, H., and Ichida, A. (1995) Development of chiral stationary phases consisting of polysaccharide derivatives, *J. Chromatogr. A* **694**, 91-100.

31 Francotte, E.J. (1992) Preparative chromatographic separation of enantiomers, *J. Chromatogr. A* **576**, 1-45.

32 Francotte, E.J. (1994) Contribution of preparative chromatographic resolution to the investigation of chiral phenomena, *J. Chromatogr. A* **666**, 565-601.

33 Schneider, M.P. and Bippi, H. (1980) Transfer of optical activity in the decomposition of (+)- and (–)-trans-3,5-diphenyl-1-pyraxoline: competing "biradical" and "cycloreversion" pathways, *J. Am. Chem. Soc.* **102**, 7363-7365.

34 Isaksson, R., Rochester, J., Sandström, J., and Wistrand, L.-G. (1985) Resolution, circular dichroism spectrum, molecular structure, and absolute configuration of cis,trans-1,3-cyclooctadiene, *J. Am. Chem. Soc.* **107**, 4074-4075.

35 Mannschreck, A., Koller, H., Stühler, B., Davies, M.A., and Traber, J. (1984) The enantiomers of methaqualone and their unequal anticonvulsive activity, *Eur. J. Med. Chem.-Chim. Ther.* **19**, 381-383.

36 Krause, N. and Hnadke, G. (1991) Enantioseparation of allenes by liquid chromatography, *Tetrahedron Lett.* **32**, 7225-7228.

37 Wittek, M., Vögtle, F., Stühler, G., Mannschreck, A., Lang, B.M., and Irngartinger, H. (1983) New helical hydrocarbons. VIII. Enantiomer separation, circular dichroism, racemization, and x-ray analysis of benzo[2.2]metacyclophane, *Chem. Ber.* **116**, 207-214.

38 Agranat, I., Suissa, M. R., Cohen, S., Isaksson, R., Sandström, J., Dale, J., and Grace, D. (1987) A novel titanium-induced aromatic dicarbonyl coupling. Synthesis of a chiral strained polynuclear aromatic hydrocarbon, *J. Chem. Soc., Chem. Commun.*, 381-383.

39 Lindsten, G., Wennerström, O., and Isaksson, R. (1987) Chiral biphenyl bis(crown ethers): synthesis and resolution, *J. Org. Chem.* **52**, 547-554.

40 Koller, H., Rimböck, K.-H., and Mannschreck, A. (1983) High-pressure liquid chromatography on triacetylcellulose, *J. Chromatogr. A* **282**, 89-94.

41 Okamoto, Y., Kawashima, M., Yamamoto, K., and Hatada, K. (1984) Useful chiral packing materials for high-performance liquid chromatographic resolution. Cellulose triacetate and tribenzoate coated on macroporous silica gel. *Chem. Lett.*, 739-742.

42 Okamoto, Y., Aburatani, R., and Hatada, K. (1987) Cellulose tribenzoate derivatives as chiral stationary phases for high-performance liquid chromatography, *J. Chromatogr.* **389**, 95-102.

43 Okamoto, Y., Kawashima, M., and Hatada, K. (1986) Controlled chiral recognition of cellulose trisphenylcarbamate derivatives supported on silica gel, *J. Chromatogr.* **363**, 173-186.

44 Okamoto, Y., Aburatani, R., Fukumoto, T., and Hatada, K. (1987) Useful chiral stationary phases for HPLC. amylose tris(3,5-dimethylphenylcarbamate) and tris(3,5-dichlorophenylcarbamate) supported on silica gel, *Chem. Lett.*, 1857-1860.

45 Okamoto, Y., Aburatani, R., and Hatada, K. (1990) Chromatographic optical resolution on 3,5-disubstituted phenylcarbamates of cellulose and amylose, *Bull. Chem. Soc. Jpn.* **63**, 955-957.

46 Okamoto, Y., Aburatani, R., Kaida, Y., and Hatada, K. (1988) Direct optical resolution of carboxylic acids by chiral HPLC on tris(3,5-dimethylphenylcarbamate)s on cellulose and amylose, *Chem. Lett.*, 1125-1128.

47 Okamoto, Y., Kaida, Y., Aburatani, R., and Hatada, K. (1989) Optical resolution of amino acid derivatives by high-performance liquid chromatography on tris(phenylcarbamate)s of cellulose and amylose, *J. Chromatogr.* **477**, 367-376.

48 Okamoto, Y., Aburatani, R., Hatano, K., and Hatada, K. (1988) Optical resolution of racemic drugs by chiral HPLC on cellulose and amylose tris(phenylcarbamate) derivatives, *J. Liq. Chromatogr.* **11**, 2147-2163.

49 Okamoto, Y., Kawashima, M., Aburatani, R., Hatada, K., Nishiyama, T., and Masuda, M. (1986) Optical resolution of β-blockers by HPLC on cellulose triphenylcarbamate derivatives, *Chem. Lett.*, 1237-2140.

50 Okamoto, Y., Kaida, Y., Hayashida, H., and Hatada, K. (1990) Tris(1-phenylethylcarbamate)s of cellulose and amylose as useful chiral stationary phases for chromatographic optical resolution, *Chem. Lett*, 909-912.

51 Kaida, Y. and Okamoto, Y. (1993) Optical resolution by high-performance liquid chromatography on benzylcarbamates of cellulose and amylose, *J. Chromatogr.* **641**, 267-278.

52 Kaida, Y. and Okamoto, Y. (1992) Optical resolution of β-lactams on 1-phenylethylcarbamates of cellulose and amylose, *Chirality* **4**, 122-124.

53 Kaida, Y. and Okamoto, Y. (1992) Efficient optical resolution of 4-hydroxy-2-cyclopentenone derivatives by HPLC on 1-phenylethylcarbamates of cellulose and amylose, *Chem. Lett.*, 85-88.

54 Kubota, T., Yamamoto, C., and Okamoto, Y. (2000) Tris(cyclohexylcarbamate)s of cellulose amylose as potential chiral stationary phases for high-performance liquid chromatography and thin-layer chromatography, *J. Am. Chem. Soc.* **122**, 4056-4059.

55 Kubota, T., Yamamoto, C., and Okamoto, Y. (2002) Chromatographic enantioseparation by cycloalkylcarbamate derivatives of cellulose and amylose, *Chirality* **14**, 372-376.

56 Okamoto, Y. Kawashima, M., and Hatada, K. (1984) Useful chiral packing materials for high-performance liquid chromatographic resolution of enantiomers: phenylcarbamates of polysaccharides coated on silica gel, *J. Am. Chem. Soc.* **106**, 5357-5359.

57 Okamoto, Y., Noguchi, J., and Yashima, E. (1998) Enantioseparation on 3,5-dichloro- and 3,5-dimethylphenycarbamates of polysaccharides as chiral stationary phases for high-performance liquid chromatography, *Reactive and Functional Polym.* **37**, 183-188.

58 Yamamoto, C., Hayashi, T., Okamoto, Y., and Kobayashi, S. (2000) Enantioseparation by using chitin phenylcarbamates as chiral stationary phases for high-performance liquid chromatography, *Chem. Lett.*, 12-13.

59 Yamamoto, C., Hayashi, T., and Okamoto, Y. (2003) High-performance liquid chromatographic enantioseparation using chitin carbamate derivatives as chiral stationary phases, *J. Chromatogr. A* **1021**, 86-91.

60 Schreiner, P.R., Fokin, A.A., Lauenstein, O., Okamoto, Y., Wakita, T., Rinderspacher, C., Robinson, G.H., Vohs, J.K., and Campana, C.F. (2002) Pseudotetrahedral polyhaloadamantanes as chirality probes: synthesis, separation, and absolute configuration, *J. Am. Chem. Soc.* **124**, 13348-13349.

61 Habaue, S., Sakamoto, H., and Okamoto, Y. (1996) Optical resolution of chiral ethylenebis(4,5,6,7-tetrahydro-1-indenyl)zirconium derivatives by high-performance liquid chromatography, *Chem. Lett.*, 383-384.

62 Okamoto, Y., Aburatani, R., and Hatada, K. (1988) Direct optical resolution of abscisic acid by high-performance liquid chromatography on cellulose tris(3,5-dimethylphenylcarbamate), *J. Chromatogr.* **448**, 454-455.

63 Tachibana, K. and Ohnishi, A. (2001) Reversed-phase liquid chromatographic separation of enantiomers on polysaccharide type chiral stationary phases, *J. Chromatogr. A* **906**, 127-154.

64 Dingene, J. (1994) Polysaccharide phases in enantioseparation, in Subramanian, G. (ed), *A Practical Approach to Chiral Separations by Liquid Chromatography*, VCH, New York, Chapter 6.

65 Ishikawa, A. and Shibata, T. (1993) Cellulosic chiral stationary phase under reversed-phase condition, *J. Liq. Chromatogr.* **16**, 859-878.

66 Francotte, E.R. (2001) Enantioselective chromatography as a powerful alternative for the preparation of drug enantiomers, *J. Chromatogr. A* **906**, 379-397.

67 Francotte, E.R. and Huynh, D. (2002) Immobilized halogenophenylcarbamate derivatives of cellulose as novel stationary phases for enantioselective drug analysis, *J. Pharm. Biomed. Anal.* **27**, 421-429.

68 Oliveros, L., Lopez, P., Minguillón, C., and Franco, P. (1995) Chiral chromatographic discrimination ability of a cellulose 3,5-dimethylphenylcarbamate/10-undecenoate mixed derivative fixed on several chromatographic matrixes, *J. Liq. Chromatogr.* **18**, 1521-1532.

69 Franco, P., Senso, A., Minguillón, C., and Oliveros, L. (1998) 3,5-Dimethylphenylcarbamates of amylose, chitosan and cellulose bonded on silica gel. Comparison of their chiral recognition abilities as high-performance liquid chromatography chiral stationary phases, *J. Chromatogr. A* **796**, 265-272.

70 Franco, P., Senso, A., Oliveros, L., and Minguillón, C. (2001) Covalently bonded polysaccharide derivatives as chiral stationary phases in high-performance liquid chromatography, *J. Chromatogr. A* **906**, 155-170.

71 Kubota, T., Kusano, T., Yamamoto, C., Yashima, E., and Okamoto, Y. (2001) Cellulose 3,5-dimethylphenylcarbamate immobilized onto silica gel via copolymerization with a vinyl monomer and its chiral recognition ability as a chiral stationary phase for HPLC, *Chem. Lett.*, 724-725.

72 Kubota, T., Yamamoto, C., and Okamoto, Y. (2003) Preparation of chiral stationary phase for HPLC based on immobilization of cellulose 3,5-dimethylphenylcarbamate derivatives on silica gel, *Chirality* **15**, 77-82.

73 Kubota, T., Yamamoto, C., and Okamoto, Y. (2003) Preparation and chiral recognition ability of cellulose 3,5-dimethylphenylcarbamate immobilized on silica gel through radical polymerization, *J. Polym. Sci. Part A, Polym. Chem.* **41**, 3703-3712.

74 Enomoto, N., Furukawa, S., Ogasawara, Y., Akano, H., Kawamura, Y., Yashima, E., and Okamoto, Y. (1996) Preparation of silica gel-bonded amylose through enzyme-catalyzed polymerization and chiral recognition ability of its phenylcarbamate derivative in HPLC, *Anal. Chem.* **68**, 2798-2804.

75 Reuter, C., Pawlittzki, G., Wörsdörfer, U., Plevoets, M., Mohry, A., Kubota, T., Okamoto, Y., and Vögtle, F. (2000) Chiral dendrophanes, dendro[2]rotaxanes, and dndro[2]catenanes: synthesis and chiroptical phenomena, *Eur. J. Org. Chem.*, 3059-3067.

76 Recker, J., Müller, W. M., Müller, U., Kubota, T., Okamoto, Y., Nieger, M., and Vögtle, F. (2002) Dendronized molecular knot: selective synthesis of various generations enantiomer separation, circular dichroism, *Chem. Eur. J.* **8**, 4434-4442.

77 Lukin, O., Recker, J., Böhmer, A., Müller, W. M., Kubota, T., Okamoto, Y., Nieger, M., Fröhlich, R., and Vögtle, F. (2003) A topologically chiral molecular dumbbell, *Angew. Chem. Int. Ed.* **42**, 442-445.

78 Yamamoto, C., Hayashi, T., Okamoto, Y., Ohkubo, S., and Kato, T. (2001) Direct resolution of C_{76} enantiomers by HPLC using an amylose-based chiral stationary phase, *Chem. Commun.* 925-926.

79 Hopf, H., Grahn, W., Barrett, D.G., Gerdes, A., Hilmer, J., Hucher, J., Okamoto, Y., and Kaida, Y. (1990) Optical resolution of [2.2]paracyclophanes by high-performance liquid chromatography on tris(3,5-dimethylphenylcarbamate) of cellulose and amylose, *Chem. Ber.* **123**, 841-845.

80 Maeda, K., Okamoto, Y., Morlender, N., Haddad, N., Eventova, I., Biali, S. E., and Rappoport, Z. (1995) Does the threshold enantiomerization route of crowded tetraarylethenes involve double bond rotation?, *J. Am. Chem. Soc.* **117**, 9686-9689.

81 Maeda, K., Okamoto, Y., Toledano, O., Becker, D., Biali, S. E., and Rappoport, Z. (1994) Multiple buttressing interactions: enantiomerization barrier of tetrakis(pentamethylphenyl)ethane, *J. Org. Chem.* **59**, 5473-5475.

82 Herges, R., Deichmann, M., Wakita, T., and Okamoto, Y. (2003) Synthesis of a chiral tube, *Angew. Chem. Int. Ed.* **42**, 1170-1172.

83 Shibata, T., Okamoto, I., and Ishii, K. (1986) Chromatographic optical resolution on polysaccharides and their derivatives, *J. Liq. Chromatogr.* **9**, 313-340.

84 Okamoto, Y., Aburatani, R., Kaida, Y., Hatada, K., Inotsume, N., and Nakano, M. (1989) Direct chromatographic separation of 2-arylpropionic acid enantiomers using tris(3,5-dimethylphenylcarbamate)s of cellulose and amylose as chiral stationary phases, *Chirality* **1**, 239-242.

85 Okamoto, Y., Cao, Z.-K., Aburatani, R., and Hatada, K. (1987) Optical resolution of alcohols as carbamates by HPLC on cellulose tris(phenylcarbamate) derivatives, *Bull. Chem. Soc. Jpn.* **60**, 3999-4003.

86 Schulte, M. and Strube, J. (2001), Preparative enantioseparation by simulated moving bed chromatography, *J. Chromatogr. A* **906**, 399-416.

87 Chankvetadze, B. and Blaschke, G. (2001) Enantioseparations in capillary electromigration techniques: recent developments and future trends, *J. Chromatogr. A* **906**, 309-363 (2001).
88 Chankvetadze, B., Yamamoto, C., and Okamoto, Y. (2003) Very fast enantioseparation in high-performance liquid chromatography using cellulose tris(3,5-dimethylphenylcarbamate) coated on monolithic silica support, *Chem. Lett.*, **32**, 240-241.
89 Yashima, E., Yamada, M., Kaida, Y., and Okamoto, Y. Computational studies on chiral discrimination mechanism of cellulose trisphenylcarbamate, *J. Chromatogr. A* **694**, 347-354.
90 Yashima, E., Yamamoto, C., and Okamoto, Y. (1996) NMR studies of chiral discrimination relevant to the liquid chromatographic enantioseparation by a cellulose phenylcarbamate derivative, *J. Am. Chem. Soc.* **118**, 4036-4048.
91 Yashima, E., Yamada, M., Yamamoto, C., Nakashima, M., and Okamoto, Y. (1997) Chromatographic enantioseparation and chiral discrimination in NMR by trisphenylcarbamate derivatives of cellulose, amylose, oligosaccharides, and cyclodextrins, *Enantiomer* **2**, 225-240.
92 Yamamoto, C., Yashima, E., and Okamoto, Y. (1999) Computational studies on chiral discrimination mechanism of phenylcarbamate derivatives of cellulose, *Bull. Chem. Soc. Jpn.* **72**, 1815-1825.
93 Yamamoto, C., Yashima, E., and Okamoto, Y. (2002) Structural analysis of amylose tris(3,5-dimethylphenylcarbamate) by NMR relevant to its chiral recognition mechanism in HPLC, *J. Am. Chem. Soc.* **124**, 12583-12589.

Index